Lecture Notes in Computer Science 13078

More information about this subseries at http://www.springer.com/series/7407

Nicholas N. Olenev · Yuri G. Evtushenko ·
Milojica Jaćimović · Michael Khachay ·
Vlasta Malkova (Eds.)

Optimization
and Applications

12th International Conference, OPTIMA 2021
Petrovac, Montenegro, September 27 – October 1, 2021
Proceedings

Springer

Editors
Nicholas N. Olenev ⓘ
Dorodnicyn Computing Centre
FRC CSC RAS
Moscow, Russia

Yuri G. Evtushenko ⓘ
Dorodnicyn Computing Centre
FRC CSC RAS
Moscow, Russia

Milojica Jaćimović ⓘ
University of Montenegro
Podgorica, Montenegro

Michael Khachay ⓘ
Krasovsky Institute of Mathematics
and Mechanics
Ekaterinburg, Russia

Vlasta Malkova ⓘ
Dorodnicyn Computing Centre
FRC CSC RAS
Moscow, Russia

ISSN 0302-9743   ISSN 1611-3349 (electronic)
Lecture Notes in Computer Science
ISBN 978-3-030-91058-7   ISBN 978-3-030-91059-4 (eBook)
https://doi.org/10.1007/978-3-030-91059-4

LNCS Sublibrary: SL1 – Theoretical Computer Science and General Issues

This Springer imprint is published by the registered company Springer Nature Switzerland AG
The registered company address is: Gewerbestrasse 11, 6330 Cham, Switzerland

Preface

This volume contains the first part of the refereed proceedings of the XII International Conference on Optimization and Applications (OPTIMA 2021)[1].

Organized annually since 2009, the conference has attracted a significant number of researchers, academics, and specialists in many fields of optimization, operations research, optimal control, game theory, and their numerous applications in practical problems of operations research, data analysis, and software development.

The broad scope of OPTIMA has made it an event where researchers involved in different domains of optimization theory and numerical methods, investigating continuous and discrete extremal problems, designing heuristics and algorithms with theoretical bounds, developing optimization software, and applying optimization techniques to highly relevant practical problems, can meet together and discuss their approaches and results. We strongly believe that this facilitates collaboration between researchers working in modern optimization theory, methods, and applications and those employing them to resolve valuable practical problems.

The conference was held during September 27 – October 1, 2021, in Petrovac, Montenegro, in the picturesque Budvanian riviera on the azure Adriatic coast. Due to the COVID-19 pandemic situation the Program Committee (PC) decided to organize online sessions for those who were not able to come to Montenegro this year. The main organizers of the conference were the Montenegrin Academy of Sciences and Arts, Montenegro, the Dorodnicyn Computing Centre, FRC CSC RAS, Russia, the Moscow Institute of Physics and Technology, Russia, Lomonosov Moscow State University, Russia, and the University of Évora, Portugal. This year, the key topics of OPTIMA were grouped into seven tracks:

- (i)   Mathematical programming
- (ii) Global optimization
- (iii) Discrete and combinatorial optimization
- (iv) Optimal control
- (v)   Optimization and data analysis
- (vi) Game theory and mathematical economics
- (vii) Applications.

The Program Committee (PC) and the reviewers of the conference included more than one hundred well-known experts in continuous and discrete optimization, optimal control and game theory, data analysis, mathematical economy, and related areas from leading institutions of 26 countries including Argentina, Australia, Austria, Belgium, China, Finland, France, Germany, Greece, India, Israel, Italy, Lithuania, Kazakhstan, Mexico, Montenegro, The Netherlands, Poland, Portugal, Russia, Serbia, Sweden, Taiwan, Ukraine, the UK, and the USA. This year we received 98 submissions, mostly from

---

[1] http://agora.guru.ru/display.php?conf=OPTIMA-2021.

Russia but also from Azerbaijan, Belarus, China, Finland, France, Germany, Kazakhstan, Moldova, Montenegro, UAE, Poland, Romania, Saudi Arabia, Serbia, and Ukraine. Each submission was reviewed by at least three PC members or invited reviewers, experts in their fields, to supply detailed and helpful comments. Out of 63 qualified submissions, the Program Committee decided to accept 22 full and 3 short papers to the first volume of the proceedings. Thus the acceptance rate for this volume was about 40%.

In addition, the Program Committee proposed about 19 papers to be included in the second volume of the proceedings after a short presentation of the candidate papers, discussion at the conference, and subsequent revision.

The conference featured five invited lecturers, and several plenary and keynote talks. The invited lectures included:

- Anton Bondarev, International Business School Suzhou, Xi'an Jiaotong-Liverpool University, China, "Optimality of sliding dynamics in hybrid control systems"
- Nenad Mladenovic, Khalifa University of Science and Technology, Abu Dhabi, UAE, "Formulation Space Search Metaheuristic"
- Yurii Nesterov, CORE/INMA, Université Catholique de Louvain, Belgium, "Inexact high-order proximal-point methods with auxiliary search procedure"
- Panos M. Pardalos, University of Florida, USA, "Artificial Intelligence, Data Sciences, and Optimization in Economics and Finance"
- Alexey Tret'yakov, Siedlce University of Natural Sciences and Humanities, Poland, "Exit from singularity. New optimization methods and the p-regularity theory applications"

We would like to thank all the authors for submitting their papers and the members of the PC for their efforts in providing exhaustive reviews. We would also like to express special gratitude to all the invited lecturers and plenary speakers.

October 2021

Nicholas N. Olenev
Yuri G. Evtushenko
Milojica Jaćimović
Michael Khachay
Vlasta Malkova

Organization

Program Committee Chairs

Milojica Jaćimović	Montenegrin Academy of Sciences and Arts, Montenegro
Yuri G. Evtushenko	Dorodnicyn Computing Centre, FRC CSC RAS, Russia
Igor G. Pospelov	Dorodnicyn Computing Centre, FRC CSC RAS, Russia

Program Committee

Majid Abbasov	St. Petersburg State University, Russia
Samir Adly	University of Limoges, France
Kamil Aida-Zade	Institute of Control Systems of ANAS, Azerbaijan
Alla Albu	Dorodnicyn Computing Centre, FRC CSC RAS, Russia
Alexander P. Afanasiev	Institute for Information Transmission Problems, RAS, Russia
Yedilkhan Amirgaliyev	Suleyman Demirel University, Kazakhstan
Anatoly S. Antipin	Dorodnicyn Computing Centre, FRC CSC RAS, Russia
Adil Bagirov	Federation University, Australia
Artem Baklanov	International Institute for Applied Systems Analysis, Austria
Evripidis Bampis	LIP6 UPMC, France
Olga Battaïa	ISAE-SUPAERO, France
Armen Beklaryan	National Research University Higher School of Economics, Russia
Vladimir Beresnev	Sobolev Institute of Mathematics, Russia
Anton Bondarev	Xi'an Jiaotong-Liverpool University, China
Sergiy Butenko	Texas A&M University, USA
Vladimir Bushenkov	University of Évora, Portugal
Igor A. Bykadorov	Sobolev Institute of Mathematics, Russia
Alexey Chernov	Moscow Institute of Physics and Technology, Russia
Duc-Cuong Dang	INESC TEC, Portugal
Tatjana Davidovic	Mathematical Institute, Serbian Academy of Sciences and Arts, Serbia
Stephan Dempe	TU Bergakademie Freiberg, Germany
Askhat Diveev	FRC CSC RAS and RUDN University, Russia
Alexandre Dolgui	IMT Atlantique, LS2N, CNRS, France
Olga Druzhinina	FRC CSC RAS, Russia
Anton Eremeev	Omsk Division of Sobolev Institute of Mathematics, SB RAS, Russia
Adil Erzin	Novosibirsk State University, Russia
Francisco Facchinei	Sapienza University of Rome, Italy

Vladimir Garanzha	Dorodnicyn Computing Centre, FRC CSC RAS, Russia
Alexander V. Gasnikov	Moscow Institute of Physics and Technology, Russia
Manlio Gaudioso	Universita della Calabria, Italy
Alexander I. Golikov	Dorodnicyn Computing Centre, FRC CSC RAS, Russia
Alexander Yu. Gornov	Institute for System Dynamics and Control Theory, SB RAS, Russia
Edward Kh. Gimadi	Sobolev Institute of Mathematics, SB RAS, Russia
Andrei Gorchakov	Dorodnicyn Computing Centre, FRC CSC RAS, Russia
Alexander Grigoriev	Maastricht University, The Netherlands
Mikhail Gusev	N.N. Krasovskii Institute of Mathematics and Mechanics, Russia
Vladimir Jaćimović	University of Montenegro, Montenegro
Vyacheslav Kalashnikov	ITESM, Monterrey, Mexico
Maksat Kalimoldayev	Institute of Information and Computational Technologies, Kazakhstan
Valeriy Kalyagin	Higher School of Economics, Russia
Igor E. Kaporin	Dorodnicyn Computing Centre, FRC CSC RAS, Russia
Alexander Kazakov	Institute for System Dynamics and Control Theory, SB RAS, Russia
Michael Khachay	Krasovsky Institute of Mathematics and Mechanics, Russia
Oleg V. Khamisov	L. A. Melentiev Energy Systems Institute, Russia
Andrey Kibzun	Moscow Aviation Institute, Russia
Donghyun Kim	Kennesaw State University, USA
Roman Kolpakov	Moscow State University, Russia
Alexander Kononov	Sobolev Institute of Mathematics, Russia
Igor Konnov	Kazan Federal University, Russia
Vera Kovacevic-Vujcic	University of Belgrade, Serbia
Yury A. Kochetov	Sobolev Institute of Mathematics, Russia
Pavlo A. Krokhmal	University of Arizona, USA
Ilya Kurochkin	Institute for Information Transmission Problems, RAS, Russia
Dmitri E. Kvasov	University of Calabria, Italy
Alexander A. Lazarev	V.A. Trapeznikov Institute of Control Sciences, Russia
Vadim Levit	Ariel University, Israel
Bertrand M. T. Lin	National Chiao Tung University, Taiwan
Alexander V. Lotov	Dorodnicyn Computing Centre, FRC CSC RAS, Russia
Olga Masina	Yelets State University, Russia
Vladimir Mazalov	Institute of Applied Mathematical Research, Karelian Research Center, Russia
Nevena Mijajlović	University of Montenegro, Montenegro
Nenad Mladenovic	Mathematical Institute, Serbian Academy of Sciences and Arts, Serbia
Mikhail Myagkov	University of Oregon, USA
Angelia Nedich	University of Illinois at Urbana Champaign, USA
Yuri Nesterov	CORE, Université Catholique de Louvain, Belgium

Yuri Nikulin	University of Turku, Finland
Evgeni Nurminski	Far Eastern Federal University, Russia
Nicholas N. Olenev	Dorodnicyn Computing Centre, FRC CSC RAS, Russia
Panos Pardalos	University of Florida, USA
Alexander V. Pesterev	V.A. Trapeznikov Institute of Control Sciences, Russia
Alexander Petunin	Ural Federal University, Russia
Stefan Pickl	Universität der Bundeswehr München, Germany
Boris T. Polyak	V.A. Trapeznikov Institute of Control Sciences, Russia
Yury S. Popkov	Institute for Systems Analysis, FRC CSC RAS, Russia
Leonid Popov	IMM UB RAS, Russia
Mikhail A. Posypkin	Dorodnicyn Computing Centre, FRC CSC RAS, Russia
Alexander N. Prokopenya	Warsaw University of Life Sciences, Poland
Oleg Prokopyev	University of Pittsburgh, USA
Artem Pyatkin	Novosibirsk State University and Sobolev Institute of Mathematics, Russia
Ioan Bot Radu	University of Vienna, Austria
Soumyendu Raha	Indian Institute of Science, India
Leonidas Sakalauskas	Institute of Mathematics and Informatics, Lithuania
Eugene Semenkin	Siberian State Aerospace University, Russia
Yaroslav D. Sergeyev	University of Calabria, Italy
Natalia Shakhlevich	University of Leeds, UK
Alexander A. Shananin	Moscow Institute of Physics and Technology, Russia
Angelo Sifaleras	University of Macedonia, Greece
Mathias Staudigl	Maastricht University, The Netherlands
Petro Stetsyuk	V.M. Glushkov Institute of Cybernetics, Ukraine
Fedor Stonyakin	V. I. Vernadsky Crimean Federal University, Russia
Alexander Strekalovskiy	Institute for System Dynamics and Control Theory, SB RAS, Russia
Vitaly Strusevich	University of Greenwich, UK
Michel Thera	University of Limoges, France
Tatiana Tchemisova	University of Aveiro, Portugal
Anna Tatarczak	Maria Curie-Skłodowska University, Poland
Alexey A. Tretyakov	Dorodnicyn Computing Centre, FRC CSC RAS, Russia
Stan Uryasev	University of Florida, USA
Frank Werner	Otto von Guericke University Magdeburg, Germany
Adrian Will	National Technological University, Argentina
Vitaly G. Zhadan	Dorodnicyn Computing Centre, FRC CSC RAS, Russia
Anatoly A. Zhigljavsky	Cardiff University, UK
Julius Žilinskas	Vilnius University, Lithuania
Yakov Zinder	University of Technology, Australia
Tatiana V. Zolotova	Financial University under the Government of the Russian Federation, Russia
Vladimir I. Zubov	Dorodnicyn Computing Centre, FRC CSC RAS, Russia
Anna V. Zykina	Omsk State Technical University, Russia

Organizing Committee Chairs

Milojica Jaćimović	Montenegrin Academy of Sciences and Arts, Montenegro
Yuri G. Evtushenko	Dorodnicyn Computing Centre, FRC CSC RAS, Russia
Nicholas N. Olenev	Dorodnicyn Computing Centre, FRC CSC RAS, Russia

Organizing Committee

Natalia Burova	Dorodnicyn Computing Centre, FRC CSC RAS, Russia
Alexander Golikov	Dorodnicyn Computing Centre, FRC CSC RAS, Russia
Alexander Gornov	Institute of System Dynamics and Control Theory, SB RAS, Russia
Vesna Dragović	Montenegrin Academy of Sciences and Arts, Montenegro
Vladimir Jaćimović	University of Montenegro, Montenegro
Michael Khachay	Krasovsky Institute of Mathematics and Mechanics, Russia
Yury Kochetov	Sobolev Institute of Mathematics, Russia
Vlasta Malkova	Dorodnicyn Computing Centre, FRC CSC RAS, Russia
Oleg Obradovic	University of Montenegro, Montenegro
Mikhail Posypkin	Dorodnicyn Computing Centre, FRC CSC RAS, Russia
Kirill Teymurazov	Dorodnicyn Computing Centre, FRC CSC RAS, Russia
Yulia Trusova	Dorodnicyn Computing Centre, FRC CSC RAS, Russia
Svetlana Vladimirova	Dorodnicyn Computing Centre, FRC CSC RAS, Russia
Victor Zakharov	FRC CSC RAS, Russia
Ivetta Zonn	Dorodnicyn Computing Centre, FRC CSC RAS, Russia

Abstracts of Invited Talks

Optimality of Sliding Dynamics in Hybrid Control Systems

Anton Bondarev

International Business School Suzhou, Xi'an Jiaotong-Liverpool University,
P. R. China
https://www.xjtlu.edu.cn/zh/departments/academic-
departments/international-business-school-
suzhou/staff/anton-bondarev

Abstract. There is growing evidence on the presence of sliding dynamics in many piecewise-smooth dynamical systems (PWS), reported in papers on population biology, renewable resources and etc. However to our best knowledge there are no studies on the optimality of such a type of dynamics. This talk will go through some recent advances in the theory of hybrid optimal control which deals with PWS dynamics and present findings on the optimality of the sliding dynamics in such systems, both in the optimal control problems and differential games. Moreover some results on the presence of hybrid limit cycles in such systems will be discussed.

In particular, hybrid control problem may have the equilibrium of the sliding flow as the only possible long-run outcome if all conventional equilibria of the PWS at hand are infeasible. Moreover, this equilibrium may be reached only from the outside of the sliding flow itself. Next, hybrid limit cycles (HLC) may be optimal or not depending on the definition of the switching manifold and the dimensionality of the problem.

At last, some further open questions of interest in the field are discussed.

Formulation Space Search Metaheuristic

Nenad Mladenovic ⓘ

Khalifa University of Science and Technology, Abu Dhabi, United Arab Emirates
http://www.mi.sanu.ac.rs/nenad/

Abstract. Many methods for solving discrete and continuous global optimization problems are based on changing one formulation to another, which is either equivalent or very close to it. These types of methods include dual, primal-dual, Lagrangian, linearization, surrogation, convexification methods, coordinate system change, discrete/continuous reformulations, to mention a few. However, in all those classes, the set of formulations of one problem are not considered as a set having some structure provided with some order relation among formulations. The main idea of Formulation Space Search (FSS) is to provide the set of formulations with some metric or quasi-metric relations, used for solving a given class or type of problem. In that way, the (quasi) distance between formulations is introduced, and the search space in solving Global optimization problems is extended to the set of formulations as well. In this talk I will present the general methodology of FSS, and give an overview of several applications taken from the literature that fall within this framework. I will also examine two of these applications in more detail.

This is joint work with J. Brimberg, R. Todosijevic and D. Urosevic.

Inexact High-Order Proximal-Point Methods with Auxiliary Search Procedure

Yurii Nesterov

CORE/INMA, Université Catholique de Louvain, Belgium
https://uclouvain.be/fr/repertoires/yurii.nesterov

Abstract. In this talk, we present new framework of Bi-Level Unconstrained Minimization based on high-order proximal-point method with the maximal convergence rate $O(1/k^{(1+3p)/2})$, where k is the iteration counter and p is the order of the scheme. Under assumption on the boundedness of the $(p + 1)$th derivative of the objective function, each iteration of the scheme can be implemented by one step of the pth order augmented tensor method. In this way, for $p = 2$, we get a new second-order method with the rate of convergence $O(1/k^{7/2})$ and logarithmic complexity of the auxiliary search at each iteration.

Another possibility is to compute the proximal-point operator by a lower-order minimization method. As an example, for $p = 3$, we consider the upper-level process convergent as $O(1/k^5)$. Assuming boundedness of the fourth derivative, an appropriate approximation of the proximal-point operator can be computed by a second-order method in a logarithmic number of iterations. This combination gives a second-order scheme with much better complexity than the existing theoretical limits.

Artificial Intelligence, Data Sciences, and Optimization in Economics and Finance

Panos M. Pardalos

University of Florida, USA
http://www.ise.ufl.edu/pardalos/
https://nnov.hse.ru/en/latna/

Abstract. Artificial Intelligence (along with data sciences and optimization) has been a fundamental component of many activities in economics and finance in recent years. In this lecture we first summarize some of the major impacts of AI tools in economics and finance and discuss future developments and limitations. In the second part of the talk we present details on neural network embeddings on corporate annual filings for portfolio selection.

Exit from Singularity. New Optimization Methods and the P-Regularity Theory Applications

Alexey Tret'yakov

Systems Research Institute, Polish Academy of Sciences, Warsaw, Poland
https://www.researchgate.net/profile/Alexey_Tretyakov

Abstract. We introduce a new nonsingular operator instead of a degenerate operator of the first derivative in a singular case for solving and describing nonregular optimization problems and some problems in calculus. Such operator is called p-factor-operator and its construction is based on the derivatives up to order p as well as on some element h, which we call the "exit from singularity".

The special variant of the method of the modified Lagrange functions for constrained optimization problems with inequality constraints is justified on the basis of the 2-factor transformation and constructions of p-regularity theory. These results are used in some classical branches of calculus: implicit function theorem is given for the singular case and is shown the existence of solutions to a boundary-valued problem for a nonlinear differential equation in the resonance case. New numerical methods are proposed including the p-factor method for solving ODEs with a small parameter.

This is joint work with Yuri Evtushenko and Vlasta Malkova.

Contents

Game Theory and Mathematical Economics

Applications

Mathematical Programming

Exit from Singularity. New Optimization Methods and the p-Regularity Theory Applications

Yuri Evtushenko[1,2] , Vlasta Malkova[1(✉)] , and Alexey Tret'yakov[1,3,4]

[1] Dorodnicyn Computing Centre, FRC CSC RAS, Vavilov Street 40,
119333 Moscow, Russia
[2] Moscow Institute of Physics and Technology, Moscow, Russia
[3] System Research Institute, Polish Academy of Sciences, Newelska 6,
01-447 Warsaw, Poland
[4] Faculty of Sciences, Siedlce University, 08-110 Siedlce, Poland
tret@ap.siedlce.pl

Abstract. In the paper, we introduce a new nonsingular operator instead of a degenerate operator of the first derivative in a singular case for solving and describing nonregular optimization problems and some problems in calculus. Such operator is called p-factor-operator and its construction is based on the derivatives up to order p as well as on some element h, which we call the "exit from singularity". The special variant of the method of the Modified Lagrangian Functions for optimization problems with inequality constraints is justified on the basis of the 2-factor transformation and constructions of p-regularity theory. These results are used in some classical branches of calculus: implicit function theorem is given for the singular case and is shown the existence of solutions to a boundary-valued problem for a nonlinear differential equation in the resonance case. New numerical methods are proposed including the p-factor method for solving ODEs with a small parameter and new formula is obtained for the solutions of such type equations.

Keywords: p-regularity · Nonlinear optimization · Modified Lagrangian Functions · Singularity

1 Introduction

Previously published papers [1–3] describe a number of applications of the p-regularity theory in various areas of mathematics. In this paper, new applications of the developed theory are considered. Namely, in the first part of the paper, the p-regularity theory is used to construct methods for solving degenerate nonlinear equations and to substantiate a special version of the method of

This work was supported in part by the Russian Foundation for Basic Research, project No. 21-71-30005.

Modified Lagrangian Functions (MLF) proposed by Yu.G. Evtushenko in [4] for solving optimization problems with inequality-type constraints in the nonregular case. In the second part of this paper, the p-regularity theory is used to analyze the existence of a solution to a degenerate nonlinear boundary value problem. In addition, a method for solving degenerate ordinary differential equations (ODE) is proposed. The method is a modification of the small parameter method.

All this became possible thanks to the introduction of a new h element into consideration, which we called "exit from singularity" and which plays a key role in the construction of an apparatus for describing and solving degenerate problems and is essentially new in comparison with traditional constructions. Its presence, as it were, guarantees in the degenerate case an exit from the singularity of the problem at the initial stage, and therefore the presence of this element in all considered schemes is a significant deference from the previously considered approaches.

2 p-Factor Operator and p-Regular Mappings

Consider the mapping $F : X \to Z$, where $F \in C^{p+1}(X, Z)$, X, and Z are Banach spaces. By x^* we denote the solution of the equation

$$F(x) = 0. \tag{1}$$

Let's assume that

$$\operatorname{Im} F'(x^*) \neq Z,$$

that is, the mapping F is degenerate at the point x^*.

In this paper, the p-factor operator is used as a tool for obtaining various kinds of results in the degenerate case. Without loss of generality, we assume that the space Z is decomposable into a direct sum of closed subspaces Z_i, $i = 1, \ldots, p$:

$$Z = Z_1 \oplus \cdots \oplus Z_p,$$

where $Z_1 = \operatorname{cl}(\operatorname{Im} F'(x^*))$, $Z_i = \operatorname{cl}(\operatorname{span} \operatorname{Im} P_{W_i} F^{(i)}(x^*)[\cdot]^i)$, $i = 2, \ldots, p-1$, $Z_p = W_p$, W_i is the closed complement of the subspace $(Z_1 \oplus \cdots \oplus Z_{i-1})$ to the space Z, $i = 2, \ldots, p$ and $P_{W_i} : Z \to W_i$ is the projection operator on W_i along $(Z_1 \oplus \cdots \oplus Z_{i-1})$. Following [5], we define the mappings

$$f_i(x) : X \to Z_i,$$
$$f_i(x) = P_{Z_i} F(x), \qquad i = 1, \ldots, p,$$

where $P_{Z_i} : Z \to Z_i$ is the operator of projection onto Z_i along $(Z_1 \oplus \cdots \oplus Z_{i-1} \oplus Z_{i+1} \oplus \cdots \oplus Z_p)$, $i = 1, \ldots, p$.

Definition 1. *The linear operator $\Psi_p(h) \in \mathcal{L}(X, Z_1 \oplus \cdots \oplus Z_p)$, $h \in X$, $h \neq 0$,*

$$\Psi_p(h) = f_1'(x^*) + f_2''(x^*)[h] + \ldots + \frac{1}{(p-1)!} f_p^{(p)}(x^*)[h]^{p-1}, \tag{2}$$

is called the p-factor operator of the mapping $F(x)$ on the element h at the point x^*.

The presence of h, which we called the "exit from singularity", yield a family of the p-factor operators "replacing" the nonregular operator $F'(x^*)$.

Definition 2. *The mapping $F(x)$ is called p-regular at the point x^* on the element h if*

$$\operatorname{Im} \Psi_p(h) = Z$$

or $\|\{\Psi_p(h)\}^{-1}\| < \infty$. Here $\|\{\Psi_p(h)\}^{-1}\| = \sup\limits_{\|z\|=1} \inf\{\|y\| \mid \Psi_p(h)[y] = z\}$.

Definition 3. *The mapping $F(x)$ is called p-regular at the point x^* if $F(x)$ is p-regular at the point x^* on the elements*

$$h \in \bigcap_{k=1}^{p} \operatorname{Ker}{}^k f_k^{(k)}(x^*),$$

where $\operatorname{Ker}{}^k f_k^{(k)}(x^*)$ is the k-kernel of the mapping $f_k^{(k)}(x^*)$ and is defined as:

$$\operatorname{Ker}{}^k f_k^{(k)}(x^*) = \{h \in X \mid f_k^{(k)}(x^*)[h]^k = 0\}.$$

3   2-Factor Methods for Solving Degenerate Equations

Following [6], consider methods for solving degenerate equations. Let in the equation (1) the mapping F acts from the space $\mathbb{R}^n$ to the space $\mathbb{R}^n$. We assume that in the solution x^* of the equation (1) the Jacobi matrix $F'(x^*)$ is degenerate, that is $\operatorname{Im} F'(x^*) \neq \mathbb{R}^n$. Consider two versions of the p-factor method for solving the degenerate system (1) for $p = 2$.

First Version of the 2-Factor Method. Due to the fact that $\ker F'(x^*) \neq \{0\}$, there is a vector $h \neq 0$ such that $F'(x^*)h = 0$. Therefore, the x^* point is also a solution to the system.

$$F(x) + F'(x)h = 0_n, \tag{3}$$

and a 2-factor-method scheme has the form:

$$x_{k+1} = x_k - \{F'(x_k) + F''(x_k)h\}^{-1} (F(x_k) + F'(x_k)h), \qquad k = 0, 1, \dots. \tag{4}$$

Here h is "exit from singularity".

Note that in this paper, taking into account the specifics of the problem, the vector h is built without using information about solving the x^* problem. We also note that the 2-factor-method scheme (4) coincides with the Newton method, but applied not to the initial Eq. (1), and to the modified system (3).

Second Version of the 2-Factor Method. Due to the fact that $\operatorname{Im} F'(x^*) \neq \mathbb{R}^n$, the projection operator P on $(\operatorname{Im} F'(x^*))^{\perp}$ non-zero, that is, $P \neq 0$. Therefore, the point x^* is also a solution of the system

$$F(x) + PF'(x)\xi = 0_n$$

for any $\xi \neq 0$. In this case, the second scheme of the 2-factor method has the form

$$x_{k+1} = x_k - \{F'(x_k) + PF''(x_k)\xi\}^{-1} (F(x_k) + PF'(x_k)\xi), \qquad k = 0, 1, \ldots . \quad (5)$$

where ξ is "exit from singularity".

The justification of the convergence of the 2-factor methods (4) and (5) are the following theorems.

Theorem 1. *Let $F \in C^3(\mathbb{R}^n)$ and for $h \in \ker F'(x^*)$, $h \neq 0_n$ there exists $(F'(x^*) + F''(x^*)h)^{-1}$. Then for the 2-factor method (4) there is an estimate of the convergence rate*

$$\|x_{k+1} - x^*\| \leq \alpha_1 \|x_k - x^*\|^2, \qquad k = 0, 1, \ldots,$$

where $x_0 \in U_\varepsilon(x^*)$, $\varepsilon > 0$ is sufficiently small, and $\alpha_1 > 0$ is independent constant.

The proof of the Theorem 1 see in [1].

Theorem 2. *Let $F \in C^3(\mathbb{R}^n)$ and for some element $\xi \neq 0_n$ there exists $(F'(x^*) + PF''(x^*)\xi)^{-1}$. Then, for the 2-factor method (5) there is an estimate of the convergence rate*

$$\|x_{k+1} - x^*\| \leq \alpha_2 \|x_k - x^*\|^2, \qquad k = 0, 1, \ldots,$$

where $x_0 \in U_\varepsilon(x^*)$, $\varepsilon > 0$ is sufficiently small, and $\alpha_2 > 0$ is independent constant.

The proof of the Theorem 2 is similar to the proof of the Theorem 1.

Remark 1. In the case of the existence of $h \in \operatorname{Ker} F'(x^*)$ such that $\operatorname{Im} F''(x^*)h = \mathbb{R}^n$, the 2-factor-method scheme may be as follows:

$$x_{k+1} = x_k - \{F''(x_k)h\}^{-1} F'(x_k)h. \quad (6)$$

In this case, for quadratic maps $F(x) = Q[x]^2 + Ax + B$, where $Q : \mathbb{R}^n \to \mathbb{R}^n$ – quadratic form, $A : \mathbb{R}^n \to \mathbb{R}^n$ – matrix, B – vector of dimension n, will be correct the exact formula for the solution point x^*:

$$x^* = -\frac{1}{2}[Qh]^{-1}(Ah), \quad (7)$$

where h satisfies the condition $(2Qx^* + A)h = 0_n$. Note that the method (6) converges for quadratic mappings in one step.

4   2-Factor Modified Lagrangian Functions Method for Solving the Problem of Constrained Optimization

Consider a nonlinear programming problem

$$\min_{x \in X} \varphi(x), \tag{8}$$

where an feasible set

$$X = \{x \in \mathbb{R}^n \mid g(x) \le 0_m\},$$

0_m is a zero vector from $\mathbb{R}^m$, $(g(x))^\top = (g_1(x), \ldots, g_m(x))$—vector function. With respect to the problem (8), we assume that the set of its solutions $X^* \in \mathbb{R}^n$ is nonempty.

We define the Lagrange function,

$$L(x, v) = \varphi(x) + \sum_{i=1}^{m} v_i g_i(x).$$

Everywhere below, it is assumed that the constraints regularity condition (RC) is satisfied, in other words, the gradients of active constraints are linearly independent. This condition guarantees that to each $x^* \in X^*$ there corresponds a unique vector of Lagrange multipliers $v^* \ge 0$ such that the relation

$$\nabla L(x^*, v^*) = \nabla \varphi(x^*) + \sum_{i=1}^{m} v_i^* \nabla g_i(x^*) = 0_n$$

holds and $v_i^* = 0$, if $g_i(x^*) > 0$, $i = 1, \ldots, m$.

Consider the non-standard version of the MLF method proposed in [4], in which the modified Lagrange function has the form

$$L_E(x, \lambda) = \varphi(x) + \frac{1}{2} \sum_{i=1}^{m} \lambda_i^2 g_i(x),$$

where $\lambda \in \mathbb{R}^m$. Combine the vectors x and λ with one symbol $w \in \mathbb{R}^{n+m}$. Similarly, the pair $[x^*, \lambda^*]$ will be denoted by w^*, therefore $L_E(x, \lambda) = L_E(w)$.

According to the Kuhn-Tucker theorem, the vector w^* satisfies the system of equations

$$G(w) = \begin{bmatrix} \nabla \varphi(x) + \dfrac{1}{2} \displaystyle\sum_{i=1}^{m} \lambda_i^2 \nabla g_i(x) \\ D(\lambda) g(x) \end{bmatrix} = 0_{n+m}. \tag{9}$$

Here $D(\lambda)$ is a diagonal matrix, the dimension of which is determined by the dimension of the vector λ, its i-th diagonal element is λ_i. Note that the system (9) can have an infinite set of solutions in a neighborhood of the point w^*.

Let $g'(x)$ be a Jacobi matrix of the mapping $g(x)$. In turn, the Jacobi matrix for the system (9) has the form

$$G'(w) = \left[\begin{array}{c|c} \nabla^2\varphi(x) + \dfrac{1}{2}\sum_{i=1}^{m}\lambda_i^2\nabla^2 g_i(x) & g'(x)D(\lambda) \\ \hline D(\lambda)[g'(x)]^\top & D(g(x)) \end{array} \right]. \tag{10}$$

For the pair $[x^*, \lambda^*]$ we define the set of active constraints $I(x^*)$, the set of weakly active constraints $I_0(x^*)$ and the set of strictly active constraints $I_+(x^*)$ using the following relations:

$$\begin{aligned} I(x^*) &= \{j = 1,\ldots,m \mid g_j(x^*) = 0\}, \\ I_0(x^*) &= \{j = 1,\ldots,m \mid \lambda_j^* = 0, \; g_j(x^*) = 0\}, \\ I_+(x^*) &= \{j = 1,\ldots,m \mid \lambda_j^* \neq 0, \; g_j(x^*) = 0\}. \end{aligned} \tag{11}$$

When justifying and analyzing the MLF method, the following conditions are usually introduced in addition to the RC:
1. the strict complementary slackness condition (SCS), i.e.

$$v_i^* g_i(x^*) = 0, \qquad i = 1,\ldots,m,$$

and, if $g_i(x^*) = 0$, then $v_i^* \neq 0$ for all $i = 1,\ldots,m$;
2. the sufficient condition for optimality of the 2nd order: there exists a number $\nu > 0$ such that

$$z^\top \nabla_{xx}^2 L(x^*, v^*)z \geq \nu\|z\|^2 \tag{12}$$

for all $z \in \mathbb{R}^n$, satisfying the conditions $[\nabla g_j(x^*)]^\top z \leq 0$ and $j \in I(x^*)$.

Suppose that at the point x^* the SCS is not satisfied. Then, for some index i, both equalities $\lambda_i^* = 0$ and $g_i(x^*) = 0$ hold, therefore the set $I_0(x^*)$ is not empty. In this case, the matrix (10) becomes degenerate at the point w^* and, therefore, Newton-type methods for solutions to the system of equations (9) are unacceptable. Let us show that in this situation it will be efficient to use the apparatus of p-factor operators.

Consider a system of nonlinear equations (9). Let the mapping G be irregular at the point w^* or, in other words, the Jacobi matrix (10) is degenerate and $\operatorname{rank} G'(w^*) = r < n + m$. In this case, w^* is called a degenerate solution of the system (9). The degeneracy condition for the matrix $G'(w^*)$ is written as $\operatorname{Im} G'(w^*) \neq \mathbb{R}^{n+m}$. It means, that there is at least one vector h ($\|h\| \neq 0$) such that $G'(w^*)h = 0_{n+m}$. Obviously, the solution of the system (9) will also be the solution of the modified system

$$\Psi(w) = G(w) + G'(w)h = 0_{n+m} \tag{13}$$

and, moreover, if the matrix $G'(w^*)$ is degenerate, then, on the contrary, the matrix $\Psi'(w^*) = G'(w^*) + G''(w^*)h$ is not degenerate and the solution w^* of the system (13) is locally unique. Here h is some vector from $\operatorname{Ker} G'(w^*)$, one of the construction examples of which is given below. The property of nondegeneracy of the matrix $\Psi'(w^*)$ underlies the construction of a 2-factor method for solving degenerate systems of nonlinear equations.

For the mapping G, we introduce the 2-factor operator

$$G'(w) + G''(w)h,$$

where the vector $h \in \operatorname{Ker} G'(w^*)$ is defined so that

$$\operatorname{rank}(G'(w^*) + G''(w^*)h) = n + m.$$

The specific form of h is formed taking into account the specifics of the system (9) and here the vector h will play role of element "exit from singularity".

Consider the first version of the 2-factor method (4) for solving the system (9):

$$w_{k+1} = w_k - \left(G'(w_k) + G''(w_k)h\right)^{-1}\left(G(w_k) + G'(w_k)h\right), \quad k = 0, 1, \ldots . \tag{14}$$

Below, we show that under sufficient optimality conditions for problem (8), the mapping G defined in (9) is 2-regular in the solution w^* on some $h \in \operatorname{Ker} G'(w^*)$. Therefore, to solve the system $G(w) = 0_{n+m}$, we can apply the 2-factor method (4) and according to Theorem 1 the resulting method will have a quadratic convergence rate, which cannot be achieved when applying Newton's method to the solution of the system (11).

Suppose that the SCS does not hold at the point x^* and without loss of generality we can assume that the set $I_0(x^*)$ consists of the first s indices, i.e. $I_0(x^*) = \{1, \ldots, s\}$. To numerically determine the set $I_0(x^*)$, the so-called procedure for identifying zero elements, first proposed in [7], can be used.

Note that Newton's method, together with a differentiable penalty, was used to solve linear programming problems with inequality-type constraints in the works [8,9], where it was shown that for penalty coefficients greater than a certain threshold value, from the results of minimizing the penalty function using simple formulas the exact solution of the problem is found.

Let's come back to substantiate the 2-factor method (14). Since $\lambda_j = 0$ and $g_j(x^*) = 0$ for all $j \in I_0(x^*) = \{1, \ldots, s\}$, the rows of the matrix $G'(w^*)$, starting from $(n + 1)$-th to $(n + s)$-th, consist of all zeros. Let's define the vector ("exit from singularity") $h \in \mathbb{R}^{n+m}$:

$$h^\top = (0_n^\top, 1_s^\top, 0_{p-s}^\top, 0_{m-p}^\top) \tag{15}$$

and consider the mapping

$$\Phi(w) = G(w) + G'(w)h, \tag{16}$$

where the vector h is defined by the formula (15).

For further consideration, we need the following auxiliary lemma.

Lemma 1. *Let the V-matrix of dimension $n \times n$ and the Q-matrix of dimension $n \times p$ be such that the columns of the Q matrix are linearly independent and*

$$\langle Vx, x \rangle > 0 \quad \text{for all} \quad x \in \{\operatorname{Ker} Q^\top\} \backslash \{0\}.$$

Suppose also that G_N is a full rank diagonal matrix of dimension $\ell \times \ell$. Then the matrix

$$\bar{A} = \begin{pmatrix} V & Q & 0 \\ Q^\top & 0 & 0 \\ 0 & 0 & G_N \end{pmatrix}$$

is not degenerate.

The Lemma 1 is similar to many results of linear algebra, therefore we omit the proof here.

Lemma 2. *Suppose that $\varphi, g_i \in C^3(\mathbb{R}^n)$, $i = 1, \ldots, m$, the RC is satisfied, the sufficient optimality conditions (12) and the mapping Φ is given by the formula (16). Then the 2-factor operator $\Phi'(w) = G'(w) + G''(w)h$ is not degenerate at the point $w^* = [x^*, \lambda^*]$.*

The proof follows from Lemma 1 if we put in it $V = \nabla^2_{xx} L_E(x^*, \lambda^*)$, $G_N = D(g_N(x^*))$, where $g_N(x) = (g_{p+1}(x), \ldots, g_m(x))^\top$, and

$$Q = \left[\nabla g_1(x^*), \ldots, \nabla g_s(x^*), \lambda^*_{s+1} \nabla g_{s+1}(x^*), \ldots, \lambda^*_p \nabla g_p(x^*) \right].$$

Then $\Phi'(w^*) = \bar{A}$.

Lemma 2 implies that the 2-factor method (14) can be used to solve the system (9), namely, the following theorem holds.

Theorem 3. *Suppose that point x^* is a solution to problem (8). Suppose that $\varphi, g_i \in C^3(\mathbb{R}^n)$, $i = 1, \ldots, m$, the RC and sufficient optimality conditions (12) are satisfied. Then there exists a sufficiently small neighborhood $U_\varepsilon(w^*)$ of the Kuhn-Tucker point $w^* = [x^*, \lambda^*]$ such that the method (14) satisfies the estimate*

$$\|w_{k+1} - w^*\| \le \beta \|w_k - w^*\|^2,$$

where $w^0 \in U_\varepsilon(w^*)$ and $\beta > 0$ is an independent constant.

Let us illustrate the application of the described method with the following example.

Example 1. Consider the problem

$$\min_{x \in \mathbb{R}^2} x_1^2 + x_2^2 + 4x_1x_2 \tag{17}$$

under the constraints

$$-x_1 \le 0, \qquad -x_2 \le 0.$$

It is easy to check that the point $x^* = (0,0)^\top$ is a solution to the problem (17) with the corresponding Lagrange multiplier $v^* = (0,0)^\top$. In this example, the set $I_0(x^*) = \{1,2\}$, and the modified Lagrange function has the form

$$L_E(x,\lambda) = x_1^2 + x_2^2 + 4x_1 x_2 - \frac{1}{2}\lambda_1^2 x_1 - \frac{1}{2}\lambda_2^2 x_2.$$

Let $h = (0,0,1,1)^\top$, then the system (9) will be written as follows:

$$G(w) = \begin{bmatrix} 2x_1 + 4x_2 - \frac{1}{2}\lambda_1^2 \\ 2x_2 + 4x_1 - \frac{1}{2}\lambda_2^2 \\ -\lambda_1 x_1 \\ -\lambda_2 x_2 \end{bmatrix} = 0_4.$$

The Jacobi matrix of the latter system has the form

$$G'(w) = \begin{bmatrix} 2 & 4 & -\lambda_1 & 0 \\ 4 & 2 & 0 & -\lambda_2 \\ -\lambda_1 & 0 & -x_1 & 0 \\ 0 & -\lambda_2 & 0 & -x_2 \end{bmatrix}.$$

This matrix is degenerate at the point $(x_1^*, x_2^*, \lambda_1^*, \lambda_2^*)^\top = (0,0,0,0)^\top$.

However, the mapping G is 2-regular at the point $(0,0,0,0)^\top$ on the introduced element h, and the scheme of the 2-factor method is written as

$$\begin{bmatrix} 2 & 4 & -\lambda_1 - 1 & 0 \\ 4 & 2 & 0 & -\lambda_2 - 1 \\ -\lambda_1 - 1 & 0 & -x_1 & 0 \\ 0 & -\lambda_2 - 1 & 0 & -x_2 \end{bmatrix} \begin{bmatrix} \bar{x}_1 - x_1 \\ \bar{x}_2 - x_2 \\ \bar{\lambda}_1 - \lambda_1 \\ \bar{\lambda}_2 - \lambda_2 \end{bmatrix} = - \begin{bmatrix} 2x_1 + 4x_2 - \frac{1}{2}\lambda_1^2 - \lambda_1 \\ 2x_2 + 4x_1 - \frac{1}{2}\lambda_2^2 - \lambda_2 \\ -\lambda_1 x_1 - x_1 \\ -\lambda_2 x_2 - x_2 \end{bmatrix},$$

where $k = 0,1,\ldots$, $(x_1, x_2, \lambda_1, \lambda_2)^\top = ((x_1)_k, (x_2)_k, (\lambda_1)_k, (\lambda_2)_k)^\top$ and $(\bar{x}_1, \bar{x}_2, \bar{\lambda}_1, \bar{\lambda}_2)^\top = ((x_1)_{k+1}, (x_2)_{k+1}, (\lambda_1)_{k+1}, (\lambda_2)_{k+1})^\top$.

In this example, the system (13) has a non-unique solution, so there is no global convergence of the specified version of the method. However, a simplified version of the 2-factor method (6) can be applied by solving the system

$$\Psi(w) = G'(w)h = 0_4, \tag{18}$$

where $h \in \mathrm{Ker} G'(w^*)$. Due to the 2-regularity of the mapping G on the element h at the point $w^* = 0_4$, the matrix $\Psi'(w^*)$ is not degenerate and, therefore, this problem has a unique solution w^*. The exact formula (7) is valid for w^*, since the system (18) will be linear relative to w and $w^* = w_0 - [G''(w_0)h]^{-1}(G'(w_0)h)$ for any $w_0 \in \mathbb{R}^4$ or $w^* = -1/2[Qh]^{-1}(Ah)$, where

$$Qh = \begin{bmatrix} 0 & 0 & -1 & 0 \\ 0 & 0 & 0 & -1 \\ -1 & 0 & 0 & 0 \\ 0 & -1 & 0 & 0 \end{bmatrix}, \quad A = \begin{bmatrix} 2 & 4 & 0 & 0 \\ 4 & 2 & 0 & 0 \\ 0 & 0 & 0 & 0 \\ 0 & 0 & 0 & 0 \end{bmatrix}, \quad h = \begin{bmatrix} 0 \\ 0 \\ 1 \\ 1 \end{bmatrix}.$$

5   The Implicit Function Theorems in the Degenerate Case

Consider the issue of the solution $y = y(x)$ existence for equation

$$F(x, y) = 0,$$

where $F \in C^{p+1}(X \times Y, Z)$, X, Y and Z – Banach spaces, $F(x^*, y^*) = 0$ and

$$\operatorname{Im} F'_y(x^*, y^*) \neq Z.$$

Determine the p-factor-operator for mapping $F(x, y)$ in the same way as in (2), replacing $F(x)$ to $F(x, y)$, derivatives to derivatives by variable y, i.e. $F^{(k)}(x^*)$ – to $F_y^{(k)}(x^*, y^*)$, and $f_i(x)$ – to $f_i(x, y)$.

We introduce a p-order mixed operator $\Psi_p : Y \to Z$:

$$\Psi_p = \left(f'_{1y}(x^*, y^*), \frac{1}{2} f''_{2y}(x^*, y^*), \dots, \frac{1}{p!} f_{py}^{(p)}(x^*, y^*) \right)$$

so that the action of Ψ_p on an element y is defined as

$$\Psi_p[y]^p = f'_{1y}(x^*, y^*)[y] + \dots + \frac{1}{p!} f_{py}^{(p)}(x^*, y^*)[y]^p$$

and the inverse (multivalued) operator Ψ_p^{-1}:

$$\Psi_p^{-1}(z) = \left\{ h \in Y \mid f'_{1y}(x^*, y^*)[h] + \dots + \frac{1}{p!} f_{py}^{(p)}(x^*, y^*)[h]^p = z \right\}.$$

Following [3], we present the following result.

Theorem 4 (the first implicit function theorem in the degenerate case). *Let X, Y and Z be Banach spaces, $U(x^*)$, $U(y^*)$ be some sufficiently small neighborhoods of the points x^* and y^*, $B(0, r)$ is a ball of radius r centered at zero in the space Y, $F \in C^{p+1}(X \times Y)$, and $F(x^*, y^*) = 0$. Suppose the following conditions are met:*

1. degeneracy condition:

$$f_{i\underbrace{x\dots x}_{q}\underbrace{y\dots y}_{r-q}}^{(r)}(x^*, y^*) = 0, \quad r = 1, \dots, i-1, \quad q = 0, \dots, r-1, \quad i = 1, \dots, p,$$

$$f_{i\underbrace{x\dots x}_{q}\underbrace{y\dots y}_{i-q}}^{(i)}(x^*, y^*) = 0, \quad q = 1, \dots, i-1, \quad i = 1, \dots, p;$$

2. p-factor approximation condition:

$$\left\| f_i(x, y^* + y_1) - f_i(x, y^* + y_2) - \frac{1}{i!} f_i^{(i)}(x^*, y^*)[y_1]^i + \frac{1}{i!} f_i^{(i)}(x^*, y^*)[y_2]^i \right\| \leq$$

$$\leq \varepsilon \left(\|y_1\|^{i-1} + \|y_2\|^{i-1} \right) \|y_1 - y_2\|, \qquad i = 1, \dots, p$$

for all $x \in U(x^*)$, $y_1, y_2 \in B(0, r)$ and $\varepsilon > 0$ sufficiently small;

3. **Banach condition**: *for $x \in U(x^*)$ there exists $h = h(x)$ such that*

$$\Psi_p[h]^p = -F(x, y^*), \quad \|h\| \le c_1 \|F(x, y^*)\|^{1/p};$$

4. **p-regularity condition with respect to the variable y:**

$$\left\| \{\Psi_p(h/\|h\|)\}^{-1} \right\| \le c_2,$$

where c_1, c_2 are independent constants.

Then for any $\delta > 0$ there exist $\sigma > 0$, $k > 0$ and a mapping $\varphi : U(x^*, \sigma) \to U(y^*, \delta)$ such that the following relations are satisfied:

$$\varphi(x) = y^* + h(x) + \vartheta(x), \quad \varphi(x^*) = y^*, \quad h(x^*) = 0, \quad \vartheta(x^*) = 0; \quad (a)$$

$$F(x, \varphi(x)) = 0, \quad \|\vartheta(x)\| = o(\|h(x)\|); \quad (b)$$

$$\|\varphi(x) - y^*\|_Y \le k \sum_{r=1}^{p} \|f_r(x, y^*)\|_{Z_r}^{1/r} \quad \forall x \in U(x^*, \sigma). \quad (c)$$

The proof of this theorem is similar to that for the case $p = 2$ (see, for example, [3]), but using the general construction from the monograph [10]. Here $h(x)$ – "exit from singularity".

Below we need one more modification of the implicit function theorem.

Theorem 5 (the second implicit function theorem in the degenerate case). *Let $F(x, y) \in C^{p+1}(X \times Y)$, $F : X \times Y \to Z$, X, Y and Z – Banach spaces, $F(x^*, y^*) = 0$ and the condition of p-regularity with respect to the variable y on the element $h \in \bigcap_{k=1}^{p} \operatorname{Ker}^k f_k^{(k)}(x^*, y^*)$, $h = (x, 0)$, that is*

$$\left\{ f_1'(x^*, y^*) + f_2''(x^*, y^*)[h] + \cdots + f_p^{(p)}(x^*, y^*)[h]^{p-1} \right\} \cdot (\{0\} \times Y) = Z.$$

Then for $x \in U(x^*)$ there exists $y = y(x)$ such that

$$F(x, y(x)) = 0$$

and

$$\|y(x) - y^*\| \le c\|x - x^*\|.$$

Here $U(x^*)$ is a sufficiently small neighborhood of the point x^*, and $c > 0$ is an independent constant.

Here element h is "exit from singularity". The proof is similar to the proof of the implicit function theorem in [10], but using the contraction map.

$$\Phi(y) = y - \left\{ f_1'(x^*, y^*) + \cdots + \frac{1}{(p-1)!} f_p^{(p)}(x^*, y^*)[h]^{p-1} \right\}_Y^{-1} \cdot F(x, y^* + y).$$

6 p-Factor Approach for Solving Degenerate Nonlinear Ordinary Differential Equations

Let us show how the apparatus of p-factor operators (or simple factor operators) can be applied to analyze and solve degenerate ordinary differential equations (ODE).

Consider an ODE of the form

$$\ddot{y}(t) + y(t) + g(y(t)) = x(t) \tag{19}$$

under the condition:

$$y(0) = y(\pi) = 0, \qquad x(0) = x(\pi) = 0. \tag{20}$$

We assume that $y(\cdot) \in C^2[0, \pi]$, $x(\cdot) \in C[0, \pi]$ and $g(\cdot) \in C^{p+1}(C^2[0, \pi])$. Moreover, we assume that

$$g(0) = g'(0) = 0. \tag{21}$$

Let us investigate the question of the solution existence of the Eq. (19). Let us introduce some notation and definitions. We define a mapping

$$F(x, y) = \ddot{y} + y + g(y) - x,$$

where $F : X \times Y \to Z$, $F \in C^{p+1}(X \times Y)$, $X = \{x \in C[0, \pi] \mid x(0) = x(\pi) = 0\}$, $Y = \{y \in C^2[0, \pi] \mid y(0) = y(\pi) = 0\}$, $Z = \{z \in C[0, 1] \mid z(0) = z(\pi) = 0\}$, and $g \in C^{p+1}(Y)$. Then the Eq. (19) can be rewritten as

$$F(x, y) = 0.$$

Without loss of generality, we assume that $x^*(t) = 0$ and $y^*(t) = 0$.

Moreover, the operator $F'_y(0, 0) = (\ddot{\cdot}) + (\cdot) + g'(0)$ is degenerate at the point $(0, 0)$ (which is why the Eq. (19) is degenerate). Indeed, the operator $F'_y(0, 0)$:

$$F'_y(0, 0)y = \ddot{y} + y$$

is not surjective, since for $z(t) = \sin t$ boundary value problem

$$\ddot{y}(t) + y(t) = \sin t, \qquad y(0) = y(\pi) = 0$$

has no solution [11]. According to the Sturm-Liouville theory, the boundary value problem

$$\ddot{y}(t) + y(t) = x(t), \qquad y(0) = y(\pi) = 0$$

has a solution only if the condition holds [12]

$$\int_0^\pi \sin \tau x(\tau) d\tau = 0.$$

Therefore, the question of the solution existence of the degenerate equation

$$F(x, y) = \ddot{y} + y + g(y) - x = 0, \qquad y(0) = y(\pi) = 0$$

cannot be investigated using the classical implicit function theorem. Let us show that for $F(x, y)$ we can apply Theorem 4 under appropriate assumptions on $g(\cdot)$ and $x(t)$. In our case

$$Z_1 = \operatorname{Im} F'_y(0, 0) = \left\{ z(\cdot) \in Z \mid \int_0^\pi \varphi(\tau) z(\tau) d\tau = 0 \right\} \neq Z, \qquad \varphi(t) = \sin t.$$

The subspace $W_2 = \operatorname{span}\{\varphi(t)\}$, therefore, according to [3], the projector P_{W_2} is introduced as follows:

$$P_{W_2} z = \frac{2}{\pi} \varphi(t) \int_0^\pi \varphi(\tau) z(\tau) d\tau, \quad z \in Z,$$

and

$$Z_2 = \operatorname{span}\left(\operatorname{Im} P_{W_2} F''_y(0, 0)[\cdot]^2\right)$$
$$= \operatorname{span}\left(z(t) \mid \exists y \in Y : z(t) = \tfrac{2}{\pi} \sin t \int_0^\pi \sin \tau g''(0)[y(\tau)]^2 d\tau\right). \tag{22}$$

The rest of the constructions – the subspaces $Z_3, \ldots, Z_p$ and the mappings $f_2(x, y), \ldots, \ldots, f_p(x, y)$ – are introduced similarly and depend only on the properties of the mapping $g(y)$. According to (21), $g'(0) = 0$ and also

$$F''_y(0, 0) = g''(0), \ldots, F_y^{(p)}(0, 0) = g^{(p)}(0).$$

Therefore, the p-factor operator has the form

$$\Psi_p(h) = (\ddot{\cdot}) + (\cdot) + \frac{1}{2} P_{Z_2} g''(0)[h] + \cdots + \frac{1}{p!} P_{Z_p} g^{(p)}(0)[h]^{p-1}.$$

Let's return to Theorem 4. Recall that in our consideration $x^*(t) = 0$, $y^*(t) = 0$. Condition 1 of Theorem 4 is fulfilled in accordance with the construction of the mappings $f_i(x, y)$ and the properties of the mappings $g(y)$. Condition 2 – p-factor-approximation – depends only on the properties of the mapping $g(y)$, i.e. if a

$$P_{Z_k}\left[\left(g(y_1) - g(y_2) - \frac{1}{k!} g^{(k)}(0)[y_1]^k + \frac{1}{k!} g^{(k)}(0)[y_2]^k\right)\right] \le \varepsilon\left(\|y_1\|_{k-1} + \|y_2\|^{k-1}\right) \|y_1 - y_2\|, \tag{23}$$

where $\varepsilon > 0$ is sufficiently small, $y_1, y_2 \in U_Y(0)$, then Condition 2 is satisfied.

Conditions 3 and 4, respectively, are equivalent to the existence of $h(t)$ such that the following relations hold

$$\ddot{h}(t) + h(t) + P_{Z_2} g''(0) h^2 + \cdots + P_{Z_p} g^{(p)}(0)[h]^p = x(t), \tag{24}$$

$$\|h(t)\| \leq c_1 \|F(x,0)\|^{1/p}, \quad x \in U_x(0);$$

$$\left\| \left\{ (\ddot{\cdot}) + (\cdot) + P_{Z_2} g''(0)[\bar{h}] + \cdots + P_{Z_p} g^{(p)}(0)[\bar{h}]^{p-1} \right\}^{-1} \right\| \leq c_2, \quad x \in U_x(0), \quad (25)$$

where $\bar{h}(t) = h(t)/\|h(t)\|$, $c_1 > 0$, $c_2 > 0$ are independent constants and $h(t)$ is "exit form singularity". Thus, we can finally formulate the following theorem.

Theorem 6. *Let for the boundary value problem (19)–(20) there exist sufficiently small neighborhoods $U_x(0)$ and $U_y(0)$ such that the conditions (23)–(25) hold. Then for $x(t) \in U_x(0)$ there exists a solution $y = y(x,t)$, and*

$$\|y(x,t)\| \leq m\|x(t)\|^{1/p},$$

where $m > 0$ is an independent constant.

To illustrate the theorem, consider the equation

$$\ddot{y}(t) + y(t) + y^2(t) = v \sin t, \qquad y(0) = y(\pi) = 0, \tag{26}$$

often found in technical applications. Here $g(y) = y^2$, $x(t) = v \sin t$, $F(x,y) = \ddot{y} + y + y^2 - v \sin t$, $F : X \times Y \to Z$, X, Y and Z were defined above. It is easy to check that the conditions of Theorem 6 are satisfied for the mapping $F(x,y)$ for $v \geq 0$ sufficiently small and $p = 2$. Thus, there exists a solution $y(x)$ of the Eq. (26) for $v \geq 0$, and

$$\|y(t)\| \leq m\|v \sin t\|^{1/2} \leq cv^{1/2}.$$

Similarly, one can illustrate the application of Theorem 5 to equations with a small parameter ε of the form

$$\varepsilon \dot{y}(t) + g(y, \varepsilon) = 0,$$

where $g(0,0) = g'(0,0) = g''(0,0) = 0$ and $y(0) = y_0$.

Put $F(\varepsilon, y) = \varepsilon \dot{y}(t) + g(y, \varepsilon)$, here the role of x will be played by ε. For example, if $g(y, \varepsilon) = y^3 + \varepsilon^3$, then for the equation $\varepsilon \dot{y}(t) + y^3(t) + \varepsilon^3 = 0$ it can be shown that $h = (\varepsilon, 0)^\top$ is "exit from singularity" and the conditions of Theorem 5 are satisfied. Therefore, there is a solution $y = y(\varepsilon, t)$ such that

$$\|y(\varepsilon, t)\| \leq c\varepsilon, \quad y(\varepsilon, 0) = y_0,$$

where $c > 0$ is a constant, $t \in U(0)$ is a small neighborhood of the point 0.

7 p-Regularity Theory and the Small Poincaré Parameter Method

When solving the equation

$$\ddot{y}(t) + a^2 y(t) + \mu y^2(t) = \sin t, \qquad y(0) = y(\pi) = 0, \tag{27}$$

where $a \neq 0$, $\mu > 0$ is a small parameter, the most popular is the search for a solution in the form of a series [11]

$$y(t) = y_0(t) + \mu y_1(t) + \mu^2 y_2(t) + \dots. \tag{28}$$

Substituting (28) in (27) and equating the coefficients at the same powers of mu, we obtain the first equation to determine $y_0(t)$

$$\ddot{y}_0(t) + a^2 y_0(t) = \sin t, \qquad y_0(0) = y_0(\pi) = 0, \tag{29}$$

from where

$$y_0(t) = \frac{c \sin t}{a^2 - 1}.$$

For $a^2 = 1$ this technique is inapplicable, because Eq. (29) has no solution. However, the original Eq. (27) has a solution for $a^2 = 1$ and $\mu > 0$ small. Note that problem (27) is equivalent to the following system:

$$\begin{aligned} P_1(y''(t) + y(t) + \mu y^2(t)) &= P_1(\sin t), \\ P_2(y''(t) + y(t) + \mu y^2(t)) &= P_2(\sin t), \end{aligned} \qquad y(0) = y(\pi) = 0, \tag{30}$$

where P_i is a projector on Z_i, $i = 1, 2$.

We are looking for a solution $y(t)$ in the form:

$$y(t) = h(t) + y_0(t) + \mu^{1/2} y_1(t) + \mu y_2(t) + \mu^{3/2} y_3(t) + \dots, \tag{31}$$

where $y_i(t)$, $i = 0, 1 \dots$ are defined as

$$y_i(t) = \tilde{y}_i(t) + \hat{y}_i(t), \qquad P_1(\hat{y}_i) = 0,$$

and $h(t)$ is "exit from singularity" and is calculated as a solution to the equation

$$P_2(\mu h^2(t)) = \sin t.$$

or

$$\frac{2\mu \sin t}{\pi} \int_0^{\pi} \sin \tau h^2(\tau) d\tau = \sin t$$

Solving this equation, we get:

$$h(t) = \sqrt{\frac{3\pi}{8\mu}} \sin t.$$

Substituting (31) into the first equation of the system (30) and equating the free terms, we obtain the following problem to determine $\tilde{y}_0$:

$$\tilde{y}_0''(t) + \tilde{y}_0(t) + \frac{3\pi}{8} \sin^2 t - \sin t = 0, \quad \tilde{y}_0(0) = \tilde{y}_0(\pi) = 0.$$

Solving, we find

$$\tilde{y}_0(t) = \frac{\pi}{4} \cos t + \sin t - \frac{t}{2} \cos t - \frac{3\pi}{16} - \frac{\pi}{16} \cos 2t. \tag{32}$$

Substitution (31) into the second equation of the system (30) and comparing the coefficients of $\mu^{1/2}$, we obtain the equation for determining the function $\hat{y}$:

$$0 = \int_0^\pi \sin(\tau)(\mu h(\tau)\tilde{y}_0(\tau) + \mu h(\tau)\hat{y}_0(\tau))d\tau$$
$$= \int_0^\pi \sin^2(\tau)(\tilde{y}_0(\tau) + \hat{y}_0(\tau))d\tau$$

We are looking for the function $\hat{y}$ in the form $\hat{y} = A\sin t$ where A is a constant defined as

$$A\int_0^\pi \sin^2(\tau)\sin(\tau)d\tau = -\int_0^\pi \sin^2(\tau)(\tilde{y}_0(\tau))d\tau$$

Substitution $\tilde{y}_0(\tau)$ defined in (32) and integrating the resulting expression, we find A and, therefore, $\hat{y}$.

Acting sequentially, we substitute $y(t)$ with the found components in the first, then in the second equation of the system (30), comparing the coefficients at the same powers of μ to find the next component in expansion of the function $y(t)$.

Finally, we get the new formula for the solution:

$$y(t) = \frac{\sqrt{3\pi}}{2\sqrt{2}\mu^{1/2}}\sin t + \tilde{y}_0(t) + \hat{y}_0(t) + \mu^{1/2}(\tilde{y}_1(t) + \hat{y}_1(t)) + \dots . \tag{33}$$

For an approximate construction of a solution to the equation (27) with $a^2 = 1$, one can also apply the 2-factor method, the scheme of which has the form

$$y_{k+1} = y_k - [F'_y(0,0) + PF''_{yy}(0,0)h]^{-1}F(x, h + y_k), \quad k = 0, 1, \dots , \tag{34}$$

where

$$y_0 = 0, \ h(t) = \sqrt{\frac{3\pi}{8\mu}}\sin t,$$

$$F(y) = \ddot{y}(t) + y(t) + \mu y^2(t) - \sin t, \quad P(\cdot) = \frac{2\sin t}{\pi}\int_0^\pi \sin\tau(\cdot)d\tau.$$

The scheme (34) is equivalent, for example, for $k = 0$, to the equation

$$\ddot{y}_1(t) + y_1(t) + \frac{\sqrt{3\mu}\sin t}{\sqrt{2\pi}}\int_0^\pi \sin^2\tau y_1(\tau)d\tau = \sin t - \frac{3\pi}{8}\sin^2 t.$$

Obviously, the last equation has a solution. The process will converge for any sufficiently small $\mu > 0$ due to the properties of the 2-factor method and the specificity of the mapping F.

References

1. Brezhneva, O.A., Tretyakov, A.A.: Methods for solving essentially nonlinear problems. CCAS of RAS, Moscow (2000).(in Russian)
2. Tret'yakov, A.A., Marsden, J.E.: Factor-analysis of nonlinear mapping: p-regularity Theory. Commun. Pure Appl. Anal. **2**, 425–445 (2003)
3. Izmailov, A.F., Tretyakov, A.A.: 2-regular solutions to nonlinear problems. Fizmatgiz, Moscow (1999). (in Russian)
4. Evtushenko, Yu.: Generalized Lagrange multiplier technique for nonlinear programming. J. Optim. Theory Appl. **21**(2), 121–135 (1977)
5. Tret'yakov, A.A.: The implicit function theorem in degenerate problems. Russ. Math. Surv. **42**, 179–180 (1987)
6. Belash, K.N., Tretyakov, A.A.: Methods for solving degenerate problems. USSR Comput. Math. Math. Phys. **28**(4), 90–94 (1988)
7. Tretyakov, A.A.: Structures of nonlinear degenerate mappings and their application to construction numerical methods. Doct. diss, Moscow (1987). (in Russian)
8. Mangasarian, O.L.: A Newton method for linear programming. J. Optimiz. Theory Appl. **121**, 1–18 (2004)
9. Golikov, A.I., Evtushenko, Y.G., Mollaverdi, N.: Application of Newton's method for solving large linear programming problems. Comput. Math. Math. Phys. **44**, 1484–1493 (2004)
10. Izmailov, A.F., Tretyakov, A.A.: Factor Analysis of Nonlinear Mappings. Nauka, Moscow (1994).(in Russian)
11. El'sgol'ts, L.E.: Ordinary Dierential Equations. Gostekhteorizdat, Moscow (1950). (in Russian)
12. Kamke, E.: Gewöhnliche Differentialgleichungen. Akademie-Verlag, Leipzig, Nauka, Moscow (1971)

Adaptive Catalyst for Smooth Convex Optimization

Anastasiya Ivanova[1,2], Dmitry Pasechnyuk[1(✉)], Dmitry Grishchenko[3], Egor Shulgin[1,4], Alexander Gasnikov[1,2], and Vladislav Matyukhin[1]

[1] Moscow Institute of Physics and Technology, Moscow, Russia
[2] National Research University Higher School of Economics, Moscow, Russia
[3] Université Grenoble Alpes, Grenoble, France
[4] King Abdullah University of Science and Technology, Thuwal, Saudi Arabia

Abstract. In this paper, we present a generic framework that allows accelerating almost arbitrary non-accelerated deterministic and randomized algorithms for smooth convex optimization problems. The major approach of our *envelope* is the same as in *Catalyst* [37]: an accelerated proximal outer gradient method, which is used as an envelope for a non-accelerated inner method for the ℓ_2 regularized auxiliary problem. Our algorithm has two key differences: 1) easily verifiable stopping condition for inner algorithm; 2) the regularization parameter can be tuned along the way. As a result, the main contribution of our work is a new framework that applies to adaptive inner algorithms: Steepest Descent, Adaptive Coordinate Descent, Alternating Minimization. Moreover, in the non-adaptive case, our approach allows obtaining Catalyst without a logarithmic factor, which appears in the standard Catalyst [37,38].

Keywords: Adaptive methods · Catalyst · Accelerated methods · Steepest descent · Coordinate descent · Alternating minimization · Distributed methods · Stochastic methods

1 Introduction

One of the main achievements in numerical methods for convex optimization is the development of accelerated methods [43]. Until 2015 acceleration schemes for different convex optimization problems seem to be quite different to unify them. But starting from the work [37] in which universal acceleration technique (*Catalyst*) was proposed, there appears a series of subsequent works [34,38,46,47] that allows generalizing Catalyst on monotone variational inequalities, non-convex problems, stochastic optimization problems. In all these works, the basic idea is to use an accelerated proximal algorithm as an outer envelope [51] with non-accelerated algorithms for inner auxiliary problems. The main practical drawback of this approach is the requirement to choose a regularization parameter

The research is supported by the Ministry of Science and Higher Education of the Russian Federation (Goszadaniye) №075-00337-20-03, project No. 0714-2020-0005.

N. N. Olenev et al. (Eds.): OPTIMA 2021, LNCS 13078, pp. 20–37, 2021.
https://doi.org/10.1007/978-3-030-91059-4_2

such that the conditional number of the auxiliary problem becomes $O(1)$. To do that, we need to know the smoothness parameters of the target that are not typically free available.

An alternative accelerated proximal envelope [48] was proposed in the paper [41]. The main difference with standard accelerated proximal envelops is the adaptability of the scheme [41]. Note, that this scheme also allows one to construct (near) optimal tensor (high-order) accelerated methods [18,19,21,43,56]. That is, the "acceleration" potential of this scheme seems to be the best known for us for the moment. So the main and rather simple idea of this paper can be formulated briefly as follows: **To develop adaptive Catalyst, we replace the accelerated proximal envelope with a fixed regularization parameter [38,48] on the adaptive accelerated proximal envelope from [41].**

In Sect. 2, we describe adaptive Catalyst envelope – Algorithm 1 and generalized Monteiro–Svaiter theorem from [41] to set out how to make this envelope work. We emphasize that the proof of the theorem contains, as a byproduct, the new theoretical analysis of the stopping condition for the inner algorithm (9). This stopping condition allows one to show that the proposed envelope in a non-adaptive mode is log-times better (see Corollary 1) in the total number of oracle calls of the inner method (here we measure the complexity of the envelope in such terms) in comparison with all other envelopes known for us.

By using this adaptive accelerated proximal envelope, we propose in Sect. 3 an accelerated variant of steepest descent [18,50] as an alternative to A. Nemirovski accelerated steepest descent (see [7,44] and references therein), adaptive accelerated variants of alternating minimization procedures [3] as an alternative to [6,25,55] and adaptive accelerated coordinate descent [42]. For the last example, as far as we know, there were no previously complete adaptive accelerated coordinate descent. The most advanced result in this direction is the work [15] that applies only to the problems with increasing smoothness parameter along the iteration process. For example, for the target function like $f(x) = x^4$, this scheme does not recognize that smoothness parameters (in particular Lipschitz gradient constant) tend to zero along the iteration process.

We expect that the proposed approach allows accelerating not only procedures that are adaptive in themselves, but also many other different non-accelerated non-adaptive randomized schemes by settings on general smoothness parameters of target function that can be difficult to analyze patently [22–24].

The first draft of this paper appeared in arXiv in November 2019. Since that time, this paper has developed (and cited) in different aspects. The main direction is a convenient (from the practical (6) and theoretical (9) point of view) condition to stop the inner algorithm that is wrapped in accelerated proximal envelope. We emphasize that our contribution in this part is not a new accelerated proximal envelope (we use the well-known envelope [41]), but we indicate that this envelope is better than the other ones due to the new theoretical analysis of its inner stopping condition that lead us from (6) to (9). Although this calculation looks simple enough, to the best of our knowledge, this was the first time when it was provably developed an accelerated proximal envelope that required to solve the auxiliary problem with prescribed relative accuracy in argument $\simeq 1/5$. Since the auxiliary problem is smooth and strongly convex, this observation eliminates the logarithmic factor (in the desired accuracy) in the complexity estimate for such an envelope in comparison with all known analogues. Note that this small observation will have a remarkable influence on the development of accelerated algorithms. A close stopping condition, for instance, arises in the following papers [8,9] that developed (sub-)optimal accelerated tensor method based on accelerated proximal envelopes. The proposed "logarithm-free" envelope allows one to improve the best known bounds [39] for strongly convex-concave saddle-point problems (with different constants of strong convexity and concavity) on logarithmic factor [11]. Composite variant of this envelope also allows one to develop "logarithm-free" gradient sliding-type methods[1] [11,12,27] and its tensor generalizations [29]. Moreover, one of the variants of the hyper-fast second-order method was also developed based on this envelope [30]. Though this envelope had known before, it seems that the original idea of this paper to use this envelope in Catalyst type procedures and new (important from the theoretical point of view) reformulation of stopping condition for inner algorithm (9) has generated a large number of applications, some of them mentioned in this paper, the others can be found in the literature cited above. As an important example, we show in Sect. 3.2 that the developed envelop e with non-accelerated coordinate descent method for auxiliary problem works much better in theory (and better in practice) than all known direct accelerated coordinate-descent algorithms for sparse soft-max type problem. Before this article, this was an open problem, how to beat standard accelerated coordinate-descent algorithms that do not allow one to take into account sparsity of the problem for soft-max type functional [16,49].

The other contribution of this paper is an adaptive choice of the smooth parameter. Since our approach requires two inputs (lower bound L_d and upper bound L_u for the unknown smoothness parameter L), it is hardly possible to call it "adaptive". Moreover, the greater the discrepancy between these two parameters, the worthier is our adaptive envelope in theory. But almost all of our experiments demonstrate low sensitivity to these parameters rather than to real smoothness parameter. But even for such a "logarithm-free" and adaptive envelope, we expect that typically the direct adaptive accelerated procedures will work better than its Catalyst type analogues. It was recently demonstrated in the following work [54] for accelerated alternating minimization procedure. But even to date, there are problems in which one can expect that firstly optimal accelerated algorithms will be developed by using Catalyst type procedures rather than direct acceleration. Recent advances in saddle-point problems [11,39,59] and decentralized distributed optimization[2] [26,35] confirm this thought. We expect that for Homogeneous Federated Learning architectures, Accelerated Local SGD

---

[1] Note, that [12] contains variance reduction [1,52] generalization (with non proximal-friendly composite) of proposed in this paper scheme.

[2] Note, that the results of these papers were further reopened by using direct acceleration [33,36].

can be developed (see [57] for the state of the art approach) by using Catalyst-type envelop e with SCAFFOLD version of local SGD algorithm [32]. As far as we know, it is still an open problem to build Accelerated local SGD as an inner algorithm. In this paper, we demonstrate some optimistic experiments in this direction.

2   The Main Scheme

Let us consider the following minimization problem

$$\min_{y \in \mathbb{R}^n} f(y), \tag{1}$$

where f is a convex function, and its gradient is Lipschitz continuous w.r.t. $\|\cdot\|_2$ with the constant L_f:

$$\|\nabla f(x) - \nabla f(y)\|_2 \leq L_f \|x - y\|_2.$$

We denote $x_\star$ a solution of (1).

To propose the main scheme of the algorithm we need to define the following functions:

$$F_{L,x}(y) = f(y) + \tfrac{L}{2}\|y - x\|_2^2,$$
$$f_L(x) = \min_{y \in \mathbb{R}^n} F_{L,x}(y) = F_{L,x}(y_L(x)),$$

then the function $F_{L,x}$ is L-strongly convex, and its gradient is Lipschitz continuous w.r.t. $\|\cdot\|_2$ with the constant $(L + L_f)$. So, the following inequality holds

$$\|\nabla F_{L,x}(y_2) - \nabla F_{L,x}(y_1)\|_2 \leq (L + L_f)\|y_1 - y_2\|_2. \tag{2}$$

Due to this definition, for all $L \geq 0$ we have that $f_L(x) \leq f(x)$ and the convex function f_L has a Lipschitz-continuous gradient with the Lipschitz constant L. Moreover, according to [50] [Theorem 5, ch. 6], since

$$x_\star \in \operatorname*{Argmin}_{x \in \mathbb{R}^n} f_L(x),$$

we obtain

$$x_\star \in \operatorname*{Argmin}_{x \in \mathbb{R}^n} f(x) \quad \text{and} \quad f_L(x_\star) = f(x_\star).$$

Thus, instead of the initial problem (1), we can consider the Moreau–Yosida regularized problem

$$\min_{x \in \mathbb{R}^n} f_L(x). \tag{3}$$

Note that the problem (3) is an ordinary problem of smooth convex optimization. Then the complexity of solving the problem (3) up to the accuracy ε with respect to the function using the Fast Gradient Method (FGM) [43] can be estimated

---

Algorithm 1. Monteiro–Svaiter algorithm

Parameters: $z^0, y^0, A_0 = 0$
for $k = 0, 1, \ldots, N - 1$ **do**
  Choose L_{k+1} and y^{k+1} such that

$$\|\nabla F_{L_{k+1}, x^{k+1}}(y^{k+1})\|_2 \leq \tfrac{L_{k+1}}{2}\|y^{k+1} - x^{k+1}\|_2,$$

  where

$$a_{k+1} = \frac{1/L_{k+1} + \sqrt{1/L_{k+1}^2 + 4A_k/L_{k+1}}}{2},$$
$$A_{k+1} = A_k + a_{k+1},$$
$$x^{k+1} = \tfrac{A_k}{A_{k+1}}y^k + \tfrac{a_{k+1}}{A_{k+1}}z^k$$

$$z^{k+1} = z^k - a_{k+1}\nabla f\left(y^{k+1}\right)$$
end for
Output: y^N

---

as follows $O\left(\sqrt{\frac{LR^2}{\varepsilon}}\right)$. The 'complexity' means here the number of oracle calls. Each oracle call means calculation of $\nabla f_L(x) = L(x - y_L(x))$, where $y_L(x)$ is the exact solution of the auxiliary problem $\min_{y \in \mathbb{R}^n} F_{L,x}(y)$.

Note that the smaller the value of the parameter L we choose, the smaller is the number of oracle calls (outer iterations). However, at the same time, this increases the complexity of solving the auxiliary problem at each iteration.

At the end of this brief introduction to standard accelerated proximal point methods, let us describe the step of ordinary (proximal) gradient descent (for more details see [48])

$$x^{k+1} = x^k - \tfrac{1}{L}\nabla f_L(x^k) = x^k - \tfrac{L}{L}(x^k - y_L(x^k)) = y_L(x^k).$$

To develop an adaptive proximal accelerated envelope, we should replace standard FGM [43] on the following adaptive variant of FGM Algorithm 1, introduced by [41] for smooth convex optimization problems.

The analysis of the algorithm is based on the following theorem.

Theorem 1. (Theorem 3.6 [41]). *Let sequence* (x^k, y^k, z^k), $k \geq 0$ *be generated by Algorithm 1 and define* $R := \|y^0 - x_\star\|_2$. *Then, for all* $N \geq 0$,

$$\tfrac{1}{2}\|z^N - x_\star\|_2^2 + A_N \cdot (f(y^N) - f(x_\star)) + \tfrac{1}{4}\sum_{k=1}^{N} A_k L_k \|y^k - x^k\|_2^2 \leq \tfrac{R^2}{2},$$

$$f(y^N) - f(x^\star) \leq \tfrac{R^2}{2A_N}, \quad \|z^N - x_\star\|_2 \leq R, \tag{4}$$

$$\sum_{k=1}^{N} A_k L_k \|y^k - x^k\|_2^2 \leq 2R^2.$$

We also need the following Lemma.

Lemma 1 (Lemma 3.7a [41]). *Let sequences $\{A_k\}, \{L_k\}$, $k \geq 0$ be generated by Algorithm 1. Then, for all $N \geq 0$,*

$$A_N \geq \tfrac{1}{4} \left(\sum_{k=1}^{N} \tfrac{1}{\sqrt{L_k}} \right)^2. \tag{5}$$

Let us define non-accelerated method $\mathcal{M}$ that we will use to solve auxiliary problem.

Assumption 1. *The convergence rate (after t iterations/oracle calls) for the method $\mathcal{M}$ for problem*

$$\min_{y \in \mathbb{R}^n} F(y)$$

can be written in the general form as follows: with probability at least $1 - \delta$ holds (for randomized algorithms, like Algorithm 4, this estimates holds true with high probability)[3]

$$F(y_t) - F(y_\star) = O\left(L_F R_y{}^2 \log \tfrac{t}{\delta}\right) \min\left\{\tfrac{C_n}{t}, \exp\left(-\tfrac{\mu_F t}{C_n L_F}\right)\right\},$$

where $y_\star$ is the solution of the problem, $R_y = \|y^0 - y_\star\|_2$, function F is μ_F-strongly convex and L_F is a constant which characterized smoothness of function F.

Typically, $C_n = O(1)$ for the standard full gradient first order methods, $C_n = O(p)$, where p is a number of blocks, for alternating minimization with p blocks and $C_n = O(n)$ for gradient-free or coordinate descent methods, where n is dimension of y. See the references in next Remark for details.

Remark 1. *Let us clarify what we mean by a constant L_F which characterized smoothness of function F. Typically for the first order methods this is just the Lipschitz constant of gradient F (see, [5, 50] for the steepest descent and [6, 31, 55] for alternating minimization); for gradient-free methods like Algorithm 4 this constant is the average value of the directional smoothness parameters, for gradient-free methods see [2, 10, 13, 14, 17, 53], for coordinate descent methods see [42, 45, 58] and for more general situations see [24].*

Remark 2. *Note that in Assumption 1 the first estimate corresponds to the estimate of the convergence rate of the method $\mathcal{M}$ for convex problems. And the second estimate corresponds to the estimate for strongly convex problems.*

Our main goal is to propose a scheme to accelerate methods of this type. But note that we apply our scheme only to degenerate convex problems since it does not take into account the strong convexity of the original problem.

Denote $F_{L,x}^{k+1} \equiv F_{L_{k+1}, x^{k+1}}$. Based on Monteiro–Svaiter accelerated proximal method we propose Algorithm 2.

---

[3] For deterministic algorithms we can skip "with probability at least $1 - \delta$" and factor $\log \tfrac{N}{\delta}$.

Algorithm 2. Adaptive Catalyst

Parameters: Starting point $x^0 = y^0 = z^0$; initial guess $L_0 > 0$; parameters $\alpha > \beta \gtrsim \gamma > 1$; optimization method $\mathcal{M}$, $A_0 = 0$.
for $k = 0, 1, \ldots, N - 1$ **do**
$\quad L_{k+1} = \beta \cdot \min\{\alpha L_k, L_u\}$
$\quad r = 0$
$\quad$**repeat**
$\quad\quad r := r + 1$
$\quad\quad L_{k+1} := \max\{L_{k+1}/\beta, L_d\}$
$\quad\quad$Compute

$$a_{k+1} = \frac{1/L_{k+1} + \sqrt{1/L_{k+1}^2 + 4A_k/L_{k+1}}}{2},$$
$$A_{k+1} = A_k + a_{k+1},$$
$$x^{k+1} = \frac{A_k}{A_{k+1}} y^k + \frac{a_{k+1}}{A_{k+1}} z^k.$$

Compute an approximate solution of the following problem with auxiliary non-accelerated method $\mathcal{M}$

$$y^{k+1} \approx \underset{y}{\operatorname{argmin}} \, F_{L,x}^{k+1}(y) :$$

By running $\mathcal{M}$ with starting point x^{k+1} and output point y^{k+1} we wait N_r iterations to fulfill adaptive stopping condition

$$\|\nabla F_{L,x}^{k+1}(y^{k+1})\|_2 \leq \frac{L_{k+1}}{2}\|y^{k+1} - x^{k+1}\|_2.$$

$\quad$**until** $r > 1$ and $N_r \geq \gamma \cdot N_{r-1}$ or $L_{k+1} = L_d$
$\quad z^{k+1} = z^k - a_{k+1}\nabla f\left(y^{k+1}\right)$
end for
Output: y^N

Now let us prove the main theorem about the convergence rate of the proposed scheme. Taking into account that $\tilde{O}(\cdot)$ means the same as $O(\cdot)$ up to a logarithmic factor, based on the Monteiro–Svaiter Theorem 1 we can introduce the following theorem:

Theorem 2. *Consider Algorithm 2 with $0 < L_d < L_u$ for solving problem (1), where $Q = \mathbb{R}^n$, with auxiliary (inner) non-accelerated algorithm (method) $\mathcal{M}$ that satisfy Assumption 1 with constants C_n and L_f such that $L_d \leq L_f \leq L_u$.*
Then the total complexity[4] of the proposed Algorithm 2 with inner method $\mathcal{M}$ is

$$\tilde{O}\left(C_n \cdot \max\left\{\sqrt{\frac{L_u}{L_f}}, \sqrt{\frac{L_f}{L_d}}\right\} \cdot \sqrt{\frac{L_f R^2}{\varepsilon}}\right)$$

with probability at least $1 - \delta$.

---

[4] The number of oracle calls (iterations) of auxiliary method $\mathcal{M}$ that required to find ε solution of (1) in terms of functions value.

Proof. Note that the Monteiro–Svaiter (M-S) condition

$$||\nabla F_{L,x}^{k+1}(y^{k+1})||_2 \leq \frac{L_{k+1}}{2}||y^{k+1} - x^{k+1}||_2 \qquad (6)$$

instead of the exact solution $y_\star^{k+1} = y_{L_{k+1}}(x^{k+1})$ of the auxiliary problem, for which

$$||\nabla F_{L,x}^{k+1}(y_\star^{k+1})||_2 = 0,$$

allows using the inexact solution that satisfies the condition (6).

Since $y_\star^{k+1}$ is the solution of the problem $\min_y F_{L,x}^{k+1}(y)$, the $\nabla F_{L,x}^{k+1}(y_\star^{k+1}) = 0$. Then, using inequality (2) we obtain

$$||\nabla F_{L,x}^{k+1}(y^{k+1})||_2 \leq (L_{k+1} + L_f)||y^{k+1} - y_\star^{k+1}||_2. \qquad (7)$$

Using the triangle inequality we have

$$||x^{k+1} - y_\star^{k+1}||_2 - ||y^{k+1} - y_\star^{k+1}||_2 \leq ||y^{k+1} - x^{k+1}||_2. \qquad (8)$$

Since r.h.s. of the inequality (8) coincide with the r.h.s. of the M-S condition and l.h.s. of the inequality (7) coincide with the l.h.s. of the M-S condition up to a multiplicative factor $L_{k+1}/2$, one can conclude that if the inequality

$$||y^{k+1} - y_\star^{k+1}||_2 \leq \frac{L_{k+1}}{3L_{k+1}+2L_f}||x^{k+1} - y_\star^{k+1}||_2 \qquad (9)$$

holds, the M-S condition holds too.

To solve the auxiliary problem $\min_y F_{L_{k+1},x^{k+1}}(y)$ we use non-accelerated method $\mathcal{M}$. Using Assumption 1 with probability $\geq 1 - \frac{\delta}{N}$ (where N is the total number of the Catalyst's steps), we obtain that the convergence rate (after t iterations of $\mathcal{M}$, see Assumption 1)

$$F_{L,x}^{k+1}(y_t^{k+1}) - F_{L,x}^{k+1}(y_\star^{k+1}) = O\left((L_f + L_{k+1})R_{k+1}^2 \log \frac{Nt}{\delta}\right) \exp\left(-\frac{L_{k+1}t}{C_n(L_f+L_{k+1})}\right).$$

Note, that $R_{k+1} = ||x^{k+1} - y_\star^{k+1}||_2$ since x^{k+1} is a starting point.

Since $F_{L,x}^{k+1}$ is L_{k+1}-strongly convex function, the following inequality holds [43]

$$\frac{L_{k+1}}{2}||y_t^{k+1} - y_\star^{k+1}||_2^2 \leq F_{L,x}^{k+1}(y_t^{k+1}) - F_{L,x}^{k+1}(y_\star^{k+1}).$$

Thus,

$$||y_t^{k+1} - y_\star^{k+1}||_2 \leq O\left(\sqrt{\frac{(L_f+L_{k+1})R_{k+1}^2}{L_{k+1}} \log \frac{Nt}{\delta}}\right) \exp\left(-\frac{L_{k+1}t}{2C_n(L_f+L_{k+1})}\right). \qquad (10)$$

From (9), (10) and the fact that we start $\mathcal{M}$ at x^{k+1}, we obtain that the complexity T (number of iterations of $\mathcal{M}$) of solving the auxiliary problem with probability at least $1 - \frac{\delta}{N}$ is determined from

$$O\left(R_{k+1}\sqrt{\frac{(L_f+L_{k+1})}{L_{k+1}}} \log \frac{NT}{\delta}\right) \exp\left(-\frac{L_{k+1}T}{2C_n(L_f+L_{k+1})}\right) \simeq \frac{L_{k+1}}{3L_{k+1}+2L_f}R_{k+1}, \qquad (11)$$

hence

$$T = \tilde{O}\left(C_n \frac{(L_{k+1} + L_f)}{L_{k+1}}\right). \tag{12}$$

Since we use in (12) $\tilde{O}(\cdot)$ notation, we can consider T to be the estimate that corresponds to the total complexity of auxiliary problem including all inner restarts on L_{k+1}.

Substituting inequality (5) into estimation (4) we obtain

$$f(y^N) - f(x_\star) \leq \frac{2R^2}{\left(\sum\limits_{k=1}^{N} \frac{1}{\sqrt{L_k}}\right)^2}.$$

Since the complexity of the auxiliary problem with probability at least $1 - \frac{\delta}{N}$ is T we assume that in the worst case all L_{k+1} are equal. Then the worst case we can estimate as the following optimization problem

$$\max_{L_d \leq L \leq L_u} \frac{L + L_f}{L} \sqrt{\frac{LR^2}{\varepsilon}},$$

Obviously, the maximum is reached at the border. So, using union bounds inequality over all N iterations of the Catalyst we can estimate the complexity in the worst two cases as follows:

- If all $L_{k+1} = L_d \leq L_f$ (at each iteration we estimate the regularization parameter as lower bound), then $\frac{(L_{k+1} + L_f)}{L_{k+1}} \approx \frac{L_f}{L_{k+1}}$ and total complexity with probability $\geq 1 - \delta$ is

$$\tilde{O}\left(C_n \frac{L_f}{L_d} \sqrt{\frac{L_d R^2}{\varepsilon}}\right) = \tilde{O}\left(C_n \sqrt{\frac{L_f}{L_d}} \cdot \sqrt{\frac{L_f R^2}{\varepsilon}}\right).$$

- If all $L_{k+1} = L_u \geq L_f$ (at each iteration we estimate the regularization parameter as upper bound), then $\frac{(L_{k+1} + L_f)}{L_{k+1}} \approx 1$ and total complexity with probability $\geq 1 - \delta$ is

$$\tilde{O}\left(C_n \sqrt{\frac{L_u R^2}{\varepsilon}}\right) = \tilde{O}\left(C_n \sqrt{\frac{L_u}{L_f}} \cdot \sqrt{\frac{L_f R^2}{\varepsilon}}\right).$$

Then, using these two estimations we obtain the result of the theorem.

Note that this result shows that such a procedure will works not worse than standard Catalyst [37,38] up to a factor $\tilde{O}\left(\max\left\{\sqrt{\frac{L_u}{L_f}}, \sqrt{\frac{L_f}{L_d}}\right\}\right)$ independent on the stopping condition in the restarts on L_{k+1}.

Since the complexity of solving the auxiliary problem is proportional to $\frac{(L_{k+1} + L_f)C_n}{L_{k+1}}$, when we reduce the parameter L_{k+1} so that $L_{k+1} < L_f$ the complexity of solving an auxiliary problem became growth exponentially. Therefore, as the stopping condition of the inner method, we select the number of iterations N_t compared to the number of iterations N_{t-1} at the previous restart $t-1$. This means that if $N_t \leq \gamma N_{t-1}$ then the complexity begins to grow exponentially and it is necessary to go to the next iteration of the external method. By using such adaptive rule we try to recognize the best possible value of $L_{k+1} \simeq L_f$. The last facts are basis of standard Catalyst approach [37,38] and have very simple explanation. To minimize the total complexity we should take parameter $L_{k+1} \equiv L$ such that

$$\min_L \sqrt{\frac{LR^2}{\varepsilon}} \cdot \tilde{O}\left(\frac{L_f + L}{L}\right).$$

This leads us to $L_{k+1} \simeq L_f$.

Note that also in non-adaptive case (if we choose all $L_{k+1} \equiv L_f$) we can obtain the following corollary from the Theorem 2.

Corollary 1. *If we consider Algorithm 2 with $L_{k+1} \equiv L_f$ for solving problem (1), then the total complexity of the proposed Algorithm 2 with inner non-randomized method $\mathcal{M}$ is*

$$O\left(C_n \sqrt{\frac{L_f R^2}{\varepsilon}}\right). \tag{13}$$

Proof. Using (11) without $\log \frac{NT}{\delta}$ factor (since $\mathcal{M}$ is non-randomized) we derive that the complexity of the auxiliary problem is (see also (12))

$$T = O\left(C_n \frac{(L_{k+1} + L_f)}{L_{k+1}} \cdot \log \frac{3L_{k+1} + 2L_f}{L_{k+1}}\right)$$

And since we choose $L_{k+1} \equiv L_f$,

$$\frac{3L_{k+1} + 2L_f}{L_{k+1}} = 5$$

Then the complexity of the auxiliary problem is $T = O(C_n)$. Using this estimate, we obtain that the total complexity is (13).

If method $\mathcal{M}$ is randomized we have the additional factor $\log \frac{NT}{\delta} \simeq \log \frac{1}{\delta\varepsilon}$. Hence, (13) changes: with probability at least $1 - \delta$

$$O\left(C_n \log \frac{1}{\delta\varepsilon} \cdot \sqrt{\frac{L_f R^2}{\varepsilon}}\right).$$

Note that in the standard Catalyst approach [37,38] the total complexity is $O\left(C_n \log \frac{1}{\delta\varepsilon} \cdot \sqrt{\frac{L_f R^2}{\varepsilon}} \cdot \log \frac{1}{\varepsilon'}\right)$, where $\varepsilon' = \text{Poly}(\varepsilon)$ is the relative accuracy of solving the auxiliary problem at each iteration. From this we get that choosing the stopping criterion for the inner method as the criterion from the Algorithm 2 we can get the Catalyst without a logarithmic cost $\log \frac{1}{\varepsilon'}$. It seems that such variant of Catalyst can be useful in many applications. For example, as universal envelope for non-accelerated asynchronized centralized distributed algorithms [40].

3   Applications

In this section, we present a few examples of algorithms that we consider as inner solvers. Most of them have an adaptive structure. It is natural to apply adaptive envelop e to adaptive algorithms since the developed methods keep adaptability.

3.1   Steepest Descent

Consider the following problem

$$\min_{x \in \mathbb{R}^n} f(x),$$

where f is a L_f-smooth convex function (its gradient is Lipschitz continuous w.r.t. $\| \cdot \|_2$ with the constant L_f).

To solve this problem, let us consider the general gradient descent update:

$$x^{k+1} = x^k - h_k \nabla f(x^k).$$

In [50] it was proposed an adaptive way to select h_k as following (see also [5] for precise rates of convergence)

$$h_k = \operatorname*{argmin}_{h \in \mathbb{R}} f(x^k - h \nabla f(x^k)).$$

Algorithm 3. Steepest descent

Parameters: Starting point x^0.
·**for** $k = 0, 1, \ldots, N - 1$ **do**
    Choose $h_k = \operatorname{argmin}_{h \in \mathbb{R}} f(x^k - h \nabla f(x^k))$
    Set $x^{k+1} = x^k - h_k \nabla f(x^k)$
end for
Output: x^N

In contrast with the standard selection $h_k = \frac{1}{L_f}$ for L_f-smooth functions f, in this method there is no need to know smoothness constant of the function. It allows to use this method for the smooth functions f when L_f is unknown (or expensive to compute) or when the global L_f is much bigger than the local ones along the trajectory.

On the other hand, as far as we concern, there is no direct acceleration of the steepest descent algorithm. Moreover, it is hard to use Catalyst with it as far as acceleration happens if L_k (κ in Catalyst article notations) is selected with respect to L_f and the scheme does not support adaptivity out of the box. Even if global L_f is known, the local smoothness constant could be significantly different from it that will lead to the worse speed of convergence.

Note that for Algorithm 3 the Assumption 1 holds with $C_n = O(1)$ and L_f is the Lipschitz constant of the gradient of function f.

3.2   Random Adaptive Coordinate Descent Method

Consider the following unconstrained problem

$$\min_{x \in \mathbb{R}^n} f(x).$$

Now we assume directional smoothness for f, that is there exists $\beta_1, \ldots, \beta_n$ such that for any $x \in \mathbb{R}^n, u \in \mathbb{R}$

$$|\nabla_i f(x + u e_i) - \nabla_i f(x)| \leq \beta_i |u|, \quad i = 1, \ldots, n,$$

where $\nabla_i f(x) = \partial f(x)/\partial x_i$. For twice differentiable f it is equivalent to $(\nabla^2 f(x))_{i,i} \leq \beta_i$. Due to the fact that we consider the situation when smoothness constants are not known, we use such a dynamic adjustment scheme from [42,58].

Algorithm 4. RACDM

Parameters: Starting point x^0;
lower bounds $\hat{\beta}_i := \beta_i^0 \in (0, \beta_i], i = 1, \ldots, n$
for $k = 0, 1, \ldots, N-1$ **do**
    Sample $i_k \sim \mathcal{U}[1, \ldots, n]$
    Set $x^{k+1} = x^k - \hat{\beta}_{i_k}^{-1} \cdot \nabla_{i_k} f(x^k) \cdot e_{i_k}$
    While $\nabla_{i_k} f(x^k) \cdot \nabla_{i_k} f(x^{k+1}) < 0$ **do**

$$\left\{ \hat{\beta}_{i_k} = 2\hat{\beta}_{i_k}, \quad x^{k+1} = x^k - \hat{\beta}_{i_k}^{-1} \cdot \nabla_{i_k} f(x^k) \cdot e_{i_k} \right\}$$

    Set $\beta_{i_k} = \frac{1}{2}\beta_{i_k}$
end for
Output: x^N

Note that for Algorithm 4 the Assumption 1 holds with $C_n = O(n)$ (for $x \in \mathbb{R}^n$) and[5] $L_f = \overline{L}_f := \frac{1}{n} \sum_{i=1}^{n} \beta_i$ (the average value of the directional smoothness parameters).

As one of the motivational example, consider the following minimization problem

$$\min_{x \in \mathbb{R}^n} \quad f(x) = \gamma \ln \left(\sum_{i=1}^{m} \exp \frac{[Ax]_i}{\gamma} \right) - \langle b, x \rangle, \tag{14}$$

where $A \in \mathbb{R}^{m \times n}$, $b \in \mathbb{R}^n$. We denote the ith row of the matrix A by A_i. A is sparse, i.e. average number of nonzero elements in A_i is less than s. f is L_f-smooth w.r.t. $\| \cdot \|_2$ with $L_f = \max_{i=1,\ldots,m} \|A_i\|_2^2$ and its gradient is component-wise β_j-continuous with $\beta_j = \max_{i=1,\ldots,m} |A_{ij}|$.

---

[5] Strictly speaking, such a constant takes place for non-adaptive variant of the CDM with specific choice of i_k [42]: $\pi(i_k = j) = \frac{\beta_j}{\sum_{j'=1}^{n} \beta_{j'}}$. For described RACDM the analysis is more difficult [49].

Fast Gradient Method (FGM) [44] requires $O\left(\sqrt{\frac{L_f R^2}{\varepsilon}}\right)$ iterations with the complexity of each iteration $O\,(ns)$. Coordinate Descent Method (CDM) [4] requires $O\left(n\frac{\overline{L}_f R^2}{\varepsilon}\right)$ iterations with the complexity of each itera-tion[6] $O\,(s)$. Accelerated Coordinate Descent Method (ACDM) [20, 45] requires $O\left(n\sqrt{\frac{\widetilde{L}_f R^2}{\varepsilon}}\right)$ iterations with the complexity of each iteration $O\,(n)$, where

$$\widetilde{L}_f = \frac{1}{n}\sum_{j=1}^{n}\sqrt{\beta_j}.$$

For proposed in this paper approach we have $O\left(n\sqrt{\frac{\overline{L}_f R^2}{\varepsilon}}\right)$ iterations of CDM with complexity of each inner iteration $O(s)$ and complexity of each outer iteration $O(ns)$. However, outer iteration executes ones per $\sim n$ inner iterations, so average-case iteration complexity is $O(s)$.

We combine all these results in the table below. From the table one can conclude that if $\overline{L}_f < L_f$, then our approach has better theoretical complexity.

Algorithm	Complexity	Reference
FGM	$O\left(ns\sqrt{\frac{L_f R^2}{\varepsilon}}\right)$	[44]
CDM	$O\left(ns\frac{L_f R^2}{\varepsilon}\right)$	[4, 42]
ACDM	$O\left(n^2\sqrt{\frac{\widetilde{L}_f R^2}{\varepsilon}}\right)$	[45]
Catalyst CDM	$O\left(ns\sqrt{\frac{\overline{L}_f R^2}{\varepsilon}}\right)$	This paper

Note that the use of Component Descent Method allows us to improve con-vergence estimate by factor $\sqrt{n}$ compared to Fast Gradient Method. Indeed, for this problem we have $L_f = \max_{i=1,\dots,m}\|A_i\|_2^2 = O(n)$, and on the other hand $\overline{L}_f = \frac{1}{n}\sum_{j=1,\dots,n}\max_{i=1,\dots,m}|A_{ij}| = O(1)$. Therefore, the total convergence estimate for Fast Gradient Method can be written as

$$O\left(ns\cdot\sqrt{n}\cdot\sqrt{\frac{R^2}{\varepsilon}}\right),$$

and for proposed in this paper method factor $\sqrt{n}$ is reduced to $O(1)$ and could be omitted:

$$O\left(ns\cdot\sqrt{\frac{R^2}{\varepsilon}}\right).$$

---

[6] Here one should use a following trick in recalculation of $\ln\left(\sum_{i=1}^{m}\exp\left([Ax]_i\right)\right)$ and its gradient (partial derivative). From the structure of the method we know that $x^{new} = x^{old} + \delta e_i$, where e_i is i-th orth. So if we've already calculate Ax^{old} then to recalculate $Ax^{new} = Ax^{old} + \delta A_i$ requires only $O(s)$ additional operations independently of n and m.

The best complexity improvement is achieved if $L_f = n$, which means there is at least one row in the matrix such that $A_i = \mathbb{1}^n$, even though all other rows can be arbitrary sparse.

3.3 Alternating Minimization

Consider the following problem

$$\min_{x=(x_1,\dots,x_p)^T \in \otimes_{i=1}^P \mathbb{R}^{n_i}} f(x),$$

where f is a L_f-smooth convex function (its gradient is Lipschitz continuous w.r.t. $\| \cdot \|_2$ with the constant L_f).

For the general case of number of blocks $p \geqslant 2$ the Alternating Minimization algorithm may be written as Algorithm 5. There are multiple common block selection rules, such as the cyclic rule or the Gauss–Southwell rule [3,6,31,55].

Algorithm 5. Alternating Minimization

Parameters: Starting point x^0.
for $k = 0, 1, \dots, N - 1$ **do**
    Choose $i \in \{1, \dots, p\}$
    Set $x^{k+1} = \underset{x_i}{\operatorname{argmin}} f(x_1^k, \dots x_{i-1}^k, x_i, x_{i+1}^k, \dots, x_p^k)$
end for
Output: x^N

Note that for Algorithm 5 the Assumption 1 holds with $C_n = O(p)$ (p is number of blocks) and L_f is the Lipschitz constant of the gradient of function f.

3.4 Theoretical Guarantees

Let us present the table that establishes the comparison of rates of convergence for the above algorithms before and after acceleration via Algorithm 2. In non-accelerated case these algorithms apply to the convex but non-strongly convex problem, therefore, we use estimates for the convex case from Assumption 1. But in the case of acceleration of these methods, we apply them to a regularized function which is strongly convex. Denote $\chi = \max\left\{\sqrt{\frac{L_u}{L_f}}, \sqrt{\frac{L_f}{L_d}}\right\}$, then we represent the following table.

The numerical experiments with the steepest descent, adaptive coordinate descent, alternating minimization and local SGD can be viewed in the full version of the article on the arXiv [28].

	Non-accelerated	M-S accelerated
Steepest descent	$\dfrac{L_f R^2}{\varepsilon}$	$\chi\sqrt{\dfrac{L_f R^2}{\varepsilon}}$
Random adaptive coordinate descent method	$n \cdot \dfrac{\overline{L}_f R^2}{\varepsilon}$	$n \cdot \chi\sqrt{\dfrac{\overline{L}_f R^2}{\varepsilon}}$
Alternating minimization	$p \cdot \dfrac{L_f R^2}{\varepsilon}$	$p \cdot \chi\sqrt{\dfrac{L_f R^2}{\varepsilon}}$

Conclusion

In this work, we present the universal framework for accelerating the non-accelerated adaptive methods such as Steepest Descent, Alternating Least Squares Minimization, and RACDM and show that acceleration works in practice (code is available online on GitHub). Moreover, we show theoretically that for the non-adaptive run proposed in this paper, acceleration has in a log-factor better rate than via Catalyst. Note, that this "fight" for the log-factor in accelerated procedure's become popular in the last time, see [33, 36] for concrete examples. In this paper, we eliminate log-factor in a rather big generality.

Acknowledgements. We would like to thank Soomin Lee (Yahoo), Erik Ordentlich (Yahoo), César A. Uribe (MIT), Pavel Dvurechensky (WIAS, Berlin) and Peter Richtarik (KAUST) for useful remarks. We also would like to thank anonymous reviewers for their fruitful comments.

References

1. Allen-Zhu, Z., Hazan, E.: Optimal black-box reductions between optimization objectives. arXiv preprint arXiv:1603.05642 (2016)
2. Bayandina, A., Gasnikov, A., Lagunovskaya, A.: Gradient-free two-points optimal method for non smooth stochastic convex optimization problem with additional small noise. Autom. Rem. Contr. **79**(7) (2018). arXiv:1701.03821
3. Beck, A.: First-order methods in optimization, vol. 25. SIAM (2017)
4. Bubeck, S.: Convex optimization: algorithms and complexity. Found. Trends® Mach. Learn. **8**(3–4), 231–357 (2015)
5. De Klerk, E., Glineur, F., Taylor, A.B.: On the worst-case complexity of the gradient method with exact line search for smooth strongly convex functions. Optim. Lett. **11**(7), 1185–1199 (2017)
6. Diakonikolas, J., Orecchia, L.: Alternating randomized block coordinate descent. arXiv preprint arXiv:1805.09185 (2018)
7. Diakonikolas, J., Orecchia, L.: Conjugate gradients and accelerated methods unified: the approximate duality gap view. arXiv preprint arXiv:1907.00289 (2019)
8. Doikov, N., Nesterov, Y.: Contracting proximal methods for smooth convex optimization. SIAM J. Optim. **30**(4), 3146–3169 (2020)
9. Doikov, N., Nesterov, Y.: Inexact tensor methods with dynamic accuracies. arXiv preprint arXiv:2002.09403 (2020)

10. Duchi, J.C., Jordan, M.I., Wainwright, M.J., Wibisono, A.: Optimal rates for zero-order convex optimization: the power of two function evaluations. IEEE Trans. Inf. Theory **61**(5), 2788–2806 (2015)
11. Dvinskikh, D., et al.: Accelerated meta-algorithm for convex optimization. Comput. Math. Math. Phys. **61**(1), 17–28 (2021)
12. Dvinskikh, D., Omelchenko, S., Gasnikov, A., Tyurin, A.: Accelerated gradient sliding for minimizing a sum of functions. Doklady Math. **101**, 244–246 (2020)
13. Dvurechensky, P., Gasnikov, A., Gorbunov, E.: An accelerated directional derivative method for smooth stochastic convex optimization. arXiv:1804.02394 (2018)
14. Dvurechensky, P., Gasnikov, A., Gorbunov, E.: An accelerated method for derivative-free smooth stochastic convex optimization. arXiv:1802.09022 (2018)
15. Fercoq, O., Richtárik, P.: Accelerated, parallel, and proximal coordinate descent. SIAM J. Optim. **25**(4), 1997–2023 (2015)
16. Gasnikov, A.: Universal Gradient Descent. MCCME, Moscow (2021)
17. Gasnikov, A., Lagunovskaya, A., Usmanova, I., Fedorenko, F.: Gradient-free proximal methods with inexact oracle for convex stochastic nonsmooth optimization problems on the simplex. Autom. Rem. Contr. **77**(11), 2018–2034 (2016). https://doi.org/10.1134/S0005117916110114. http://dx.doi.org/10.1134/S0005117916110114. arXiv:1412.3890
18. Gasnikov, A.: Universal gradient descent. arXiv preprint arXiv:1711.00394 (2017)
19. Gasnikov, A., et al.: Near optimal methods for minimizing convex functions with lipschitz p-th derivatives. In: Conference on Learning Theory, pp. 1392–1393 (2019)
20. Gasnikov, A., Dvurechensky, P., Usmanova, I.: On accelerated randomized methods. Proc. Moscow Inst. Phys. Technol. **8**(2), 67–100 (2016). (in Russian), first appeared in arXiv:1508.02182
21. Gasnikov, A., Gorbunov, E., Kovalev, D., Mokhammed, A., Chernousova, E.: Reachability of optimal convergence rate estimates for high-order numerical convex optimization methods. Doklady Math. **99**, 91–94 (2019)
22. Gazagnadou, N., Gower, R.M., Salmon, J.: Optimal mini-batch and step sizes for saga. arXiv preprint arXiv:1902.00071 (2019)
23. Gorbunov, E., Hanzely, F., Richtarik, P.: A unified theory of SGD: variance reduction, sampling, quantization and coordinate descent (2019)
24. Gower, R.M., Loizou, N., Qian, X., Sailanbayev, A., Shulgin, E., Richtárik, P.: SGD: general analysis and improved rates. arXiv preprint arXiv:1901.09401 (2019)
25. Guminov, S., Dvurechensky, P., Gasnikov, A.: Accelerated alternating minimization. arXiv preprint arXiv:1906.03622 (2019)
26. Hendrikx, H., Bach, F., Massoulié, L.: Dual-free stochastic decentralized optimization with variance reduction. In: Advances in Neural Information Processing Systems, vol. 33 (2020)
27. Ivanova, A., et al.: Oracle complexity separation in convex optimization. arXiv preprint arXiv:2002.02706 (2020)
28. Ivanova, A., Pasechnyuk, D., Grishchenko, D., Shulgin, E., Gasnikov, A., Matyukhin, V.: Adaptive catalyst for smooth convex optimization. arXiv preprint arXiv:1911.11271 (2019)
29. Kamzolov, D., Gasnikov, A., Dvurechensky, P.: Optimal combination of tensor optimization methods. In: Olenev, N., Evtushenko, Y., Khachay, M., Malkova, V. (eds.) OPTIMA 2020. LNCS, vol. 12422, pp. 166–183. Springer, Cham (2020). https://doi.org/10.1007/978-3-030-62867-3_13
30. Kamzolov, D., Gasnikov, A.: Near-optimal hyperfast second-order method for convex optimization and its sliding. arXiv preprint arXiv:2002.09050 (2020)

31. Karimi, H., Nutini, J., Schmidt, M.: Linear convergence of gradient and proximal-gradient methods under the Polyak-Łojasiewicz condition. In: Frasconi, P., Landwehr, N., Manco, G., Vreeken, J. (eds.) ECML PKDD 2016. LNCS (LNAI), vol. 9851, pp. 795–811. Springer, Cham (2016). https://doi.org/10.1007/978-3-319-46128-1_50

32. Karimireddy, S.P., Kale, S., Mohri, M., Reddi, S.J., Stich, S.U., Suresh, A.T.: Scaffold: stochastic controlled averaging for federated learning. arXiv preprint arXiv:1910.06378 (2019)

33. Kovalev, D., Salim, A., Richtárik, P.: Optimal and practical algorithms for smooth and strongly convex decentralized optimization. In: Advances in Neural Information Processing Systems, vol. 33 (2020)

34. Kulunchakov, A., Mairal, J.: A generic acceleration framework for stochastic composite optimization. arXiv preprint arXiv:1906.01164 (2019)

35. Li, H., Lin, Z.: Revisiting extra for smooth distributed optimization. arXiv preprint arXiv:2002.10110 (2020)

36. Li, H., Lin, Z., Fang, Y.: Optimal accelerated variance reduced extra and diging for strongly convex and smooth decentralized optimization. arXiv preprint arXiv:2009.04373 (2020)

37. Lin, H., Mairal, J., Harchaoui, Z.: A universal catalyst for first-order optimization. In: Advances in Neural Information Processing Systems, pp. 3384–3392 (2015)

38. Lin, H., Mairal, J., Harchaoui, Z.: Catalyst acceleration for first-order convex optimization: from theory to practice. arXiv preprint arXiv:1712.05654 (2018)

39. Lin, T., Jin, C., Jordan, M.: On gradient descent ascent for nonconvex-concave minimax problems. In: International Conference on Machine Learning, pp. 6083–6093. PMLR (2020)

40. Mishchenko, K., Iutzeler, F., Malick, J., Amini, M.R.: A delay-tolerant proximal-gradient algorithm for distributed learning. In: International Conference on Machine Learning, pp. 3587–3595 (2018)

41. Monteiro, R.D., Svaiter, B.F.: An accelerated hybrid proximal extragradient method for convex optimization and its implications to second-order methods. SIAM J. Optim. 23(2), 1092–1125 (2013)

42. Nesterov, Y.: Efficiency of coordinate descent methods on huge-scale optimization problems. SIAM J. Optim. 22(2), 341–362 (2012)

43. Nesterov, Y.: Lectures on Convex Optimization, vol. 137. Springer, Cham (2018). https://doi.org/10.1007/978-3-319-91578-4

44. Nesterov, Y., Gasnikov, A., Guminov, S., Dvurechensky, P.: Primal-dual accelerated gradient descent with line search for convex and nonconvex optimization problems. arXiv preprint arXiv:1809.05895 (2018)

45. Nesterov, Y., Stich, S.U.: Efficiency of the accelerated coordinate descent method on structured optimization problems. SIAM J. Optim. 27(1), 110–123 (2017)

46. Palaniappan, B., Bach, F.: Stochastic variance reduction methods for saddle-point problems. In: Advances in Neural Information Processing Systems, pp. 1416–1424 (2016)

47. Paquette, C., Lin, H., Drusvyatskiy, D., Mairal, J., Harchaoui, Z.: Catalyst acceleration for gradient-based non-convex optimization. arXiv preprint arXiv:1703.10993 (2017)

48. Parikh, N., Boyd, S., et al.: Proximal algorithms. Found. Trends® Optim. 1(3), 127–239 (2014)

49. Pasechnyuk, D., Anikin, A., Matyukhin, V.: Accelerated proximal envelopes: application to the coordinate descent method. arXiv preprint arXiv:2101.04706 (2021)

50. Polyak, B.T.: Introduction to optimization. Optimization Software (1987)
51. Rockafellar, R.T.: Monotone operators and the proximal point algorithm. SIAM J. Control. Optim. **14**(5), 877–898 (1976)
52. Shalev-Shwartz, S., Zhang, T.: Accelerated proximal stochastic dual coordinate ascent for regularized loss minimization. In: International Conference on Machine Learning, pp. 64–72 (2014)
53. Shamir, O.: An optimal algorithm for bandit and zero-order convex optimization with two-point feedback. J. Mach. Learn. Res. **18**, 52:1–52:11 (2017)
54. Tupitsa, N.: Accelerated alternating minimization and adaptability to strong convexity. arXiv preprint arXiv:2006.09097 (2020)
55. Tupitsa, N., Dvurechensky, P., Gasnikov, A.: Alternating minimization methods for strongly convex optimization. arXiv preprint arXiv:1911.08987 (2019)
56. Wilson, A.C., Mackey, L., Wibisono, A.: Accelerating rescaled gradient descent: Fast optimization of smooth functions. In: Advances in Neural Information Processing Systems, pp. 13533–13543 (2019)
57. Woodworth, B., et al.: Is local SGD better than minibatch SGD? arXiv preprint arXiv:2002.07839 (2020)
58. Wright, S.J.: Coordinate descent algorithms. Math. Program. **151**(1), 3–34 (2015)
59. Yang, J., Zhang, S., Kiyavash, N., He, N.: A catalyst framework for minimax optimization. In: Advances in Neural Information Processing Systems, vol. 33 (2020)

On Local Error Bound in Nonlinear Programs

L. I. Minchenko and S.I. Sirotko$^{(\boxtimes)}$

Belarusian State University of Informatics and Radioelectronics, Minsk, Belarus
leonidm@insoftgroup.com, sergeyis@bsuir.by

Abstract. Numerous efforts in the literature are devoted to studying error bounds in optimization problems. The existence of local error bounds is closely related with constraint qualifications. It is well-known that some constraint qualifications imply error bounds. We consider constraint qualifications, which do not impose as strong requirements on the structure of the optimization problem as traditional conditions do. On their base necessary and sufficient conditions of error bounds are derived.

Keywords: Error bound · Necessary conditions · Sufficient conditions · Nonlinear programming

1 Introduction

The concept of error bound are studied in numerous papers [2–4,9,11,15–17, 20,21,24,26], etc. It has been proved to be extremely useful in analyzing the convergence of many algorithms for solving optimization problems, as well as serving as a constraint qualification for optimality conditions [2,3,5,11,16,25, 26]. The study of error bounds turns also out to be of great importance in stability and sensitivity issues, subdifferential calculus [4,5,9,17]. Results in the framework of the theory of error bounds also include the investigation of the local upper Lipschitz stability for the solutions of parametric Karush – Kuhn – Tucker systems [10], the relation between the equivalence of an error bound and a quadratic growth condition [7] necessary and sufficient criteria for metric subregularity of set-valued mappings [12]. Subdifferential conditions for Hölder error bounds and some new estimates for the corresponding modulus are treated in [13].

For more details, we refer the reader to the surveys by Pang [24], Lewis and Pang [15], to the paper by Fabian et al. [9], and their references.

Let us consider a set

$$C = \{y \in R^m \mid h_i(y) \leq 0 \quad i \in I, \ h_i(y) = 0 \quad i \in I_0\} \tag{1}$$

Supported by Belarusian State Program for Scientific Research "Mathematical Models and Methods".

N. N. Olenev et al. (Eds.): OPTIMA 2021, LNCS 13078, pp. 38–49, 2021.
https://doi.org/10.1007/978-3-030-91059-4_3

where $I = \{1, \ldots, s\}$ and $I_0 = \{s + 1, \ldots, p\}$, h_i $i = 1, 2, \ldots, p$ are continuously differentiable functions from R^m to R.

Let $d_C(y)$ be the Euclidean distance from the point y to C.

We say the local error bound (LEB) holds at a point $y^0 \in C$ if there exist a number $M > 0$ and a neighborhood $V(y^0)$ of this point such that $d_C(y) \leq M \max\{0, h_i(y)\ i \in I, |h_i(y)|\ i \in I\}$ for all $y \in V(y^0)$.

Constraint qualifications (see [1–3, 5, 16–21, 25] and other numerous publications) play an important role in studying optimization problems. This concerns an analytical description of tangent cones to sets of feasible points, in derivations of duality relations, in sensitivity analysis with respect to parameter perturbations, as well as in the investigations of convergence of numerical optimization algorithms. A fundamental necessary optimality condition in nonlinear programming is the Karush – Kuhn – Tucker condition (KKT) [14]. However, KKT holds only under constraint qualifications and it loses validity if they are not satisfied.

Since the local error bound itself is a constraint qualification and implies KKT, the concept of error bound is closely related with some other constraint qualifications. Beginning with the work of Robinson [26] there is a question whether some constraint qualification implies an error bound.

One of such constraint qualifications is the well-known *Mangasarian – Fromovitz constraint qualification* (MFCQ) [18].

Denote $I(y) = \{i \in I \mid h_i(y) = 0\}$.

MFCQ holds at a point $y \in C$ if the vectors $\nabla h_i(y)$ $i \in I$ are linearly independent and there exists a vector $\hat{y}$ such that

$$\langle \nabla h_i(y), \hat{y} \rangle = 0 \ i \in I_0, \ \langle \nabla h_i(y), \hat{y} \rangle < 0 \ i \in I(y).$$

It is known that MFCQ at a point $y \in C$ is equivalent to the condition $\Lambda_0(y) = 0$, where

$$\Lambda_0(y) = \{\lambda \in R^p \mid \sum_{i=1}^{p} \lambda_i \nabla h_i(y) = 0, \lambda_i \geq 0 \text{ and } \lambda_i h_i(y) = 0 \ i \in I\}.$$

Despite the wide applicability of MFCQ and the fact that it is quite effective, there are some sets (1) for which this condition is not valid, so other constraint qualifications are needed. The *constant rank condition* (CRCQ) [8] and its generalization, the *relaxed constant rank condition* (RCRCQ) [21], are constraint qualifications of different nature than that of MFCQ and they are independent of MFCQ.

It is proved [21] that RCRCQ implies LEB.

It is interesting to have weaker constraint qualifications which generalize the both MFCQ and RCRCQ.

Consider the tangent cone $T_C(y)$ and the Clarke tangent cone $\hat{T}_C(y)$ to the set C at a point $y \in C$, given by, respectively,

$$T_C(y) = \{\bar{y} \in R^m \mid \exists t_k \downarrow 0 \text{ and } \bar{y}^k \to \bar{y} \text{ such that } y + t_k \bar{y}^k \in C, k = 1, 2, \ldots\},$$

$$\hat{T}_C(y) = \{\bar{y} \in R^m \mid \forall t_k \downarrow 0 \text{ and } \forall y^k \xrightarrow{C} y, \exists \bar{y}^k \to \bar{y}$$
$$\text{such that } y^k + t_k \bar{y}^k \in C, k = 1, 2, \ldots\}.$$

Introduce also the linearized tangent cone to C at a point $y \in C$:

$$\Gamma_C(y) = \{\bar{y} \in R^m \mid \langle \nabla h_i(y), \bar{y} \rangle \leq 0 \ i \in I(y), \ \langle \nabla h_i(y), \bar{y} \rangle = 0 \ i \in I_0\}.$$

One of the weakest constraint qualifications is the *Abadie condition* (ACQ) [1] *which holds at* $y \in C$ *if* $T_C(y) = \Gamma_C(y)$. ACQ ensures KKT but counterexamples show that ACQ doesn't imply error bound.

The *relaxed Mangasarian – Fromovitz constraint qualification* (RMFCQ) was introduced in [11,19] and in the work [2] (under the name CRSC, i.e. constant rank of the subspace component condition). Following [21,22] we introduce the set of indices of all essentially active inequality constraints for the set $\Gamma(C, y)$:

$$I^a(y) = \{i \in I(y) \mid \langle \nabla h_i(y), \bar{y} \rangle = 0 \ \text{for all} \ \bar{y} \in \Gamma(C, y)\}$$

We say that RMFCQ holds at a point $y^0 \in C$ *if there exists a neighborhood of* $V(y^0)$ *such that the system of vectors* $\{\nabla h_i(y) \ i \in I_0 \cup I^a(y^0)\}$ *is of constant rank for all points* $y \in V(y^0)$.

Due to [2,11,19] *RMFCQ is a constraint qualification implying ACQ.*

Note that, if $I^a(y) = \emptyset$ and the gradients $\nabla h_i(y) \ i \in I_0$ are linearly independent, RMFCQ reduces to MFCQ. Moreover, in [22] it has been proved that RMFCQ was the MFCQ-like constraint qualification and implied MFCQ to hold in some alternative parametrization of the set C. Since RMFCQ is implied by some other constraint qualifications (CRCQ [8], RCRCQ [21], CPLD [25], RCPLD [3]), they all are MFCQ-like too.

Andreani et al. [2] proved that, under the twice continuous differentiability of constraint functions, CRSC (or, equivalently, RMFCQ) implies the existence of local error bound. In [16] this result was extended to the systems with continuously differentiable constraint functions.

The main goal of our paper is to find sufficient conditions for LEB.

We denote by $|y|$ the Euclidean norm of a vector $y \in R^m$. Denote also by $V(y^0)$ a neighborhood of the point y^0.

Let $v \in R^m$. Denote by $\Pi_C(v)$ the set of points from C closest to v and introduce the Mordukhovich normal cone [23] $N_C(y) = \limsup_{v \to y}[cone(v - \Pi_C(v))]$ to the set C at a point $y \in C$. It is known [23,27] that $N_C(y)^* = \hat{T}_C(y)$ where $K^* = \{y^* \in R^m \mid \langle y^*, y \rangle \leq 0 \ \ \forall y \in K\}$ for a set $K \subset R^m$.

2 Necessary Conditions for Error Bound

Propositions below give necessary conditions for the local error bound holds at a feasible point y^0.

Proposition 1. *Let LEB hold at* $y^0 \in C$. *Then* $\hat{T}_C(y^0) = \Gamma_C(y^0)$.

Proof. If $\Gamma_C(y^0) = \{0\}$, the assertion is evident. Let $\bar{y} \in \Gamma_C(y^0)$ and $\bar{y} \neq 0$. It is known that LEB at $y^0 \in C$ implies LEB at all feasible points $y \in y^0 + \delta_0 B$, where δ_0 is some small positive number. Take $\delta > 0$ such that $\delta < 2^{-1}\delta_0$ and such that

$h_i(y) < 0$ for all $y \in y^0 + 2\delta B$ and all $i \in (I \backslash I(y^0))$. Denote $t_0 = \delta |\bar{y}|^{-1}$. Then $h_i(y + t\bar{y})$ for all $y \in y^0 + \delta B$, $t \in [0, t_0]$, $i \in (I \backslash I(y^0))$.

For any $y \in C \cap (y^0 + \delta B)$ and for all $t \in [0, t_0]$ from LEB at y follows

$$d_C(y + t\bar{y}) - d_C(y)$$
$$\leq M \max\{0, h_i(y + t\bar{y}) \ i \in I, |h_i(y + t\bar{y})| \ i \in I_0\}$$
$$= M \max\{0, h_i(y + t\bar{y}) \ i \in I(y^0), |h_i(y + t\bar{y})| \ i \in I_0\}$$
$$= M \max\{0, h_i(y) + t\langle \nabla h_i(y), \bar{y}\rangle + t\gamma_i \ i \in I(y^0),$$
$$|h_i(y) + t\langle \nabla h_i(y), \bar{y}\rangle + t\gamma_i| \ i \in I_0\}$$
$$\leq tM \max\{0, \langle \nabla h_i(y), \bar{y}\rangle \ i \in I(y^0), |\langle \nabla h_i(y), \bar{y}\rangle| \ i \in I_0\} + t\gamma,$$

where $\tau_i \in (0,1)$, $\gamma_i = \langle \nabla h_i(y + \tau_i t\bar{y}) - \langle \nabla h_i(y), \bar{y}\rangle$, $\gamma = \max\{|\gamma_i| \ \tau_i \in (0,1), i \in I_0 \cup I(y^0)\}$.

The latter inequality implies

$$d_C^0(y^0; \bar{y}) = \limsup_{y \xrightarrow{C} y^0, y \downarrow 0} t^{-1}[d_C(y + t\bar{y}) - d_C(y)]$$

$$\leq M \max\{0, \langle \nabla h_i(y), \bar{y}\rangle \ i \in I(y^0), |\langle \nabla h_i(y), \bar{y}\rangle| \ i \in I_0\} = 0.$$

This means that $\bar{y} \in \hat{T}_C(y^0)$ (see, [27]) and, hence, $\Gamma_C(y^0) \subset \hat{T}_C(y^0)$. Since the opposite inclusion always holds, we obtain $\Gamma_C(y^0) = \hat{T}_C(y^0)$.

Corollary 1. *Let LEB hold at $y^0 \in C$. Then there exists a neighborhood $V(y^0)$ such that $\hat{T}(y) = \Gamma_C(y)$ for all $y \in V(y^0) \cap C$ and $\liminf_{y \xrightarrow{C} y^0} \Gamma_C(y) = \Gamma_C(y^0)$.*

Proposition 2. *Let LEB hold at $y^0 \in C$. Then there exists a neighborhood $V(y^0)$ such that $I^a(y) \subset I^a(y^0)$ for all $y \in V(y^0) \cap C$.*

Proof. Suppose the opposite. Then there is a sequence $\{y^k\} \subset C$ such that $y^k \to y^0$ and for any k there exists $i_k \in I^a(y^k)$ such that $i_k \notin I^a(y^0)$. Since the set I is finite, without loss of generality one may assume that $I(y^k) = K \subset I(y^0)$ and $I^a(y^k) = K^a$, where K and K^a don't depend of k. Moreover, we can assume that $i_k = j$ and doesn't depend of k. Thus, there exists some index $j \in K^a$ such that $j \notin I^a(y^0)$. Take an arbitrary vector $\bar{y} \in \Gamma_C(y^0)$. From Corollary 1 follows that there exists a sequence $\bar{y}^k \in \Gamma_C(y^k)$ such as $\bar{y}^k \to \bar{y}$ as $k \to \infty$. Pass to the limit in the following equalities and inequalities

$$\langle \nabla h_i(y^k), \bar{y}^k\rangle = 0 \ i \in I_0 \cup K^a, \langle \nabla h_i(y^k), \bar{y}^k\rangle \leq 0 \ i \in K \backslash K^a \text{ as } k \to \infty.$$

In result we have

$$\langle \nabla h_i(y^0), \bar{y}\rangle = 0 \ i \in I_0 \cup K^a, \langle \nabla h_i(y^0), \bar{y}\rangle \leq 0 \ i \in K \backslash K^a$$

for any $\bar{y} \in \Gamma_C(y^0)$. Thus, $\langle \nabla h_j(y^0), \bar{y}\rangle = 0$ for all $\bar{y} \in \Gamma_C(y^0)$ and, hence, $j \in I^a(y^0)$. This contradicts to our assumption.

Example 1. Let $C = \{y \in R^3 \mid y_1 + y_2 = 0, y_3^2 \leq 0\}$, $y^0 = (0,0,0)^T$. In this example $\Gamma_C(y^0) = \{\bar{y} \in R^3 \mid \bar{y}_1 + \bar{y}_2 = 0\} \neq \hat{T}_C(y^0) = \{\bar{y} \in R^3 \mid \bar{y}_1 + \bar{y}_2 = 0, \bar{y}_3 = 0\}$. Therefore, due to Proposition 1 LEB can't hold at y^0.

An example below shows that the condition $\hat{T}_C(y^0) = \Gamma_C(y^0)$ is not sufficient for LEB.

Example 2. Let $C = \{y \in R^2 \mid y_1^4 - y_2^2 \leq 0, -y_2 \leq 0\}$, $y^0 = (0,0)^T$. Then $\Gamma_C(y^0) = \{\bar{y} \in R^2 \mid \bar{y}_1 \in R, \bar{y}_2 \geq 0\}$, $N_C(y^0) = \{v \in R^2 \mid v_1 = 0, v_2 \leq 0\}$. Therefore, $N_C(y^0) = \Gamma_C^*(y^0)$, i.e. $T_C(y^0) = \Gamma_C(y^0)$. On the other hand, for $v = (\varepsilon, 0)^T$, where $\varepsilon \downarrow 0$, we obtain $d_C(v) \geq \varepsilon^2$, $\max\{0, h_1(v), h_2(v)\} = \varepsilon^4$, i.e. LEB doesn't hold at the given point.

Example 3. Let $C = \{y \in R^2 \mid y_2 - y_1^2 \leq 0, -y_2 \leq 0\}$, $y^0 = (0,0)^T$. Here $h_1(y) = y_2 - y_1^2$, $h_2(y) = -y_2$. It is not difficult to see that $rank\{\nabla h_1(y), \nabla h_2(y)\} \neq const$ in $V(y^0)$. One can check that $I^a(y^0) = \{1,2\}$ but $I^a(y) = \emptyset$ for $y \in V(y^0)\backslash y^0$.

3   Sufficient Conditions for Local Error Bound

Let $y^0 \in C$ and $\varepsilon > 0$. Consider all subsets $K \subset I(y^0)$ such that for every K there is a continuous arc $g : [0, \varepsilon] \to R^m$ such that $g(0) = y^0$, $g(t) \in C$ for all $t \in [0, \varepsilon]$, $h_i(g(t)) = 0$ $i \in K$ and $h_i(g(t)) < 0$ $i \in I\backslash K$ for all $t \in [0, \varepsilon]$. Introduce the sets

$$C(K) = \{y \in R^m \mid h_i(y) \leq 0 \ i \in K, h_i(y) = 0 \ i \in I_0\}$$

and consider

$$\Gamma_{C(K)} = \{\bar{y} \in R^m \mid \langle \nabla h_i(y^0), \bar{y}\rangle \leq 0 \ i \in K, \langle \nabla h_i(y^0), \bar{y}\rangle = 0 \ i \in I_0\}.$$

Let $v \in R^m$, $v \notin C$. Consider the set $\Pi_C(v)$ of points from C closest to v. Evidently these points are the solutions of the following problem

$$|y - v| \to \min, y \in C. \tag{2}$$

Theorem 1. *Let $y^0 \in C$ and let $\hat{T}_{C(K)}(y^0) = \Gamma_{C(K)}(y^0)$ (or, equivalently, $N_{C(K)}(y^0) = \Gamma_{C(K)}(y^0)^*$) hold for every $K \subset I(y^0)$. Then LEB holds at y^0.*

Proof. The assertion of the theorem is valid, if $y^0 \in intC$. Assume that y^0 is a boundary point for C. Suppose that C doesn't satisfy LEB at $y^0 \in C$. Then there exists a sequence $v^k \to y^0$ such that $v^k \notin C$ and

$$d_C(v^k) > k \max\{0, h_i(v^k) \ i \in I, |h_i(v^k)| \ i \in I_0\}, \quad k = 1, 2, \ldots. \tag{3}$$

Let $y^k = y(v^k) \in \Pi_C(v^k)$, where $\Pi_C(v)$ is the set of points from C closest to a point v. Denote $\bar{v}^k = (v^k - y^k)|v^k - y^k|^{-1}$, $k = 1, 2, \ldots$. Then $|v^k - y^k| \leq |v^k - y^0|$ and, hence, $y^k \to y^0$.

In view of finiteness of the set I one can extract from the sequences $\{v^k\}$ and $\{y^k\}$ such subsequences (for the simplicity we keep the notation $\{v^k\}$ and $\{y^k\}$ and for them) that $I(y^k) = K \subset I(y^0)$ and doesn't depend of k. Then, $h_i(y^k) = 0$ for $i \in K$ and $h_i(y^k) < 0$ for $i \in I \setminus K$. Due to the continuity of functions h_i there is a neighborhood $V(y^k)$ such that $h_i(y) < 0$ for all $i \in I \setminus K$ and all $y \in V(y^k)$, i.e., $V(y^k) \cap C(K)$ and $\Pi_C(v^k) = \Pi_{C(k) \cap V(y^k)}(v^k)$.

Without loss of generality we can also assume that $\bar{v}^k \to \bar{v}$. Then, from (3) follows

$$|v^k - y^k| > k \max\{0, \langle \nabla h_i(\tilde{v}_i^k), v^k - y^k \rangle \ i \in K, |\langle \nabla h_i(\tilde{v}_i^k), v^k - y^k \rangle| \ i \in I_0, \}$$

and

$$\frac{1}{k} > \max\{0, \langle \nabla h_i(\tilde{v}_i^k), \bar{v}^k \rangle \ i \in K, |\langle \nabla h_i(\tilde{v}_i^k), \bar{v}^k \rangle| \ i \in I_0\}$$

where $\tilde{v}_i^k = y^k + \tau_{ki}(v^k - y^k)$, $0 \le \tau_{ki} \le 1$.

Therefore, $\max\{0, \langle \nabla h_i(y^0), \bar{v} \rangle \ i \in K, |\langle \nabla h_i(y^0), \bar{v} \rangle| \ i \in I_0\} \le 0$ and, therefore,

$$\bar{v} \in \Gamma_{C(K)}(y^0) \tag{4}$$

From the definition of $N_{C(K)}(y^0)$ follows that $\bar{v} \in N_{C(K)}(y^0)$ if $y^k \in \Pi_{C(K)}(v^k)$. Otherwise there exist small numbers $\tau_k > 0$ such that $\Pi_{C(K)}(y^k + \tau_k(v^k - y^k)) = \{y^k\}$ and $\bar{v}^k$ is a proximal normal (see, 6.16 [27]) to $C(K)$ at a point y^k. Then, $\dfrac{\tau_k(v^k - y^k)}{|\tau_k(v^k - y^k)|} = \bar{v}^k \to \bar{v}$, $\bar{v} \in N_{C(K)}(y^0)$. Consequently, $\bar{v} \in N_{C(K)}(y^0)$ and taking into account of (4) $\bar{v} \in N_{C(K)}(y^0) \cap \Gamma_{C(K)}(y^0) = \Gamma^*_{C(K)}(y^0) \cap \Gamma_{C(K)}(y^0)$. Since $\bar{v} \ne 0$, the latter inclusion is impossible. Thus, C satisfies LEB at y^0.

The following examples show that Theorem 1 is effective.

Example 4. Let $h_1(y) = 1 - y_1^2 - (y_2 - 1)^2 \le 0$, $h_2(y) = 1 - y_1^2 - (y_2 + 1)^2 \le 0$, $y^0 = (0,0)^T$. It's easy to check that MFCQ and RMFCQ don't hold for C at y^0. However, the sets C, $C(K_1) = \{y \in R^2 \mid 1 - y_1^2 - (y_2 - 1)^2 \le 0\}$ under $K_1 = \{1\}$ and $C(K_2) = \{y \in R^2 \mid 1 - y_1^2 - (y_2 + 1)^2 \le 0\}$ under $K_2 = \{2\}$ satisfy the condition $\hat{T}_{C(K)}(y^0) = \Gamma_{C(K)}(y^0)$. Thus, LEB holds at y^0 according to Theorem 1.

Example 5. Let $C = \{y \in R^2 \mid y_2 - y_1^2 \le 0, -y_2 \le 0, -y_2 - y_1^2 \le 0\}$, $y^0 = (0,0)^T$. Then $h_1(y) = y_2 - y_1^2$, $h_2(y) = -y_2$, $h_3(y) = -y_2 - y_1^2$. Since $\Gamma_C(y^0) = \{\bar{y} \in R^2 \mid \bar{y}_2 = 0\}$, all constraints are essentially active at the point y^0, i.e. $I^a(y^0) = \{1, 2, 3\}$. We have

$$rank\{\nabla h_1(y), \nabla h_2(y), \nabla h_3(y)\} = rank \begin{pmatrix} -2y_1 & 0 & -2y_1 \\ 1 & -1 & -1 \end{pmatrix} \ne const$$

Thus, RMFCQ does not hold at y^0. In this case stronger constraint qualifications MFCQ, RCRCQ, RCPLD do not hold too. On the other hand, it is easy to check that the conditions of Theorem 1 are fulfilled. This means in virtue of Theorem 1 LEB holds at the given point.

We can check it immediately. For all y such that $y_2 < 0$, we obtain $d_C(y) = |y_2|$ and $\max\{0, y_2 - y_1^2, -y_2, -y_2 - y_1^2\} = |y_2|$. Also, for y such that $y_2 > 0$ and $y \notin C$ we have $d_C(y) \leq y_2 - y_1^2$ and $\max\{0, y_2 - y_1^2, -y_2, -y_2 - y_1^2\} = y_2 - y_1^2$. Thus, $d_C(y) \leq \max\{0, h_1(y), h_2(y), h_3(y)\}$ near y^0.

This means LEB holds at the point under consideration.

4   Essentially Active Inequality Constraints and Error Bound Properties

In this section we need the following lemmas about essentially active inequality constraints.

Lemma 1. *([22]) Let $y \in C$. Then there exists a vector $\bar{y} \in \Gamma_C(y)$ such that $\langle \nabla h_i(y), \bar{y} \rangle = 0$ for all $i \in I_0 \cup I^a(y)$, $\langle \nabla h_i(y), \bar{y} \rangle < 0$ for all $i \in I(y) \backslash I^a(y)$.*

Lemma 2. *([6, 22]) Let $y \in C$. Then $I^a(y) = \{i \in I(y) \mid \exists \lambda \in \Lambda_0(y)$ such that $\lambda_i > 0\}$.*

It is easy to see that from Lemma 2 follows that there exists $\lambda \in \Lambda_0(y)$ such that $\lambda_i > 0$ for all $i \in I^a(y)$ and $\lambda_i = 0$ for all $i \in I \backslash I^a(y)$.

We also need the following Caratheodory's Lemma [3].

Lemma 3. *Let $0 \neq y = \sum\limits_{i \in J} \alpha_i v^i + \sum\limits_{i \in K} \alpha_i v^i$, $\alpha_i \neq 0$ for all $i \in K$, where the vectors $\{v^i \; i \in J\}$ are linearly independent. Then there exist a set $S \subset K$ and numbers $\beta_i \; i \in J \cup S$ such that the vectors $\{v^i \; i \in J \cup S\}$ are linearly independent and $y = \sum\limits_{i \in J} \beta_i v^i + \sum\limits_{i \in S} \beta_i v^i$, where $\alpha_i \beta_i > 0$ for $i \in S$.*

Let $y^0 \in C$ and let subsets $K \subset I(y^0)$ be defined as at the beginning of the Sect. 3.

The following definition is motivated by the definition of RCPLD [3], considering positive-linearly dependent families of gradients.

Let $S_0 \subset I_0$, $S \subset I$. A vector family $\{\nabla h_i(y) \; i \in S_0 \cup S\}$ is called *positive-linearly dependent* if there exist numbers $\lambda_i \in R$ for all $i \in S_0$ and $\lambda_i \geq 0$ for all $i \in S$ such that

$$\sum_{i \in S_0 \cup S} \lambda_i \nabla h_i(y) = 0.$$

Let I_{00} be such that $\{\nabla h_i(y^0) \; i \in I_{00}\}$ is a basis for the vector family $\{\nabla h_i(y^0) \; i \in I_0\}$. We say that *a feasible point y^0 satisfies the positive linear dependence condition (PLD) if there exists a neighborhood $V(y^0)$ of y^0 such that*

1) *$\{\nabla h_i(y) \; i \in I_0\}$ has the same rank for every $y \in V(y^0)$;*
2) *for every $K \subset I(y^0)$, if $\{\nabla h_i(y^0) \; i \in I_{00} \cup (I^a(y^0) \cap K)\}$ is positive-linearly dependent, then $\{\nabla h_i(y) \; i \in I_{00} \cup (I^a(y^0) \cap K)\}$ is linearly dependent for every $y \in V(y^0)$.*

We say that a feasible point y^0 satisfies the positive linear dependence condition on C (PLD$_C$) if the both requirements in the definition of PLD are fulfilled for $y \in V(y^0) \cap C$.

Theorem 2. *Let $y^0 \in C$. Suppose that there is a neighborhood $V(y^0)$ such that*

(i) there is a constraint qualification which holds at any feasible point near y^0;
(ii) y^0 satisfies PLD$_C$.

Then LEB holds at the given point.

Proof. Suppose that LEB doesn't hold at $y^0 \in C$. Then, repeating the first part of the proof Theorem 1 we obtain that there exist sequences $v^k \to y^0$, $v^k \notin C$, and $y^k \in \Pi_C(v^k)$ such that $y^k \to y^0$, $\bar{v}^k = (v^k - y^k)|v^k - y^k|^{-1} \to \bar{v}$, $I(y^k) = K \subset I(y^0)$, and (4) holds.

Due to KKT there exist Lagrange multipliers $\lambda^k \in R^p$ such that

$$\frac{v^k - y^k}{|v^k - y^k|} = \sum_{i=1}^{p} \lambda_i^k \nabla h_i(y^k), \ \lambda_i^k \geq 0 \ i \in I, \text{ and } \lambda_i^k = 0 \text{ for } i \in I\backslash K.$$

From the latter equality follows

$$\bar{v}^k = \sum_{i \in I_0 \cup K} \lambda_i^k \nabla h_i(y^k), \ \lambda_i^k \geq 0 \ i \in K, \ \lambda_i^k = 0 \ i \in I\backslash K. \tag{5}$$

We can rewrite (5) in the following way

$$\bar{v}^k - \sum_{i \in K\backslash I^a(y^0)} \lambda_i^k \nabla h_i(y^k) = \sum_{i \in I_{00}} \alpha_i^k \nabla h_i(y^k) + \sum_{i \in I^a(y^0) \cap K} \lambda_i^k \nabla h_i(y^k), \tag{6}$$

where $\lambda_i^k \geq 0 \ i \in K$, $\alpha_i^k \in R \ i \in I_{00}$.
1) Suppose that

$$\bar{v}^k - \sum_{i \in K\backslash I^a(y^0)} \lambda_i^k \nabla h_i(y^k) \neq 0$$

for all $k = 1, 2, \ldots$ beginning with some k_0. Then due to Lemma 3 for any k there exists a set $J(k) \subset I^a(y^0) \cap K$ such that

$$\bar{v}^k - \sum_{i \in K\backslash I^a(y^0)} \lambda_i^k \nabla h_i(y^k) = \sum_{i \in I_{00}} \alpha_i^k \nabla h_i(y^k) + \sum_{i \in J(k)} \alpha_i^k \nabla h_i(y^k),$$

where $\alpha_i^k > 0$ for all $i \in J(k)$ and the vectors $\{\nabla h_i(y^k) \ i \in I_{00} \cup J(k)\}$ are linearly independent.

Without loss of generality one may assume that the set $J(k)$ is the same for all k, i.e. $J(k) = J$.

Then

$$\bar{v}^k = \sum_{i \in K\backslash I^a(y^0)} \lambda_i^k \nabla h_i(y^k) + \sum_{i \in I_{00}} \alpha_i^k \nabla h_i(y^k) + \sum_{i \in J} \alpha_i^k \nabla h_i(y^k), \tag{7}$$

where vectors $\{\nabla h_i(y^k)\ i \in I_{00} \cup J\}$ are linear independent, and $\alpha_i^k > 0$ for all $i \in J$.

Let $M_k = \max\{\lambda_i^k\ i \in K\backslash I^a(y^0), \alpha_i^k\ i \in J, |\alpha_i^k|\ i \in I_{00}\} \to \infty$. Then dividing (7) by M_k, passing to a subsequence, if necessary, and taking the limit we obtain

$$\sum_{i \in K\backslash I^a(y^0)} \lambda_i \nabla h_i(y^0) + \sum_{i \in I_{00}} \lambda_i \nabla h_i(y^0) + \sum_{i \in J} \lambda_i \nabla h_i(y^0) = 0, \qquad (8)$$

where $\lambda_i \geq 0\ i \in J \cup (K\backslash I^a(y^0))$, $\lambda_i \in R\ i \in I_{00}$, and there are some $\lambda_i \neq 0$.

Multiply (8) by a vector $\bar{y}$ from Lemma 1. Then we have

$$\sum_{i \in K\backslash I^a(y^0)} \lambda_i \langle \nabla h_i(y^0), \bar{y} \rangle = 0.$$

On the other hand, according to Lemma 2 $\langle \nabla h_i(y^0), \bar{y} \rangle$ for all $i \in K\backslash I^a(y^0)$. This means that $\lambda_i = 0$ for all $i \in K\backslash I^a(y^0)$ and, therefore, (8) implies

$$\sum_{i \in I_{00}} \lambda_i \nabla h_i(y^0) + \sum_{i \in J} \lambda_i \nabla h_i(y^0) = 0,$$

where there are $\lambda_i \neq 0$.

Since vectors $\{\nabla h_i(y^k)\ i \in I_{00} \cup J\}$ are linear independent, this equality contradicts PLD$_C$.

Suppose that $\{M_k\}$ is bounded. Then without loss of generality one may assume that $\lambda_i^k \to \lambda_i \geq 0\ i \in K\backslash I^a(y^0)$, $\alpha_i^k \to \lambda_i \geq 0\ i \in J$, $\alpha_i^k \to \lambda_i \in R\ i \in I_{00}$.

Taking the limit in (7) we have

$$\bar{v} = \sum_{i \in K\backslash I^a(y^0)} \lambda_i \nabla h_i(y^0) + \sum_{i \in I_{00}} \lambda_i \nabla h_i(y^0) + \sum_{i \in J} \lambda_i \nabla h_i(y^0) \in [\Gamma_{C(K)}(y^0)]^*.$$

This contradicts (4).

2) Suppose that

$$\bar{v}^k = \sum_{i \in K\backslash I^a(y^0)} \lambda_i^k \nabla h_i(y^k) \qquad (9)$$

for some infinite number of k (for the simplicity assume that for all k). If $\{\lambda_i^k\}$ is unbounded, then dividing (9) by $\max\{\lambda_i^k\ i \in K\backslash I^a(y^0)\}$ and passing to the limit we have

$$\sum_{i \in K\backslash I^a(y^0)} \lambda_i \nabla h_i(y^0) = 0,$$

where there are some $\lambda_i \neq 0$.

However, due to Lemma 2 and the definition of the set $K\backslash I^a(y^0)$ all λ_i should be null. This means that under (9) the sequence $\{\lambda_i^k\}$ cannot be unbounded. Thus, the sequence $\{\lambda_i^k\}$ is bounded. Then without loss of generality $\lambda_i^k \to \lambda_i \geq 0$ and, hence,

$$\bar{v} = \sum_{i \in K\backslash I^a(y^0)} \lambda_i \nabla h_i(y^0) \in [\Gamma_{C(K)}(y^0)]^*.$$

This contradicts (4).

Thus, LEB holds at y^0.

Remark. Remind that the set C satisfies the weakest constraint qualification at a point $y \in C$ if, independently of the objective function f having a local minimizer y on the set C, there exists vector $\lambda \in R^p$ such that the KKT condition holds:

$$\nabla f(y) + \sum_{i=1}^{p} \lambda_i \nabla h_i(y) = 0, \quad \lambda_i \in R \text{ for } i \in I_0, \quad \lambda_i \geq 0 \text{ and } \lambda_i h_i(y) = 0 \text{ for } i \in I.$$

The following example shows that Theorem 2 can be useful in the case where RMFCQ (hence, also MFCQ, RCRCQ and RCPLD) is not effective.

Example 6. Let $C = \{y \in R^2 \mid y_2 - y_1^2 \leq 0, -y_2 \leq 0, -y_2 - y_1^2 \leq 0\}$, $y^0 = (0,0)^T$. Then $h_1(y) = y_2 - y_1^2$, $h_2(y) = -y_2$, $h_3(y) = -y_2 - y_1^2$. Since $\Gamma_C(y^0) = \{\bar{y} \in R^2 \mid \bar{y}_2 = 0\}$, all constraints are essentially active at the point y^0, i.e. $I^a(y^0) = \{1,2,3\}$. We have

$$rank\{\nabla h_1(y), \nabla h_2(y), \nabla h_3(y)\} = rank\begin{pmatrix} -2y_1 & 0 & -2y_1 \\ 1 & -1 & -1 \end{pmatrix} \neq const$$

Hence, RMFCQ (hence, MFCQ, RCPLD, RCRCQ too) does not hold at y^0.

Check the condition of Theorem 2. It is easy to check that ACQ holds at all feasible points near y^0. In the example the sets K from Theorem 2 are the following:

$$K = K_1 = \{1\}, \quad K = K_2 = \{2\}, \quad K = K_{123} = \{1,2,3\}.$$

In this case the vector families $\{\nabla h_i(y^0) \ i \in K_1\}$ and $\{\nabla h_i(y^0) \ i \in K_2\}$ are not positively-linear dependent. At the same time

$$\{\nabla h_i(y^0) \ i \in K_{123}\} = \{\begin{pmatrix} 0 \\ 1 \end{pmatrix}, \begin{pmatrix} 0 \\ -1 \end{pmatrix}, \begin{pmatrix} 0 \\ -1 \end{pmatrix}\}$$

is positively-linear dependent, and

$$\{\nabla h_i(y) \ i \in K_{123}\} = \{\begin{pmatrix} -2y_1 \\ 1 \end{pmatrix}, \begin{pmatrix} 0 \\ -1 \end{pmatrix}, \begin{pmatrix} -2y_1 \\ -1 \end{pmatrix}\}$$

is linearly dependent in $V(y^0)$.

Thus y^0 satisfies PLD$_C$ and, hence, in virtue of Theorem 2 LEB holds at the point y^0.

Check this directly. For all y such that $y_2 < 0$, we obtain $d_C(y) = |y_2|$ and $\max\{0, y_2 - y_1^2, -y_2, -y_2 - y_1^2\} = |y_2|$. Also, for y such that $y_2 > 0$ and $y \notin C$ we have $d_C(y) \leq y_2 - y_1^2$ and $\max\{0, y_2 - y_1^2, -y_2, -y_2 - y_1^2\} = y_2 - y_1^2$. Thus, $d_C(y) \leq \max\{0, h_1(y), h_2(y), h_3(y)\}$ near y^0.

This means LEB holds at the point under consideration.

5   Final Remarks

In the paper we have investigated necessary conditions for the local error bound. We also proved sufficient conditions which imply the existence of local error bounds for the systems of inequalities and equalities whenever constraints functions are continuously differentiable. The obtained sufficient conditions extend some known results devoted to the local error bound in nonlinear programming.

One of the results derives sufficient conditions for the local error bound on the base of normal and tangent cones. On the other hand, it is well known that some constraint qualifications (in particular, RCPLD) guarantee the existence of the local error bound. In the paper we prove sufficient conditions involving weaker assumptions than in RCPLD on the behavior of the gradients for the active inequality constraints.

References

1. Abadie, J.: On the Kuhn - Tucker theorem. In: Abadie, J. (ed.) Nonlinear Programming, pp. 21–36. John Wiley, New York (1967)
2. Andreani, R., Haeser, G., Schuverdt, M.L., Silva, P.J.S.: Two new weak constraint qualifications and applications. SIAM J. Optim. **22**(3), 1109–1135 (2012)
3. Andreani, R., Haeser, G., Schuverdt, M.L., Silva, P.J.S.: A relaxed constant positive linear dependence constraint qualification and applications. Math. Program. **135**, 255–273 (2012)
4. Bosch, P., Jourani, A., Henrion, R.: Sufficient conditions for error bounds and applications. Appl. Math. Appl. **50**, 161–181 (2004)
5. Giannessi, F.: Constrained Optimization and Image Space Analysis, Separation of Sets and Optimality Conditions, vol. 1. Springer, New York (2005). https://doi.org/10.1007/0-387-28020-0
6. Gorokhovik, V.V.: Finite-Dimensional Optimization Problems. BSU Publ, Minsk (2007)
7. Drusvyatskiy, D., Lewis, A.S.: Error bounds, quadratic growth, and linear convergence of proximal methods. Math. Oper. Res. **43**(3), 919–948 (2018)
8. Janin, R.: Directional derivative of the marginal function in nonlinear programming. In: Fiacco, A.V. (ed.) Sensitivity, Stability and Parametric Analysis. Mathematical Programming Studies, vol. 21, pp. 110–126. Springer, Heidelberg (1984). https://doi.org/10.1007/BFb0121214
9. Fabian, M.J., Henrion, R., Kruger, A.Y., Outrata, J.V.: Error bounds: necessary and sufficient conditions. Set-Valued Var. Anal. **18**(2), 121–149 (2010)
10. Izmailov, A.F., Kurennoy, A.S., Solodov, M.V.: A note on upper Lipschitz stability, error bounds, and critical multipliers for Lipschitz-continuous KKT systems. Math. Program., 591–604 (2012). https://doi.org/10.1007/s10107-012-0586-z
11. Kruger, A.Y., Minchenko, L.I., Outrata, J.V.: On relaxing the Mangasarian - Fromovitz constraint qualification. Positivity **18**, 171–189 (2014)
12. Kruger, A.Y.: Error bounds and metric subregularity. Optimization **64**(1), 49–79 (2015)
13. Kruger, A.Y., López, M.A., Yang, X., Zhu, J.: Hölder error bounds and Hölder calmness with applications to convex semi-infinite optimization. Set-Valued Var. Anal. **27**(4), 995–1023 (2019). https://doi.org/10.1007/s11228-019-0504-0

14. Kuhn, H.W., Tucker, A.W.: Nonlinear programming. In: Neyman, J. (ed.) Proceedings of the 2nd Berkeley Symposium on Mathematical Statistics and Probability, vol. 2, pp. 481–492. University of California Press, Berkeley (1951)
15. Lewis A.S., Pang J.S.: Error bounds for convex inequality systems. In: Crouzeix, J.P., Martinez-Legaz, J.E., Volle, M. (eds.) Generalized Convexity, Generalized Monotonicity: Recent Results. Nonconvex Optimization and Its Applications, vol. 27, pp. 75–110. Springer, Boston (1998). https://doi.org/10.1007/978-1-4613-3341-8_3
16. Guo, L., Zhang, J., Lin, G.-H.: New results on constraint qualifications for nonlinear extremum problems and extensions. J. Opt. Theory Appl. **163**, 737–754 (2014)
17. Luderer, B., Minchenko, L., Satsura, T.: Multivalued Analysis and Nonlinear Programming Problems with Perturbations. Kluwer Academic Publisher, Dordrecht (2002)
18. Mangasarian, O.L., Fromovitz, S.: The Fritz-John necessary optimality conditions in presence of equality and inequality constraints. J. Math. Anal. Appl. **17**, 37–47 (1967)
19. Minchenko, L., Stakhovski, S.: About generalizing the Mangasarian - Fromovitz regularity condition. Doklady BGUIR **8**, 104–109 (2010)
20. Minchenko, L.I., Tarakanov, A.N.: On error bounds for quasinormal programs. J. Opt. Theory Appl. **148**, 571–579 (2011)
21. Minchenko, L.I., Stakhovski, S.M.: On relaxed constant rank regularity condition in mathematical programming. Optimization **60**, 429–440 (2011)
22. Minchenko, L.I.: Note on MFCQ-like constraint qualifications. J. Opt. Theory Appl. **3**(182), 1199–1204 (2019)
23. Mordukhovich, B.S.: Variational Analysis and Generalized Differentiation. Vol. I: Basic Theory. Springer, Heidelberg (2005). https://doi.org/10.1007/3-540-31246-3
24. Pang, J.-S.: Error bounds in mathematical programming. Math. Program. **79**, 299–332 (1997)
25. Qi, L., Wei, Z.: On the constant positive linear independence condition and its application to SQP methods. SIAM J. Optim. **10**(4), 963–981 (2000)
26. Robinson, S.M.: An application of error bounds for convex programming in a linear space. SIAM J. Control **13**(2), 271–273 (1975)
27. Rockafellar, R.T., Wets, R.J.-B.: Variational Analysis. Springer, Heidelberg (1998). https://doi.org/10.1007/978-3-642-02431-3

Row-Oriented Decomposition in Large-Scale Linear Optimization

Evgeni Nurminski[1] and Natalia Shamray[2]

[1] Center for Research and Education in Mathematics, Far Eastern Federal
University, Vladivostok, Russia
nurminski.ea@dvfu.ru
[2] Institute of Automation and Control Processes, Vladivostok, Russia
shamray@dvo.ru

Abstract. The single-projection linear optimization method (Nurminski, E.A.: Single-projection procedure for linear optimization. Journal of Global Optimization. 66(1), 95–110 (2016)) demonstrated a promising computational performance on the series of giga-scale academic and practical problems shown in this talk. Another attractive feature of this method is its potential for row-wise decomposition of large-scale problems. It can be applied irrelevant to the problem structure but also can make use of it if present. This decomposition technique might take different forms as well, and a few variants will be presented.

Keywords: Large-scale linear optimization · Projection · Decomposition

Introduction

This article deals with the esteemed linear optimization problem

$$\min_{x \in X} cx = cx^{\star} \tag{1}$$

where X is a polyhedral set defined by the different combinations of equalities and inequalities. It was shown in [1] that (1) can be solved by a single projection operation, which might be an interesting theoretical result but its practical significance was difficult to estimate. Since that we implemented several versions of projection procedures and accumulated specific experience with this idea which raised some hopes for the competitive advantage of this approach even in the well-established area of linear optimization. Apart from rather interesting computational results obtained by numerical experiments with this approach, this idea presents new possibilities for the decomposition of large problems which

This work is supported by RF Ministry of Education and Science, project 1.7658.2017/6.7 and RFBR grant 18-29-03071.

have a high degree of sparseness only without any specific structure or have a mixture of these which are difficult to separate.

The computational core of this approach is a projection operation that is widely used in theoretical mathematics, computational algorithms and numerous applications: image processing, machine learning, automatic classification, to name just a few. So projection toolbox contains many procedures, the most notable and related to the aims of this paper is Von Neumann alternating projection algorithm [2] and its extensions (see f.i. [3]) and polytope projection method of Ph. Wolfe [4] which we adapted for projection onto a polyhedral convex cone.

1 Notations and Preliminaries

Let E be a finite-dimensional vector space of the primal variables with the standard inner product and the norm $\|x\|^2 = xx$. This space is then self-conjugate with the duality relation induced by the inner product. The dimensionality of this space, if needed, is determined as $\dim(E)$ and the space of dimensionality n when necessary is denoted as E^n.

The non-negative part of any space E will be denoted as E_+. Among the others special vectors and sets we mention the null vector $\mathbf{0}$, vector of ones $\mathbf{1} = (1, 1, \ldots, 1)$, and the standard simplex $\Delta = E_+ \cap \{x : \mathbf{1}x = 1\}$. Convex and conical hulls of set X are denoted in more or less standard notation: $\mathrm{co}(X)$ and $\mathrm{Co}(X)$. By using the inner product we can define a notion of the orthogonal complement of a linear subspace $L \subset E$: $L^\perp = \{x : xz = 0 \text{ for all } z \in L\}$.

We define linear operators, acting from E into E' with $\dim(E') = m$ as collections of vectors $\mathcal{A} = \{a^1, a^2, \ldots, a^m\}$ with $a^i \in E$ which produce vector $y = (y_1, y_2, \ldots, y_m) \in E^m$ according to following relations $y_i = a^i x, i = 1, 2, \ldots, m$. In the classical matrix-vector notation vectors $\mathcal{A}$ form the rows of the matrix A and $y = Ax$. At the same time we will consider the row subspace E' as the linear envelope of $\mathcal{A}$:

$$E' = \lin(\mathcal{A}) = \{x = \sum_{i=1}^{m} a^i z_i = A^T z, z \in E^m\} \subset E.$$

The projection operator of a point p onto a closed convex set X in E is defined as

$$p \downarrow X = \operatorname*{argmin}_{x \in X} \min \|p - x\|,$$

that is $\min_{x \in X} \|p - x\| = \|p - p \downarrow X\|$. For closed convex X, this operator is well-defined and Lipschitz-continuous with the Lipschitz constant less or equal 1. We will also notice that this operator is idempotent: $(p \downarrow X) \downarrow X = p \downarrow X$ and linear for projection on linear subspace L of E: $\alpha p \downarrow L = \alpha(p \downarrow L)$ for $\alpha \in \mathbb{R}$ and $(p + q) \downarrow L = p \downarrow L + q \downarrow L$. Of course $p = p \downarrow L + p \downarrow L^\perp$. The point-to-set projection operation is naturally generalized for sets: $X \downarrow \mathcal{A} = \{z = x \downarrow \mathcal{A}, x \in X\}$. Using this generalization it is easy to prove the following result which will be used further on.

Lemma 1. *For any set X and linear space L with orthogonal complement $L^\perp$*

$$X \downarrow L^\perp = (X + L) \cap L^\perp. \tag{2}$$

Proof. If $x^\perp \in X \downarrow L^\perp$ then $x^\perp \in L^\perp$, and there exists $x \in X$ such that $(x^\perp - x)z = 0$ for any $z \in L^\perp$ by optimality condition for $x^\perp$. But that means that $x^\perp - x \in L$, and hence $x^\perp \in x + L \subset X + L$ which proves $X \downarrow L^\perp \subset (X+L) \cap L^\perp$.

The reverse holds true as well. If $x^\perp \in (X + L) \cap L^\perp$ then there exists $x \in X$ such that $(x + L) \cap L^\perp = x^\perp$ and therefore $x^\perp L^\perp$, and $x^\perp \in x + L$ or $x - x^\perp \in L$. Hence $(x - x^\perp)z = 0$ for any $z \in L^\perp$.

By optimality conditions which are in this case necessary and sufficient and uniqness arguments $x^\perp = x \downarrow L^\perp \subset X \downarrow L^\perp$ which complete the proof.

For a closed convex set X denote as $(X)_z$ its support function

$$(X)_z = \min_{x \in X} xz. \tag{3}$$

With the help of this function we can transform the least-norm problem for polyhedron X into unconditional non-smooth problem and vice versa.

Lemma 2 [10]. *Let X is a bounded convex polyhedron. Then*

$$\min_{x \in X} \frac{1}{2}\|x\|^2 = -\min_x \{\frac{1}{2}\|x\|^2 + (X)_x\}. \tag{4}$$

Proof. Indeed, X can be represented as a convex hull of its extreme points: $X = \mathrm{co}(x^i, i = 1, 2, \ldots, N)$. Then

$$\min_{x \in X} \frac{1}{2}\|x\|^2 = \min_{x = Xu, u \in \Delta} \frac{1}{2}\|x\|^2 = \min_{x, u \in \Delta} \max_w \{\frac{1}{2}\|x\|^2 + w(x - Xu)\}$$

$$= \max_w \min_x \{\frac{1}{2}\|x\|^2 + wx\} + \min_{u \in \Delta} (-wXu) = \max_w \{-\frac{1}{2}\|w\|^2 - \max_{z \in X} wz\}$$

$$= -\min_w \{\frac{1}{2}\|w\|^2 + (X)_w\}.$$

It is interesting to note that the initial form of the problem and the final expression do not make any use of the finite representation of X. Nevertheless it is required for the proof.

2   Projection and Linear Optimization

The main result, which is going to be used here, is the fact that the projection operation can be used to solve the linear optimization problem (1) where X is a polyhedral set defined by the different combinations of equalities and inequalities. It was shown in [1] that under not very restrictive assumptions (1) can be solved by a single projection operation, that is $x^\star = (x^0 - \tau c) \downarrow X$ for arbitrary x^0

and $\tau > 0$ large enough. This result is an interesting fact by itself was difficult to estimate from the practical point of view and this note aims to show its computational perspectives.

To begin with we consider (1) in more details as

$$\begin{array}{c} \min \quad cx = cx^\star \\ Ax \leq b \end{array} \tag{5}$$

with n-dimensional $x \in E^n$ and m-dimensional right-hand side $b \in E^m$.

The above-mentioned result claims that

$$\begin{array}{c} \min \quad \|x - (x^0 - \tau c)\|^2 = \|x^\star - (x^0 - \tau c)\|^2 \\ Ax \leq b \end{array} \tag{6}$$

for arbitrary x^0 and $\tau > 0$ large enough. After simple shift in variables $y = x - (x^0 - \tau c)$ the latter problem becomes

$$\begin{array}{c} \min \quad \|y\|^2 = \|y^\star\|^2 \\ Ay \leq b' \end{array} \tag{7}$$

where $b' = b - A(x^0 - \tau c)$. It means that without any loss of generality, we may consider the least-norm problem for the feasible set described by the same linear operator and recalculated right-hand sides of constraints.

The following lemma justifies the use of convenient exact penalty function based on Chebyshev-like cost of constraint violation for solving problems like (7).

Lemma 3. *There exists $\Gamma > 0$ such that for all $\gamma \geq \Gamma$ the problem (5) and*

$$\min\{\tfrac{1}{2}\|x\|^2 + \gamma|Ax - b|^+\} = \min\{\tfrac{1}{2}\|x\|^2 + \gamma\pi_{A,b}(x)\} \tag{8}$$

where

$$\pi_{A,b}(x) = |Ax - b|^+ = \max\{0, \max_{i=1,2,\ldots,n} (Ax - b)_i\},$$

and $(Ax - b)_i$—i-th coordinate of the vector $Ax - b$ are equivalent.

For further developments we complement the vector x with one additional coordinate, i.e. define $\bar{x} = (x, \chi) \in E^{n+1}$ and extended $m \times (n + 1)$ matrix $\bar{A} = \|A \vdots - b\|$. Then the penalty term in (5) can be rewritten as

$$\pi_{A,b}(x) = \gamma|Ax - b|^+ = \gamma\max\{0, \max_{i=1,2,\ldots,m} (\bar{A}\bar{x})_i\}$$

$$= \gamma\max\{0, \max_{u \in \Delta_m} u\bar{A}\bar{x}\} = \gamma \max_{z \in \mathrm{co}(\{0, \bar{A}^T \Delta_m\})} z\bar{x} \tag{9}$$

$$= \gamma(\mathrm{co}(\{0, \bar{A}^T \Delta_m\}))_{\bar{x}} = (\mathrm{co}(\{0, \gamma\bar{A}^T \Delta_m\}))_{\bar{x}}$$

with $\bar{x} = (x, 1)$.

Combining Lemmas 2, 3, and the penalty formula (9) we can obtain the following theorem.

Theorem 1 [11]. *Solution of the problem* $\min \|x\|^2, Ax \le b$ *can be derived from the solution* $p_e \downarrow K$ *of the projection problem*

$$\min_{\bar{x} \in K} \|\bar{x} - p_e\|^2 = \|p_e \downarrow K - p_e\|^2 \tag{10}$$

where $\bar{x} = (x, \chi) \in E^{n+1}$, $p_e = (0, 0, \ldots, 0, 1) \in E^{n+1}$, *and* $K = \bar{A}^T E_+^m$ *is a cone, generated by rows of the extended matrix* $\bar{A}$.

Proof. Referring to Eq. 8 consider the problem

$$\begin{aligned}
\min_{\bar{x} p_e = 1} & \{\frac{1}{2}\|\bar{x}\|^2 + \gamma(D)_{\bar{x}}\} \\
= \max_{\omega} \min_{\bar{x}} & \{\frac{1}{2}\|\bar{x}\|^2 + \gamma(D)_{\bar{x}} + \omega(\bar{x} p_e - 1)\} \\
= \max_{\omega} & \{-\omega + \min_{\bar{x}} \{\frac{1}{2}\|\bar{x}\|^2 + \gamma(D)_{\bar{x}} + \omega \bar{x} p_e\}\} \\
= \max_{\omega} & \{-\omega + \min_{\bar{x}} \{\frac{1}{2}\|\bar{x}\|^2 + (\gamma D + \omega p_e)_{\bar{x}}\}\} \tag{11} \\
= \max_{\omega} & \{-\omega - \min_{\bar{x} \in \gamma D + \omega p_e} \frac{1}{2}\|\bar{x}\|^2\} \\
= \max_{\omega} & \{-\omega - \min_{\bar{x} \in \gamma D} \frac{1}{2}\|\bar{x} - \omega p_e\|^2\} \\
= -\omega_\star & - \frac{1}{2} \min_{\bar{x} \in \gamma D} \|\bar{x} - \omega_\star p_e\|^2,
\end{aligned}$$

where ω is the Lagrange multiplier for the constraint $\bar{x} p_e = 1$ and $\omega_\star$ is its unique optimal value.

Together with γD we consider the cone $K = \mathrm{Co}(D)$ and prove that

$$\min_{\bar{x} \in \gamma D} \|\bar{x} - \omega_\star p_e\|^2 = \min_{\bar{x} \in K} \|\bar{x} - \omega_\star p_e\|^2 = \delta(\omega_\star)^2, \tag{12}$$

for $\gamma > 0$ large enough.

Indeed, let

$$\min_{\bar{x} \in K} \|\bar{x} - \omega_\star p_e\|^2 = \|\bar{x}_\star - \omega_\star p_e\|^2, \tag{13}$$

where $\bar{x}_\star = \sum_{i=1}^{N} \mu_i \bar{A}_i$. Of course

$$\|\bar{x}_\star - \omega_\star p_e\|^2 \le \min_{\bar{x} \in \gamma D} \|\bar{x} - \omega_\star p_e\|^2 \tag{14}$$

as $\gamma D \subset K$ for any γ.

On the other hand

$$\bar{x}_\star = \sum_{i=1}^{N} \mu_i \bar{A}_i \in \gamma D = \{z = \sum_{i=1}^{N} \lambda_i \bar{A}_i, \lambda_i \in [0, \gamma]\}$$

for $\gamma > \max_i \mu_i$, so

$$\|\bar{x}_\star - \omega_\star p_e\|^2 \geq \min_{\bar{x} \in \gamma D} \|\bar{x} - \omega_\star p_e\|^2 \tag{15}$$

for γ large enough. Combination of (14) and (15) proves (12).

To complete we specify the dependence of $\delta(\omega_\star)$ upon $\omega_\star$ in (12). By definition

$$
\begin{aligned}
\delta(\omega)^2 &= \min_{\bar{x} \in K} \|\bar{x} - \omega p_e\|^2 = \omega^2 \min_{\bar{x} \in K} \|\bar{x}/\omega - p_e\|^2 \\
&= \omega^2 \min_{\bar{x} \in K/\omega} \|\bar{x} - p_e\|^2 = \omega^2 \min_{\bar{x} \in K} \|\bar{x} - p_e\|^2 = \omega^2 \delta^2
\end{aligned}
\tag{16}
$$

and the dual problem with respect to ω becomes $\min_\omega \{\omega + \frac{1}{2}\delta^2 \omega^2\}$ with the solution $\omega_\star = -\delta^{-2}$.

From the computational point of view, we replaced the linear problem with many constraints (we need at least $n + 1$ constraints to obtain a bounded polyhedron) with the cone projection problem of less dimensionality but many cone generators. The correspondence between (1) and the projection problem (10) is graphically demonstrated by Fig. 1.

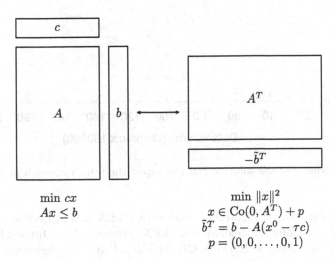

Fig. 1. Transformation of the linear optimization problem (1) into projection problem (10)

As seen from Fig. 1, this transformation adds one additional variable into the projection problem, and uses rows-constraints of the original problem as the code generators. There is a hope that the projection algorithm may effectively select active generators which actually determine solution of (10) and thus have good computational behavior.

These hopes were confirmed by the computational experiments, where we compared run times for the solution of large-scale dense linear optimization problems by projection algorithm and state-of-art optimization solver CPLEX. This set of experiments was conducted with linear optimization problems (5) with random dense matrices A generated according to the suggestion of Ph. Wolfe [4] which produces sufficiently different random rows of constraints. The results of these experiments are shown in Fig. 2, where you can see the dependence of run-time for the solver CPLEX [5] (dotted line) and the projection algorithm (two kinds of crosses) upon the total data volume measured in units of 100 000 double-precision numbers. To test the stability of the projection algorithm, we run it twice, which is shown in the Figure.

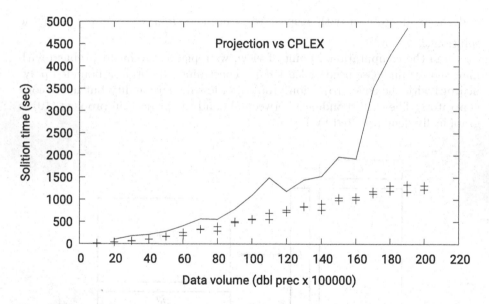

Fig. 2. Dependence of the solution run-time upon the total volume of problem data

For technical and administrative reasons, CPLEX and projection algorithms were executed on different machines. CPLEX (version 12.6.1.0) used Intel Xeon E7-4870 (1 socket, 8 cores, RAM 96 Gb 2.4 GHz). Projector algorithm used Intel Core i3-3220 CPU 3.30 GHz, 4 cores computer with 4 Gb RAM packed into HP nettop.

For CPLEX, each test problem was presented in MPS format. We allowed CPLEX to choose the best solving algorithm and presolver automatically for the input problem and to use the maximum possible number of threads for any of the parallel CPLEX optimizers. The analysis of the CPLEX protocols demonstrated first that the automatic presolve procedure excluded no rows or columns from any test problem. It means that the problems were complex enough. Secondly, the conducted experiments showed that CPLEX used the dual simplex method for the number of nonzeros elements of the constraint matrix less than $3 \cdot 10^6$, and the barrier algorithm otherwise. So it can be said that CPLEX tried its best.

The linear optimization problems used 100% dense random matrices with elements in the range $[-5, +5]$ and the rows-to-columns ratio of about $3 : 1$. All in all, twenty test problems (4) were generated in this experiment if disregard repetition of the test computations for projection algorithm. The number of nonzero elements of the constraint matrix varied from 10^6 to $2 \cdot 10^7$ with the step size 10^6, so the pick memory usage for input data only was about 160 Mb. It can be seen from Fig. 2 that projection routine steadily outperform CPLEX and its advantage becomes large when problems grow. Further details of the experiment, test generator, etc., are available upon request.

Of course, there are other linear optimization solvers, but they are hardly free accessible for the problems of this side. For instance, the popular NEOS platform provides free access to such solvers as MINOS, GUROBI, but for the problems which can be written into the file up to 16 MB. Our smallest test problem that has been generated for numerical experiments occupies 21,25 MB in MPS format.

3   Canonical Linear Optimization Problems

In this section, we consider the standard formulation of the problem (1) where contrains are divided into two sets X_A and X_P accoding to the textbook formulation of the linear optimization problems. There the feasible set X is considered as the intersection of affine subspace X_A, described by the set of linear equations $X_A = \{x : Ax = b\}$ and polyheadron $X_P = \{x : Px \leq q\}$, which is mostly often just the nonnegativity constraints.

This representation provides a convenient framework for the simplex method and often really takes place in practical problems where these parts are very different in their structure and size. The affine subspace X_A can be responsible for something like material an/or flow balances and has essentially lesser rows than columns, which represent processes itself. Polyhedron X_P describes technological and physical constraints for process variables and typically has the number of rows of the order of variables, by the necessity for instance, including non-negativity or boundary constraints on variables. However, it may include much more rows of additional restrictions.

The constraines may be subdivided in additional subgroupes, related for instance to diffrent time periods in dynamic linear programming. If these groupes have a few linking variables it may be used to construct effective numerical methods [9].

Taking into account this decomposition the feasible set X becomes in algebraic term the intersection

$$X = X_A \cap X_P = \{x : Ax = b\} \cap \{x : Px \leq q\}. \tag{17}$$

After transformation of (1) into dual projection problem, described early in the Sect. 2 the cone $Co(X)$ will look like following:

$$K = \{z : z = A^T u + P^T v, u \in E_u, v \in E_v^+\} = A + P \tag{18}$$

where we misuse the same notations for the different things:

$A = A^T E_u$—the linear row space of the matrix A;
$P = P^T E_v^+$—convex closed cone, generated by rows of the matrix P

Without going into finer decomposition the main projection problem becomes

$$\min_{z \in Z} \|p - z\|^2 = \|p - p \downarrow Z\|^2, \qquad (19)$$

where $Z = A + P$.

The problem (19) in its turn can be rewritten as the least distance problem between two sets: $A_\alpha = A - \alpha p$ and $P_\alpha = (1 - \alpha)p - P$ where α is an arbitrary constant which can be used to split p between these two sets. Consequently (19) becomes

$$\min_{\substack{z_A \in A_\alpha \\ z_P \in P_\alpha}} \|z_A - z_P\|^2 = \|z_A^\star - z_P^\star\|^2 \qquad (20)$$

the notable problem for computational mathematics, which dates back to [2].

3.1   Alternative Projection Algorithms

As we are specifically interested in large-scale problems, we are looking for the most straightforward methods for solving (20). A very natural candidate for this is the alternating projection method of von Neumann [2] which is described in Algorithm 1. This method is most often considered for the solution of convex feasibility problems, but fortunately, for the case of nonoverlapping convex sets, it solves namely the least distance problem (20).

Data: Sets $\mathrm{Co}(X), \mathrm{lin}(A)$, the vector p.
Result: $z_A^\star \in \mathrm{lin}(A)$ and $z_X^\star \in \mathrm{Co}(X)$ which solve (20).
Initialization: Set $k = 0$, $z_X^k = p, z_A^k = 0$;
while z_A^k and z_X^k do not solve (20) **do**
    Compute $z_A^k = (p + z_X^k) \downarrow A$;
    Compute $z_X^k = z_A^k \downarrow \mathrm{Co}(X)$;
    Increment iteration counter $k \to k + 1$, etc.
end

Algorithm 1: Alternating projection algorithm

Figure 3 demonstrates not very impressive convergence of this algorithm for computing the minimal distance between 40-dimensional linear subspace $\mathcal{A} = A^T E^{40}$ of 400-dimensional non-negative ortant E_+^{400} shifted by a random vector to avoid intersection of these two sets. The algorithm demonstrates the expected linear convergence with the multiplier of the order of 0.985 which implies the 10-fold accuracy improvement in 150 iterations. This convergence was accelerated using many ideas, the earliest is probably [6], and see, for instance, [3,7] and references therein for further developments. However, these ideas mainly focused either on the convex feasibility problem or on the case of linear subspaces. At the moment, we have no experience with these suggestions but we intend to study these possibilities.

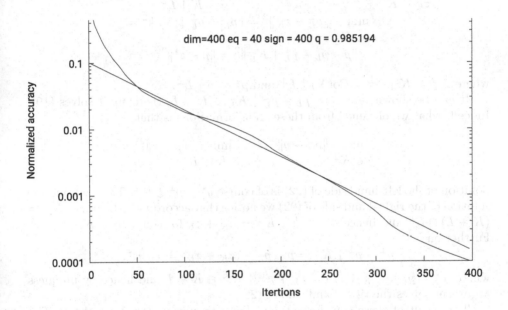

Fig. 3. Von Neumann alternating projection algorithm

For a moment, we consider the case of two sets, one of them being a linear subspace and the other being a convex polyhedral cone with a finite number of generators. The problem of two sets is a special case anyway, see [8], and subspace-cone is also a special one which allows to transform (19) into reduced projection problem which can be solved in a finite number of elementary projection operations. There are suggestions for accelerating the convergence of this method; see f.i. [7] and references therein.

3.2   Single Projection Algorithm

The problem (19) can be solved in two-stage manner:

$$
\begin{aligned}
\min_{\substack{z = z_L + z_K \\ z_L \in L,\ z_K \in K}} \|p - z\|^2 &= \min_{z_K \in K} \min_{z_L \in A} \|p - z_L - z_K\|^2 = \\
&\min_{z_K \in K} \|p - z_K - (p - z_K) \downarrow L\|^2 = \\
&\min_{z_K \in K} \|p - p \downarrow L - (z_K - z_K \downarrow L)\|^2 = \\
\min_{z_K \in K} \|p \downarrow L^\perp - z_K \downarrow L^\perp)\|^2 &= \min_{z_K \in K \downarrow L^\perp} p_L^\perp - z_K\|^2 = \\
\min_{z_K \in K_L^\perp} \|p_L^\perp - z_K\|^2 &= \|p_L^\perp - p_L^\perp \downarrow K_A^\perp\|^2 = \\
\|p - p_L - p_L^\perp \downarrow K_L^\perp\|^2 &= \|p - z^\star\|^2 .
\end{aligned}
\tag{21}
$$

where $K_L^\perp = K \downarrow L^\perp = \mathrm{Co}(X) \downarrow L^\perp$ and $p_L^\perp = p \downarrow L^\perp$.

It can be shown that $z^\star = p_L + p_L^\perp \downarrow K_L^\perp \in K + L$, therefore it solves (19). Indeed, what we obtained from these transformations is that

$$
\min_{y \in K^\perp} \|p^\perp - y\|^2 = \min_{z \in K + L} \|p - z\|^2
\tag{22}
$$

Solution of the left-hand side of (22) is of course $y^\star = p^\perp \downarrow K^\perp$. To determine the solution of the right-hand side of (22) we notice that according to Lemma 1 $K^\perp = (K + L) \cap L^\perp$ and hence $y^\star = p^\perp \downarrow K^\perp = z_K + z_L$ for some $z_K \in K$, $z_L \in L$. Further on

$$
p^\perp - p^\perp \downarrow K^\perp = p - p_L - (z_K + z_L) = p - w^\star,
$$

where $w^\star = p_L + (z_K + z_L) = p_L + p^\perp \downarrow K^\perp \in K + L$ and hence by uniqness argument solves the right-hand side of (22).

The result of these transformations allows to suggest the Algorithm 2. The main computational burden of Algorithm 2 consists in computing $p_A^\perp \downarrow K_A^\perp$ which in turn involves the projection of the cone K onto $A^\perp$.

Data: Sets $\mathrm{Co}(X), \mathrm{lin}(A)$, the vector p.
Result: $z^\star \in \mathrm{lin}(A) + \mathrm{Co}(X)$ which solves (20).
Step 1. Compute $p_A = p \downarrow A$ and set $p_A^\perp = p - p_A$;
Step 2. Compute $K_A = K \downarrow A$ and set $K_A^\perp = K - K_A$ component-wise, that is by generators;
Step 3. Compute $p_A^\perp \downarrow K_A^\perp$;
Step 4. Set $z^\star = p_A + p_A^\perp \downarrow K_A^\perp$.

Algorithm 2: Cone projection algorithm

The computational performance of this algorithm can be estimated from the experiments presented on Fig. 4. The exciting feature of these experiments is that with the growing problem, the dependence is becoming closer to linear and so the algorithm may outperform the conventional algorithms, which generally demonstrate polynomial growth.

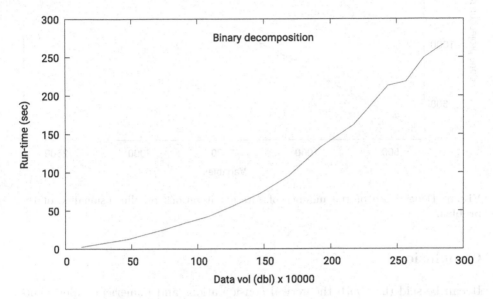

Fig. 4. Dependence of the solution run-time upon the total volume of problem data

Another set of experiments was performed with the randomly generated linear optimization problems, where the numbers of the equality constrains consisted 10% of the number of variables, so the peak number of nonzero elements was about 600 ths, and the run-time for the largest test was about 210 s.

The detailed analysis of the performance of this algorithm reveals that the number of internal iterations in the procedure for computing $p_A^\perp \downarrow K_A^\perp$ grows practically linear with the dimensionality of the problem, as presented on Fig. 5.

It means that the algorithm pracically does not err in selecting active generators in the conical part of constraints, so the major improvements which can be done consist in speeding up the linear algebra operations and selections of candidats to enter the basis. Great part of these operations can be, for instance, parallelized.

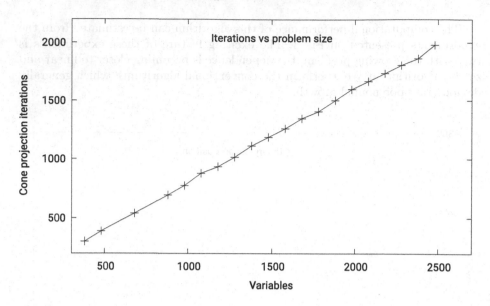

Fig. 5. Dependence of the number of internal iterations on dimensionality of the problem

Conclusion

It can be said that both theoretical considerations and numerical experiments allow saying that the idea of using projection operation for solving large linear optimization problems has certain computational advantages. However, this algorithm's key operation–projection of a point onto a polyhedral cone needs substantial speedup, connected, for instance, with the effective use of sparsity features and/or parallel computations.

Acknowledgements. The research contribution of E. Nurminski was supported by the Ministry of Science and Higher Education of the Russian Federation, agreement No. 075-02-2021-1395. N. Shamray acknowledges the support from the grant 18-29-03071 of the Russian Foundation for Basic Research.

References

1. Nurminski, E.A.: Single-projection procedure for linear optimization. J. Glob. Optim. **66**(1), 95–110 (2015). https://doi.org/10.1007/s10898-015-0337-9
2. von Neumann, J.: Functional Operators, vol. II. Princeton University Press, Princeton (1950)
3. Escalante, R., Raydan, M.: Alternating Projection Methods. SIAM, Philadelphia (2011)
4. Wolfe, P.: Finding the nearest point in a polytope. Math. Program. **11**, 128–149 (1976). https://doi.org/10.1007/BF01580381

5. IBM ILOG CPLEX Optimization Studio. https://www.ibm.com/products/ilog-cplex-optimization-studio. Accessed 18 May 2021
6. Gurin, L.G., Polyak, B.T., Raik, E.V.: The method of projections for finding the common point of convex sets. USSR Comput. Math. Math. Phys. **7**(6), 1–24 (1967)
7. Bauschke, H.H., Deutsch, F., Hundal, H., Park, S.-H.: Accelerating the convergence of the method of alternating projections. Trans. Am. Math. Soc. **355**(9), 3433–3461 (2003)
8. Bauschke, H.H., Borwein, J.M.: On the convergence of von Neumann's alternating projection algorithm for two sets. Set-Valued Anal. **1**, 185–212 (1993). https://doi.org/10.1007/BF01027691
9. Censor, Y.: Row-action methods for huge and sparse systems and their applications. SIAM Rev. **23**(4), 444–466 (1981)
10. Nurminski, E.A.: Acceleration of iterative methods for projection onto a polyhedron. Far Eastern Math. Digest **1**, 51–62 (1995). (in Russian)
11. Nurminski, E.A.: Projection onto polyhedra in outer representation. Comput. Math. Math. Phys. **148**(3), 367–375 (2008). https://doi.org/10.1134/S0965542508030044

The Variant of Primal Simplex-Type Method for Linear Second-Order Cone Programming

Vitaly Zhadan[✉][iD]

Federal Research Center "Computer Science and Control" of the Russian Academy of Sciences, 40, Vavilova Street, Moscow 119333, Russia
zhadan@ccas.ru

Abstract. The linear second-order cone programming problem is considered. For its solution the variant of the primal simplex-type method is proposed. The notion of the **S**-extreme point of the feasible set is introduced. Among all **S**-extreme points the regular **S**-extreme points are regarded separately. The passage from the regular **S**-extreme point to another one is described. The method can be treated as the generalization of the primal simplex-type method for linear programming. At each iteration the dual variable together with the dual weak variable are defined. As in linear programming, the basis of the extreme point is used. Among all basic variables the facet basic variables and the interior basic variable are selected.

Keywords: Linear second-order cone programming · Primal simplex-type method · Extreme and **S**-extreme points

1 Introduction

The standard linear second-order cone programming problem (SOCP) is the problem of minimizing the linear goal function on the intersection of the linear manyfold with the direct product some second-order cones [1,2]. Many others optimization problems, including combinatorial optimization problems, can be reformulated as SOCP [1,3].

The main approach to constructing numerical techniques for solving the SOCP is based on the generalization of the corresponding methods from linear programming. The interior point methods are especially well generalized for the SOCP [4,5]. Unlike to interior point methods, the generalization of the simplex-type algorithms for SOCP is encountered with many difficulties. Certainly, this is because the second-order cone is not polyhedral. The other difficulty is connected with the possible different dimensions of the partial second-order cones. Nevertheless, there are some variants of the primal simplex-type methods for

The research was supported by Russian Science Foundation (project No. 21-71-30005).

N. N. Olenev et al. (Eds.): OPTIMA 2021, LNCS 13078, pp. 64–75, 2021.
https://doi.org/10.1007/978-3-030-91059-4_5

partial cases of the SOCP [6,7]. The principal simplex-type method for cone programming is given in [8] too.

In [9] the variant of the primal simplex-type method for SOCP had been proposed. In this method the passage from one extreme point to another one was described under assumption that all extreme points in iterative process are regular. In other words, all these extreme points are non-degenerate. Due to this regularity the system of linear equations for determining the dual variable has the single solution. But in many cases we have the non-regular extreme points in which the system of linear equations for computing the dual variable is undetermined. In order to overcome this difficulty in present paper the special notion of the **S**-extreme point is introduced. This notion is based on consideration of sub-cones of less dimensions with respect to the initial second-order cones.

The paper is organized as follows. In Sect. 2, the statement of SOCP is given. In Sect. 3, the notion of the **S**-extreme points is introduced. The pivoting in the case of regular **S**-extreme point is described in Sect. 4. Here the definitions of basis of the **S**-extreme point and the definitions of facet and interior variables are given too.

2   The Problem Statement

Let K^n denote the second order cone (the Lorentz cone) in $\mathbb{R}^n$. According to its definition

$$K^n = \left\{ x = [x^1; \bar{x}] \in \mathbb{R} \times \mathbb{R}^{n-1} : \ x^1 \geq \|\bar{x}\| \right\},$$

where $\| \cdot \|$ is the standard Euclidean norm. Here and in what follows the point with comma in enumerations of vectors or components of the vector means that these vectors or components are placed one under another. The cone K^n is self-dual, i.e. $(K^n)^* = K^n$. It induces in $\mathbb{R}^n$ the partial order, namely, $x_1 \succeq_{K^n} x_2$, if $x_1 - x_2 \in K^n$.

Consider the cone programming problem

$$\min \sum_{i=1}^{r} \langle c_i, x_i \rangle,$$
$$\sum_{i=1}^{r} A_i x_i = b, \quad x_1 \succeq_{K^{n_1}} 0_{n_1}, \ \ldots, x_r \succeq_{K^{n_r}} 0_{n_r}. \tag{1}$$

Here $c_i \in \mathbb{R}^{n_i}$, $1 \leq i \leq r$, and $b \in \mathbb{R}^m$. The matrix A_i has the dimension $m \times n_i$, $1 \leq i \leq r$. Angle brackets denote the Euclidean inner product in $\mathbb{R}^{n_i}$.

The following problem is dual to (1)

$$\max \langle b, u \rangle,$$
$$A_i^T u + y_i = c_i, \ 1 \leq i \leq r, \quad y_1 \succeq_{K^{n_1}} 0_{n_1}, \ \ldots, \ y_r \succeq_{K^{n_r}} 0_{n_r}, \tag{2}$$

where $u \in \mathbb{R}^m$.

Let $n = n_1 + \cdots + n_r$. Using the notations

$$c = [c_1; \ldots; c_r] \in \mathbb{R}^n, \quad x = [x_1; \ldots; x_r] \in \mathbb{R}^n; \quad y = [y_1; \ldots; y_r] \in \mathbb{R}^n,$$

and also

$$\mathcal{A} = [A_1, \ \ldots A_r, \ ], \quad \mathcal{K} = K^{n_1} \times \cdots \times K^{n_r},$$

the pair of problems (1) and (2) may be written in the form:

$$\min \langle c, x \rangle, \quad Ax = b, \quad x \succeq_{\mathcal{K}} 0_n, \tag{3}$$

$$\max \langle b, u \rangle, \quad A^T u + y = c, \quad y \succeq_{\mathcal{K}} 0_n. \tag{4}$$

We assume that solutions of problems (3) and (4) exist, and the rows of the matrix $\mathcal{A}$ are linear independent. The feasible set of problem (3) denote by $\mathcal{F}_P$.

The necessary optimality conditions for problems (3) and (4) are following:

$$\langle x, y \rangle = 0, \quad Ax = b, \quad A^T u + y = c, \quad x \in \mathcal{K}, \quad y \in \mathcal{K}. \tag{5}$$

The simplex-method is one of the ways for solving this system of equations and inclusions,

3   The Extreme and S-Extreme Points of $\mathcal{F}_P$

Let $x \in \mathcal{K}$. Choose among all components x_i of the vector x the null components and the non-null components. In addition, we choose among non-null components the interior components x_i, which belong to the interior of the cone K^{n_i}, and the boundary components x_i, which belong to the boundary of K^{n_i} (more exactly, which belong to any non-zero face of this cone). We compose from boundary, interior and zero components of x three blocks: x_F, x_I and x_O. Without loss of generality suppose that at the given point x these blocks are located at the indicate order, that is $x = [x_F; x_I; x_O]$. Suppose also that

$$x_F = [x_1; \ldots; x_{r_F}], \quad x_I = [x_{r_F+1}; \ldots; x_{r_F+r_I}], \quad x_N = [x_{r_F+r_I+1}; \ldots; x_{r_F+r_I+r_O}].$$

Thus, the first block consists of the r_F components. The second and the third blocks consist of r_I and r_O components, respectfully. Some blocks may be empty, in this case the corresponding numbers r_F, r_I or r_O are zero. We have $r_F + r_I + r_O = r$.

According to introduced partition of x onto three blocks of components there exists the partition of the index set $J^r = [1 : r]$ onto three subsets

$$J_F(x) = [1, \ldots, r_F], \quad J_I(x) = [r_F + 1, \ldots, r_B], \quad J_O(x) = [r_B + 1, \ldots, r],$$

where $r_B = r_F + r_I$. Below the notation $J_B(x) = J_F(x) \cup J_I(x)$ is also used.

For each non-zero component x_i, $i \in J_B(x)$, the following decomposition

$$x_i = \eta_{i,1} \, \mathbf{e}_{i,1} + \eta_{i,n_i} \, \mathbf{e}_{i,n_i}, \tag{6}$$

takes place (see [1]). Here $\mathbf{e}_{i,1}$ and $\mathbf{e}_{i,n_i}$ are the vectors having the form

$$\mathbf{e}_{i,1} = \frac{1}{\sqrt{2}} \left[1; \frac{\bar{x}_i}{\|\bar{x}_i\|} \right], \quad \mathbf{e}_{i,n_i} = \frac{1}{\sqrt{2}} \left[1; -\frac{\bar{x}_i}{\|\bar{x}_i\|} \right],$$

and

$$\eta_{i,1} = \frac{1}{\sqrt{2}} \left(x_i^1 + \|\bar{x}_i\| \right), \qquad \eta_{i,n_i} = \frac{1}{\sqrt{2}} \left(x_i^1 - \|\bar{x}_i\| \right).$$

Both vectors $\mathbf{e}_{i,1}$ and $\mathbf{e}_{i,n_i}$ have the unit length and are orthogonal each to other. If $x_i \in K^{n_i}$, then $\eta_{i,1} \geq 0$ and $\eta_{i,n_i} \geq 0$. Moreover, if $i \in J_F(x)$, then $\eta_{i,1} = \sqrt{2} x_i^1$, $\eta_{i,n_i} = 0$.

With each non-zero point $x_i \in \mathbb{R}^{n_i}$ the following symmetric matrix

$$\mathrm{Arr}\,(x_i) = \begin{bmatrix} x_i^1 & \bar{x}_i^T \\ \bar{x}_i & x_i^1 I_{n_i-1} \end{bmatrix}.$$

is associated. Here and in what follows I_k is a unite matrix of order k. The point x_i belongs to the cone K^{n_i} if and only if the matrix $\mathrm{Arr}\,(x_i)$ is positive semi-definite. Moreover, if x_i is an interior point of K^{n_i}, the matrix $\mathrm{Arr}\,(x_i)$ is positive definite.

Denote by $\mathbf{E}_i$ an orthogonal matrix with columns being the eigenvectors of the matrix $\mathrm{Arr}\,(x_i)$. The vectors $\mathbf{e}_{i,1}$ and $\mathbf{e}_{i,n_i}$ are the eigenvectors of $\mathrm{Arr}\,(x_i)$. The matrix $\mathbf{E}_i$ can be represented in the form

$$\mathbf{E}_i = \begin{bmatrix} \mathbf{e}_{i,1}, & \mathbf{E}_i', & \mathbf{e}_{i,n_i} \end{bmatrix},$$

where $\mathbf{E}_i'$ is the matrix with dimension $n_i \times (n_i - 2)$. All columns of $\mathbf{E}_i$ are orthogonal each to others and have the unit length. Moreover, columns of $\mathbf{E}_i$ are orthogonal to the vectors $\mathbf{e}_{i,1}$ and $\mathbf{e}_{i,n_i}$.

In what follows, we need in the definition of *an extreme point* of the feasible set $\mathcal{F}_P$ of problem (3). Let $F_{\min}(x|\mathcal{K})$ be the minimal face of the cone $\mathcal{K}$ containing the point $x \in \mathcal{K}$. Moreover, let $\mathcal{N}(\mathcal{A})$ be the null space of the matrix $\mathcal{A}$. According to [8] $x \in \mathcal{F}_P$ be *an extreme point* of the set $\mathcal{F}_P$, if

$$\mathrm{lin}\,(F_{\min}(x\,|\,\mathcal{K})) \cap \mathcal{N}(\mathcal{A}) = \{0_n\},$$

where $\mathrm{lin}\,(F_{\min}(x|\mathcal{K}))$ is a liner hull of the face $F_{\min}(x|\mathcal{K})$. Therefore, the dimension of the minimal face $F_{\min}(x|\mathcal{K})$ and the dimension of the null-space $\mathcal{N}(\mathcal{A})$, which is equal to $n - m$, does not exceed the dimension of the whole space $\mathbb{R}^n$. This circumstance imply the inequality: $\dim F_{\min}(x|\mathcal{K}) \leq m$.

We have

$$\dim F_{\min}(x_i \,|\, K^{n_i}) = \begin{cases} 1, & i \in J_F(x), \\ n_i, & i \in J_I(x). \end{cases}$$

Therefore, for the dimension of the minimal face, to which the extreme point x belongs, the inequality $\dim F_{\min}(x\,|\,\mathcal{F}_P) \leq m$ must hold, where

$$\dim F_{\min}(x\,|\,\mathcal{F}_P) = r_F + n_I, \qquad n_I = n_I(x) = \sum_{i \in J_I(x)} n_i.$$

An extreme point $x \in \mathcal{F}_P$ is called *regular*, if $\dim F_{\min}(x\,|\,\mathcal{F}_P) = m$. In the case, where $\dim F_{\min}(x\,|\,\mathcal{F}_P) < m$, an extreme point $x \in \mathcal{F}_P$ is called *irregular*.

Let $x = [x_F; x_I; x_O]$ be the extreme point of the set $\mathcal{F}_P$. Let, in addition, $x_i = \eta_{i,1}\mathbf{e}_{i,1}$, where $i \in J_F(x)$, and

$$\eta_i = [\eta_{i,1}; \; 0; \; \ldots; 0; \; \eta_{i,n_i}] \in \mathbb{R}_+^{n_i}, \quad i \in J_I(x).$$

We set

$$\eta_F = [\eta_{1,1}, \ldots, \eta_{r_F,1}] \in \mathbb{R}_+^{r_F}, \qquad \eta_I = [\eta_{r_F+1}; \; \ldots; \; \eta_{r_B}] \in \mathbb{R}_+^{n_I}.$$

Assume also that $x_i = \mathbf{E}_i\eta_i$, $i \in J_I(x)$.

Denote

$$\mathcal{A}_F^{\mathbf{E}} = \left[A_1\mathbf{e}_{1,1}, \ldots, A_{r_F}\mathbf{e}_{r_F,1}\right], \quad \mathcal{A}_I^{\mathbf{E}} = \left[A_{r_F+1}\mathbf{E}_{r_F+1}, \ldots, A_{r_B}\mathbf{E}_{r_B}\right], \quad (7)$$

and join together both matrices in the unique matrix

$$\mathcal{A}_B^{\mathbf{E}} = [\mathcal{A}_F^{\mathbf{E}}, \mathcal{A}_I^{\mathbf{E}}] = \left[A_1\mathbf{e}_{1,1}, \ldots, A_{r_F}\mathbf{e}_{r_F,1}, \; A_{r_F+1}\mathbf{E}_{r_F+1}, \ldots, A_{r_B}\mathbf{E}_{r_B}\right].$$

The matrix $\mathcal{A}_B^{\mathbf{E}}$ is consisted from $n_B = r_F + n_I$ columns. If the extreme point x is irregular, then $n_B < m$. If x is a regular extreme point, then $n_B = m$. The following equality $\mathcal{A}_B^{\mathbf{E}}\eta_B = b$ takes place, where $\eta_B = [\eta_F; \eta_I]$. In the case, where x is a regular extreme point, the matrix $\mathcal{A}_B^{\mathbf{E}}$ is called *the basis* of this point.

Proposition 1 (The criterion of an extreme point [1]). *The point $x \in \mathcal{F}_P$ is an extreme point, if and only if the columns of the matrix $\mathcal{A}_B^{\mathbf{E}}$ are linear independent.*

Let us introduce the notion of **S**-*extreme point* of the set $\mathcal{F}_P$. This notion is based on replacement of the cone K^{n_i} by the cone $K_S^{n_i} \subseteq K^{n_i}$ of the lesser dimension.

Assume that we have a point $x_i \in \mathbb{R}^{n_i}$. Assume also that the corresponding matrix $\mathrm{Arr}\,(x_i)$ is determined. Moreover, the orthogonal matrix $\mathbf{E}_i$, consisting from eigenvectors of $\mathrm{Arr}\,(x_i)$, is determined too.

Suppose that in the index set $J^{n_i} = [1 : n_i]$, $i \in J_I(x)$, by some manner the subset $J_S^{n_i} \subseteq J^{n_i}$ is selected. The total number of indexes containing in this subset is equal to n_i^S. Suppose that $n_i^S \geq 2$, and that the indexes 1 and n_i be a part of $J_S^{n_i}$.

Further, introduce into consideration the square matrix $\mathbf{S}_i$ of order n_i with columns being unite vectors e_j of the space $\mathbb{R}^{n_i}$ or null vectors. More precisely, we assume $\mathbf{S}_{i,j} = e_j$, if $j \in J_S^{n_i}$, and we assume $\mathbf{S}_{i,j} = 0_j$ at the opposite case. Here $\mathbf{S}_{i,j}$ is the jth column of the matrix $\mathbf{S}_i$.

The following matrix

$$\mathbf{S}_i = \begin{bmatrix} 1 & 0 & \ldots & 0 & 0 \\ 0 & 0 & \ldots & 0 & 0 \\ & & \vdots & & \\ 0 & 0 & \ldots & 0 & 1 \end{bmatrix}, \quad (8)$$

is an example of the matrix $\mathbf{S}_i$. It corresponds to the case, where $J_S^{n_i} = \{1, n_i\}$. Thus, in this case the variable $x_i = \mathbf{E}_i \mathbf{S}_i \eta_i$, where $\eta_i \in \mathbb{R}^{n_i}$, must belong to the two-dimensional linear subspace, determining by vectors $\mathbf{e}_{i,1}$ and $\mathbf{e}_{i,n_i}$. If $J_S^{n_i} = J^{n_i}$, then $\mathbf{S}_i = I_{n_i}$.

Assume that at the point $x \in \mathcal{F}_P$ the matrices $\mathbf{S}_i$, $i \in J_I(x)$, are specified. In the matrix $\mathbf{S}_i$ we extract the sub-matrix $\mathbf{S}_i'$ with dimension $n_i \times n_i^S$, removing from $\mathbf{S}_i$ all null columns and leaving only unit columns. Replace the matrix $\mathcal{A}_I^{\mathbf{E}}$ in (7) on the following one

$$\mathcal{A}_I^{\mathbf{E},\mathbf{S}'} = \left[A_{r_F+1} \mathbf{E}_{r_F+1} \mathbf{S}_{r_F+1}', \ \ldots, \ A_{r_B} \mathbf{E}_{r_B} \mathbf{S}_{r_B}' \right]. \tag{9}$$

Join together the matrix $\mathcal{A}_F$ from (7) and the matrix $\mathcal{A}_I^{\mathbf{E},\mathbf{S}'}$ from (9) to the unique matrix $\mathcal{A}_B^{\mathbf{E},\mathbf{S}'} = \left[\mathcal{A}_F^{\mathbf{E}}, \ \mathcal{A}_I^{\mathbf{E},\mathbf{S}'} \right]$.

Definition 1. *The point $x \in \mathcal{F}_P$ is called $\mathbf{S}$-extreme, if columns of the matrix $\mathcal{A}_B^{\mathbf{E},\mathbf{S}'}$ are linear independent.*

It is clear that, if the point $x \in \mathcal{F}_P$ is $\mathbf{S}$-extreme, then the number of columns n_B^S in the matrix $\mathcal{A}_B^{\mathbf{E},\mathbf{S}'}$ does not exceed the number m. In the case, where the number of columns is equal to m, such $\mathbf{S}$-extreme point x we will call *regular*. In the opposite case we will call the $\mathbf{S}$-extreme point x *irregular*. The matrix $\mathcal{A}_B^{\mathbf{E},\mathbf{S}'}$ is regarded as $\mathbf{S}$-*basis* of the $\mathbf{S}$-extreme point $x \in \mathcal{F}_P$.

At regular $\mathbf{S}$-extreme point $x \in \mathcal{F}_P$ we have: $n_B^S = m$. If all $\mathbf{S}_i$, $i \in J_I(x)$, are identity matrices, then for $\mathbf{S}$-extreme point the criterion 1 is fulfilled, as for the usual extreme point.

4   Pivoting at the Regular S-Extreme Point

Assume that the regular $\mathbf{S}$-extreme point $x \in \mathcal{F}_P$ is given. Assume also that the $\mathbf{S}$-basis of this $\mathbf{S}$-extreme point $\mathcal{A}_B^{\mathbf{E},\mathbf{S}'}$ is known. Our aim is to make passage from this $\mathbf{S}$-extreme point to another $\mathbf{S}$-extreme point (possibly, to the usual extreme point). This transfer we will make preserving the complementarity condition from (5).

Firstly, compute the weak dual variable y, taking into consideration the condition $\langle x, y \rangle = 0$. This condition can be rewritten in the form $\sum_{i=1}^r \langle x_i, y_i \rangle = 0$. Since $x_i = 0_{n_i}$, where $i \in J_O(x)$, it is sufficient to require that

$$\sum_{i \in J_B(x)} \langle x_i, y_i \rangle = 0. \tag{10}$$

This condition holds, if $\langle x_i, y_i \rangle = 0$ for each $i \in J_B(x)$. Therefore, it is worth to take $y_i = 0_{n_i}$ for $i \in J_B(x)$.

Take into account that $y_i = c_i - A_i^T u$ for all $i \in J^r$. Denote $c_i^{\mathbf{E}} = \mathbf{e}_{i,1}^T c_i$, where $i \in J_F(x)$, and denote $c_i^{\mathbf{E},\mathbf{S}'} = (\mathbf{S}_i')^T \mathbf{E}_i^T c_i$, where $i \in J_I(x)$. Denote also

$$c_B^{\mathbf{E},\mathbf{S}'} = \left[c_1^{\mathbf{E}}; \ \ldots; \ c_{r_F}^{\mathbf{E}}; \ c_{r_F+1}^{\mathbf{E},\mathbf{S}'}; \ \ldots; \ c_{r_B}^{\mathbf{E},\mathbf{S}'} \right].$$

Setting equal $y_i = 0_{n_i}$, $i \in J_B(x)$, we obtain the system of linear equations with respect to the dual variable u:

$$c_B^{\mathbf{E},\mathbf{S}'} - \left(A_B^{\mathbf{E},\mathbf{S}'} \right)^T u = 0_m. \tag{11}$$

This system consists of m equations.

Since, by assumption, x is a regular **S**-extreme point, the matrix $A_B^{\mathbf{E},\mathbf{S}'}$ of system (11) is nonsingular

Solving system (11), we get

$$u = \left(A_B^{\mathbf{E},\mathbf{S}'} \right)^{-T} c_B^{\mathbf{E},\mathbf{S}'}.$$

Here the notation M^{-T} is used instead of $\left(M^T \right)^{-1}$.

Compute y_i, $i \in J^r$. Let $i \in J_F(x)$. Define the linear hyperplane

$$\Gamma(\mathbf{e}_{i,1}) = \{ z_i \in \mathbb{R}^{n_i} : \ \langle \mathbf{e}_{i,1}, z_i \rangle = 0 \}$$

with the directing vector $\mathbf{e}_{i,1}$. The ray

$$l(\mathbf{e}_{i,1}) = \{ z_i \in \mathbb{R}^{n_i} : \ z_i = \lambda_i \mathbf{e}_{i,1}, \ \lambda_i \geq 0 \},$$

is a minimal face of the cone K^{n_i} containing the point x_i. The ray

$$l(\mathbf{e}_{i,n_i}) = \{ z_i \in \mathbb{R}^{n_i} : \ z_i = \lambda_i \mathbf{e}_{i,n_i}, \ \lambda_i \geq 0 \}$$

is a conjugate face to the face $l(\mathbf{e}_{i,1})$ of the cone K^{n_i}.

Let now $i \in J_I(x)$. Denote by $\overline{\mathbf{S}}_i'$ the complement of the matrix $\mathbf{S}_i'$ in the sense that columns of $\overline{\mathbf{S}}_i'$ are unit vectors e_j, which do not contain in $\mathbf{S}_i'$. Introduce into considerations the linear manyfold

$$\Gamma_i(\mathbf{E}_i, \mathbf{S}_i) = \left\{ y_i \in \mathbb{R}^{n_i} : \ y_i = \mathbf{E}_i \overline{\mathbf{S}}_i' \overline{\theta}_i' \right\}, \quad i \in J_I(x),$$

where $\overline{\theta}_i' \in \mathbb{R}^{\bar{n}_i^S}$, and $\bar{n}_i^S$ is the number of columns in the matrix $\overline{\mathbf{S}}_i'$.

It follows from system (11) that $y_i \in \Gamma(\mathbf{e}_{i,1})$ for $i \in J_F(x)$, and $y_i \in \Gamma_i(\mathbf{E}_i, \mathbf{S}_i)$ for $i \in J_I(x)$. If $\mathbf{S}' = I_{n_i}$, then $y_i = 0_{n_i}$, $i \in J_I(x)$. In the case, where $i \in J_O(x)$, the vector y_i may be arbitrary.

Proposition 2. Let $i \in J_F(x)$. Then $y_i \in K^{n_i}$ if and only if $y_i \in l(\mathbf{e}_{i,n_i})$.

Proposition 3. *Let $i \in J_I(x)$. Then $y_i \in K^{n_i}$ if and only if*

$$y_i \in \Gamma_{i,+}(\mathbf{E}_i, \mathbf{S}_i) = \left\{ y_i \in^{n_i}: \ y_i = \mathbf{E}_i \overline{\mathbf{S}}_i' \overline{\theta}_i', \ \overline{\theta}_i' \in \mathbb{R}_+^{\bar{n}_i^S} \right\}.$$

Analyzing y_i, $i \in J^r$, we derive to the following possible cases:

1. The optimal solution. This case is realized under the following assertion.

Proposition 4. *Let $y_i \in K^{n_i}$ for all $i \in J^r$. Then the point $x \in \mathcal{F}_P$ is the solution of problem* (1).

Proof. In this case due to (10) and made assumptions all optimality conditions (5) are satisfied.

2. The current point x is not optimal. Assume that there is the index $k \in J^r$ such that $y_k \notin K^{n_i}$. If x is a regular S-extreme point, it is possible only when $k \in J_O(x)$ or $k \in J_F(x)$. If x is a non-regular extreme point, it is possible also that $k \in J_I(x)$.

Similar to (6) decompose the vector y_k:

$$y_k = \theta_{k,1} \mathbf{d}_{k,1} + \theta_{k,n_k} \mathbf{d}_{k,n_k},$$

where $\mathbf{d}_{k,1} \ \mathbf{d}_{k,n_k}$ are vectors of the form

$$\mathbf{d}_{k,1} = \frac{1}{\sqrt{2}} \left[1; \frac{\bar{y}_k}{\|\bar{y}_k\|} \right], \quad \mathbf{d}_{k,n_k} = \frac{1}{\sqrt{2}} \left[1; -\frac{\bar{y}_k}{\|\bar{y}_k\|} \right].$$

Both vectors $\mathbf{d}_{k,1}$ and $\mathbf{d}_{k,n_k}$ have the unit length. The coefficients $\theta_{k,1}$ and θ_{k,n_k} are following:

$$\theta_{k,1} = \frac{1}{\sqrt{2}} \left(y_k^0 + \|\bar{y}_k\| \right), \quad \theta_{k,n_k} = \frac{1}{\sqrt{2}} \left(y_k^0 - \|\bar{y}_k\| \right).$$

Since by assumption $y_k \notin K^{n_k}$, at least one of two coefficients $\theta_{k,1}$ or θ_{k,n_k} is negative. Assume without loss of generality that $\theta_{k,1} < 0$.

Let $\Delta x_B = [\Delta x_1; \ \ldots; \ \Delta x_{r_B}]$. Moreover, let $\Delta x_i = \mathbf{e}_{i,1} \Delta \eta_{i,1}$, $i \in J_F(x)$, and $\Delta x_i = \mathbf{E}_i \mathbf{S}_i' \Delta \eta_i'$, $i \in J_I(x)$. Consider the system of equations

$$\mathcal{A}_B \Delta x_B + A_k \mathbf{d}_k = 0_m, \tag{12}$$

where $\mathcal{A}_B = [A_1, \ ,\ldots, \ A_{r_B}]$. By setting

$$\Delta \eta_B = \left[\Delta \eta_{1,1}; \ldots; \Delta \eta_{r_F,1}; \ \Delta \eta_{r_F+1}'; \ldots; \Delta \eta_{r_B}' \right],$$

system (12) may be written in the form

$$\mathcal{A}_B^{\mathbf{E},\mathbf{S}'} \Delta \eta_B + A_k \mathbf{d}_k = 0_m. \tag{13}$$

As, by assumption, x is a regular **S**-extreme point of $\mathcal{F}_P$, the matrix $\mathcal{A}_B^{\mathbf{E},\mathbf{S}'}$ is nonsingular. Solving system (13), we get

$$\Delta\eta_B = -\left(\mathcal{A}_B^{\mathbf{E},\mathbf{S}'}\right)^{-1} A_k \mathbf{d}_k. \tag{14}$$

Therefore, $\Delta x_B = \mathbb{E}\,\Delta\eta_B$, where $\mathbb{E}$ is a block diagonal matrix with the vectors $\mathbf{e}_{1,1}, \ldots, \mathbf{e}_{r_F,1}$ and with blocks $\mathbf{E}_{r_F+1}\mathbf{S}'_{r_F+1}, \ldots, \mathbf{E}_{r_B}\mathbf{S}'_{r_B}$ on its diagonal.

Denote $\Delta x_N = \left[\Delta x_{r_B+1}; \ldots; \Delta x_r\right]$, $\Delta x = \left[\Delta x_B; \Delta x_N\right]$. The following assertion is valid.

Proposition 5. *Let all components of the vector Δx_N equal to zero. Moreover, let the kth component be taken equal to $\Delta x_k := \Delta x_k + \mathbf{d}_{k,1}$. Then $A\Delta x = 0_m$.*

Proof. We have

$$A\Delta x = A_B\Delta x_B + A_k\Delta x_k. \tag{15}$$

Substituting Δx_k and Δx_B, we conclude that equality (15) holds.

Proposition 6. *Let x be a regular **S**-extreme point of the set $\mathcal{F}_P$. Then Δx is the feasible direction with respect to the cone $\mathcal{K}$.*

Proof. The cone of feasible directions $\mathcal{C}_{\mathcal{K}}(x)$ with respect to $\mathcal{K}$ at the point $x \in \mathcal{K}$ is the direct product of the cones of feasible directions $\mathcal{C}_{K^{n_i}}(x_i)$, $i \in J^r$. But if $h \in \mathcal{C}_{K^{n_i}}(x_i)$, then $h = h_1 + h_2$, where h_1 belongs to the linear hull of the minimal face containing the point x_i, and h_2 is a direction belonging to the same cone K^{n_i}.

Assume for simplicity, that $k \in J_O(x)$. At the case, where $i \in J_F(x)$, the Δx_i belongs to the linear hull of minimal face of the cone K^{n_i}, coinciding with the linear hull of the vector $\mathbf{e}_{i,1}$. If $i \in J_I(x)$, then x_i is an interior point of the cone K^{n_i}. Hence, the cone $\mathcal{C}_{K^{n_i}}(x_i)$ is the whole space $\mathbb{R}^{n_i}$. All the more, we obtain that $\Delta x_i = \mathbf{E}_i\mathbf{S}'_i\Delta\eta'$ belongs to $\mathcal{C}_{K^{n_i}}(x_i)$. If $k \in J_B(x)$, then as far as $\mathbf{d}_{k,1} \in \partial K^{n_k}$, we have $\Delta x_k + \mathbf{d}_{k,1} \in \mathcal{C}_{K^{n_k}}(x_k)$. The inclusions $\Delta x_i \in \mathcal{C}_{K^{n_i}}$ for the rest $i \in J_B(x)$ are preserved.

Corollary 1. *Taking into account Propositions 5 and 6, we conclude that at the regular **S**-extreme point $x \in \mathcal{F}_P$ the direction Δx is feasible with respect to the set $\mathcal{F}_P$.*

Proposition 7. *Let assumptions of Propositions 5 and 6 hold. Then Δx is the decreasing direction of the goal function in problem (3).*

Proof. According to (14)

$$\langle c, \Delta x\rangle = \langle c_B, \Delta x_B\rangle + \langle c_k, \mathbf{d}_{k,1}\rangle = -\langle c_B^{\mathbf{E},\mathbf{S}'}, \left(\mathcal{A}_B^{\mathbf{E},\mathbf{S}'}\right)^{-1} A_k\mathbf{d}_k\rangle + \langle c_k, \mathbf{d}_{k,1}\rangle$$
$$= -\langle (\mathcal{A}_B^{\mathbf{E},\mathbf{S}'})^{-T}c_B^{\mathbf{E},\mathbf{S}'}, A_k\mathbf{d}_k\rangle + \langle c_k, \mathbf{d}_{k,1}\rangle = -\langle u, A_k\mathbf{d}_k\rangle + \langle c_k, \mathbf{d}_{k,1}\rangle = \langle c_k - A_k^T u, \mathbf{d}_k\rangle$$
$$= \langle y_k, \mathbf{d}_k\rangle = \langle \theta_{k,1}\mathbf{d}_{k,1} + \theta_{k,n_k}\mathbf{d}_{k,n_k}, \mathbf{d}_{k1}\rangle = \theta_{k,1} < 0.$$

From here the required inequality $\langle c, \Delta x\rangle < 0$ is followed.

3. The solution absence. The solution absence is possible only in the case when the feasible set $\mathcal{F}_P$ in problem (3) is unbounded. Let $\hat{x}(\alpha) = x + \alpha \Delta x$.

Proposition 8. *Let the solution of the system* (13) *be such that* $\Delta x_i \in K^{n_i}$, $i \in J_B(x)$, *and* $\Delta x_N = 0$. *Besides, let* $\Delta x_k := \Delta x_k + \mathbf{d}_k$. *Then the feasible set* $\mathcal{F}_P$ *in problem* (3) *is unbounded, and* $\hat{x}(\alpha) \in \mathcal{K}$ *under* $\alpha \geq 0$. *Moreover,*

$$\langle c, \hat{x}(\alpha) \rangle \to -\infty \tag{16}$$

when $\alpha \to +\infty$.

Proof. According to made assumptions we obtain that $x_i + \alpha \Delta x_i \in K^{n_i}$ for all $i \in J^r$ $\alpha \geq 0$. Moreover, by Proposition 5 the equality $\mathcal{A}\Delta x = 0_m$ takes place. Therefore, $\hat{x}(\alpha) \in \mathcal{F}_P$, where $\alpha \geq 0$. Since by Proposition 7 the inequality $\langle c, \Delta x \rangle < 0$ holds, the limit equality (16) is valid.

4. Passage to the new extreme point. Assume that the conditions of Propositions 4 or 8 do not hold. Then, in principle, the passage from **S**-extreme point $x \in \mathcal{F}_P$ to the new **S**-extreme point $\hat{x} \in \mathcal{F}_P$ with lesser value of the goal function is possible.

Suppose that x is the regular **S**-extreme point. Describe the algorithm of the passage.

For each $i \in J_B(x)$ compute the maximal possible step size:

$$\alpha_i = \max\{\alpha \geq 0: \ \eta_{i,1} + \alpha\, \Delta\eta_{i,1} \geq 0\}, \quad i \in J_F(x),$$

$$\alpha_i = \max\{\alpha \geq 0: \ x_i + \alpha\, \Delta x_i \in K^{n_i}\}, \quad i \in J_I(x).$$

Denote $\hat{\alpha}_F = \min_{i \in J_F(x)} \alpha_i$, $\hat{\alpha}_I = \min_{i \in J_I(x)} \alpha_i$ and $\hat{\alpha}_B = \min\{\hat{\alpha}_F, \hat{\alpha}_I\}$. Determine the index sets

$$J_F^U(x) = \{i \in J_F(x): \ \alpha_i = \hat{\alpha}_B\}, \quad J_I^U(x) = \{i \in J_I(x): \ \alpha_i = \hat{\alpha}_B\},$$

and also the set $J_B^U(x) = J_F^U(x) \cup J_I^U(x)$. The set $J_B^U(x)$ contains at least one index. By $\sigma_B^U(x)$ denote the number if indexes in the set $J_B^U(x)$, i.e. $\sigma_B^U(x) = |J_B^U(x)|$. Below for simplicity suppose, that the set $\sigma_B^U(x)$ consists of *the single index* $l \in J_B^U(x)$.

If $l \in J_F^U(x)$, we take the step size equal to $\hat{\alpha}$. If $l \in J_I^U(x)$, we set $\hat{\alpha} := \hat{\alpha} - \varepsilon$, where $\varepsilon > 0$ is a sufficiently small number. Taking $\Delta x_i = 0_{n_i}$, $i \in J_O(x)$, we make the passage to the new **S**-extreme point

$$\hat{x} = x + \hat{\alpha}\Delta x. \tag{17}$$

In addition we make the passage

$$\hat{x}_k := \hat{x}_k + \hat{\alpha}\,\mathbf{d}_{k,1}. \tag{18}$$

Since $\mathbf{d}_{k,1}$ belongs to the boundary of the cone K^{n_k}, we obtain at any case that the point $\hat{x}_k$ is turned out in the cone K^{n_k}.

The variable x_i, $i \in J_B(x)$, we call *basic*. Among all basic variables we distinguish *facet basic variables*, when $i \in J_F(x)$, and *interior basic variables*, when $i \in J_I(x)$. Introduce the notion of **S**-index of the basic variable x_i. The **S**-index of the facet basic variable is set equal to one. By the **S**-index of the interior basic variable x_i we call the number of columns in the matrix $\mathbf{S}_i'$ and denote it $\gamma_{\mathbf{S}}(x_i)$. The **S**-index of the interior basic variable can not be less than two. In the case, where $\gamma_{\mathbf{S}}(x_i) = n_i$, the interior basic variable x_i we call by *the full interior basic variable*. If for the interior basic variable x_i the equality $\gamma_{\mathbf{S}}(x_i) = 2$ holds, such variable x_i we call *the principal interior basic variable*.

The following number

$$\varGamma_{\mathbf{S}}(x) = \sum_{i \in J_B(x)} \gamma_{\mathbf{S}}(x_i)$$

we call *the general* **S**-*index* of the point x. The general **S**-index of the regular **S**-extreme point x is equal to m.

Consider possible cases of belonging indexes k and l to the various index sets, when passage from the **S**-extreme point to updated **S**-extreme point is realized by formulas (17) and (18). At the same time, we will describe **the passage rules**, when the index l is unique.

1. Let $k \in J_O(x)$. If $l \in J_F^U(x)$, we obtain that the variable $\hat{x}_l$ ceased to be facet basic variable. But the variable $\hat{x}_k$ becomes the new facet basic variable. Taking the $\gamma_{\mathbf{S}}$-indexes of other variables $\hat{x}_i$ former, we obtain that the general $\varGamma_{\mathbf{S}}$-index of $\hat{x}$ is equal to m, i.e. $\hat{x}$ is turned out to be regular **S**-extreme point.

   Assume now that $l \in J_I^U(x)$. At this case $\hat{x}_i$, $i \in J_F^U(x)$, remain the facet basic variables. All variables $\hat{x}_i$, where $i \in J_I^U(x)$ and $i \neq l$, remain interior basic. Moreover, their $\gamma_{\mathbf{S}}$-indexes we remain equal to $\gamma_{\mathbf{S}}$-indexes of the corresponding variables x_i. If the $\gamma_{\mathbf{S}}$-index of the variable x_l greater than two, we decrease the $\gamma_{\mathbf{S}}$-index of the variable $\hat{x}_l$ at one. The general $\varGamma_{\mathbf{S}}$-index of the point $\hat{x}$ does not change with respect to the point x and remains equal to m.

2. Suppose now that $k \in J_F(x)$ and consider separately the cases, where $l \in J_F(x)$ and $l \in J_I(x)$. Let $l \in J_F(x)$. If $l = k$, the $\hat{x}_l$ remains by facet basic variable. If $l \neq k$, the variable $\hat{x}_l$ becomes zero, i.e. $l \in J_O(\hat{x})$. The $\hat{x}_k$ is the interior basic variable . We set $\gamma_{\mathbf{S}}(\hat{x}_k) = 2$. Then in any case, regardless $l = k$ or $l \neq k$, the equality $\varGamma_{\mathbf{S}}(\hat{x}) = m$ holds, that is $\hat{x}$ turns out to be the regular **S**-extreme point.

   Let $l \in J_I(x)$. Then $\hat{x}_k$ we turn into the principal interior basic variable. The variable $\hat{x}_l$ is remained by the interior basic variable, but its **S**-index $\gamma_{\mathbf{S}}(\hat{x}_l)$ we decrease at one with respect to **S**-index of x_l, i.e. $\gamma_{\mathbf{S}}(x_l) = \gamma_{\mathbf{S}}(x_l) - 1$. If x_l is a principal interior variable, then $\hat{x}_l$ turns out to be the facet basic variable.

3. At last, consider the case, where $k \in J_I(x)$. This is possible only, when among interior basic variables there are the variables, which are not full interior basic

variables. The variable $\hat{x}_k$ we remain by the interior basic variable with the same $\mathbf{S}$-index, which the variable x_k has. It is possible to make, because the weak dual variable y_k belongs to the boundary of the cone K^{n_k}.

Proposition 9. *Let x be the $\mathbf{S}$-extreme point of $\mathcal{F}_P$, and let the step size $\hat{\alpha}$ be finite. Moreover, let the set $J_B^U(x)$ consist of the single index l. Then $\hat{x}$ is also the regular $\mathbf{S}$-extreme point of $\mathcal{F}_P$.*

Proof. This assertion is followed from the described above rules of passage from the $\mathbf{S}$-extreme point x at the new $\mathbf{S}$-extreme point $\hat{x}$.

Remark 1. If the index l at the point x is not unique, the point $\hat{x}$ turns out to be non-regular, that is $\Gamma_{\mathbf{S}}(\hat{x}) < m$. In order to choose l, it should use any technique for overcoming the degeneracy. It may be the approach similar to using in the linear programming, for example, ε-technique.

5   Conclusion

We have proposed the variants of the primal simplex-type method for solving the linear second-order cone programming problem. This variant of the method has preference with respect to other variants since the pivoting is possible both in regular and non-regular extreme points.

References

1. Alizadeh, F., Goldfarb, D.: Second-order cone programming. Math. Program. Ser. B. **95**, 3–51 (2003)
2. Anjos, M.F., Lasserre, J. (eds.): Handbook of Semidefinite, Cone and Polynomial Optimization: Theory, Algorithms, Software and Applications. ISOR, vol. 166, p. 915. Springer, New York (2011). https://doi.org/10.1007/978-1-4614-0769-0
3. Lobo, M.S., Vandenberghe, L., Boyd, S., Lebret, H.: Applications of second order cone programming. Linear Algebra Appl. **284**, 193–228 (1998)
4. Nesterov, Y.E., Todd, M.J.: Primal-dual interior-point methods for self-scaled cones. SIAM J. Optim. **8**, 324–364 (1998)
5. Monteiro, R.D.C., Tsuchiya, T.: Polynomial convergence of primal-dual algorithms for second-order cone program based on the MZ-family of directions. Math. Program. **88**(1), 61–83 (2000)
6. Muramatsu, M.: A pivoting procedure for a class of second-order cone programming. Optim. Methods Softw. **21**(2), 295–314 (2006)
7. Hayashi, S., Okuno, T., Ito, Y.: Simplex-type algorithm for second-order cone programming via semi-infinite programming reformulation. Optim. Methods Softw. **31**(6), 1272–1297 (2016)
8. Pataki, G.: Cone-LP's and semidefinite programs: geometry and a simplex-type method. In: Cunningham, W.H., McCormick, S.T., Queyranne, M. (eds.) IPCO 1996. LNCS, vol. 1084, pp. 162–174. Springer, Heidelberg (1996). https://doi.org/10.1007/3-540-61310-2_13
9. Zhadan, V.: A variant of the simplex method for second-order cone programming. In: Khachay, M., Kochetov, Y., Pardalos, P. (eds.) MOTOR 2019. LNCS, vol. 11548, pp. 115–129. Springer, Cham (2019). https://doi.org/10.1007/978-3-030-22629-9_9

Global Optimization

A Computational Study of the DC Minimization Global Optimality Conditions Applied to K-Means Clustering

Tatiana V. Gruzdeva[(✉)] [iD] and Anton V. Ushakov [iD]

Matrosov Institute for System Dynamics and Control Theory of SB RAS,
134 Lermontov Street, 664033 Irkutsk, Russia
{gruzdeva,aushakov}@icc.ru

Abstract. Clustering is traditionally one of the basic tools of data analysis widely applied in diverse fields. By now, one of the most common clustering models is the Euclidean minimum-sum-of-squares clustering problem (MSSC). Consequently, Lloyd's algorithm, often referred to as k-means, is probably the most popular clustering algorithm. Despite its popularity, Lloyd's algorithm is a local search heuristic for MSSC that in general converges to local optima only. In this paper, we aim at enhancing k-means by employing the global optimality conditions for MSSC represented as a problem with DC (difference of convex) functions. We then embed the k-means algorithm into the so-called global search framework for DC minimization problems where it is employed to find local optimal solutions. We tested such an improved implementation of k-means in a series of computation experiments on well-known test library of medium-size datasets and compared it with the conventional k-means and k-means++ algorithms.

Keywords: Clustering · k-means · Nonconvex optimization · Global optimality conditions · Machine learning

1 Introduction

Cluster analysis is one of the traditional and basic subroutines in unsupervised machine learning. At the same time, it is one of the oldest machine learning problems. Clustering is to partition a set of data items or objects into groups (clusters) based on their similarity, i.e. the data items of the same cluster are of high similarity, whereas the items of different clusters are of low similarity. Clustering is traditionally one of the basic tools of data analysis widely applied in diverse fields. Since there are no strict mathematical definitions of similarity and quality partitions, there is a wide range of clustering approaches proposed in the

The research was funded by the Ministry of Education and Science of the Russian Federation (state registration No. 121041300065-9).

N. N. Olenev et al. (Eds.): OPTIMA 2021, LNCS 13078, pp. 79–93, 2021.
https://doi.org/10.1007/978-3-030-91059-4_6

fields ranging from mathematics to bioinformatics and computational medicine. All the approaches can be categorized in many ways according to the principles of organizing data into groups. However, one of the most popular and widespread approaches to cluster analysis is the so-called center-based clustering. It assumes that a partition of data items is defined by a set of cluster representatives (or cluster centers), and clusters are obtained by assigning data items to the "closest" (most similar) cluster center.

The problem of finding such a set of cluster centers can be formulated as an optimization problem. The number of clusters can also be integrated as a variable in the problem [30]. Unfortunately, most of such problems are NP-hard. By now, the most popular center-based clustering models are k-medoids (minimum sum of stars clustering), k-center, and k-means (minimum sum of squares) clustering. All these models are to find a set of cluster representatives to minimize the total sum of dissimilarities between data items and the closest cluster centers. Probably, the most popular clustering model applied in plenty of applications is the minimum sum of squares clustering model (MSSC), whereas the most popular and well-known clustering algorithm is Lloyd's algorithm, which is often referred to as k-means [25]. The latter is a local search heuristic for MSSC.

Besides being a basic data analysis tool, cluster analysis is a useful approach to image segmentation [33] or optimal location of service and distribution centers. Indeed, the center-based clustering models can be considered as facility location problems, e.g., k-medoids problem from machine learning is the well known p-median facility location problem, and MSSC is a particular case of the multi-source Weber problem, in which the locations of customers are given in a multidimensional space, the distances between customers and facilities are computed according to the squared Euclidean distance (instead of the Euclidean distance in the Weber problem), and the customers' demands are equal to 1. Finally, center-based clustering is often utilized as a compression or vector quantization method that aims at representing each data item by the center it is assigned to or by the distances to all cluster centers.

The problem MSSC can be formulated as the following non-smooth optimization problem. Given a set $J = \{1, \ldots, m\}$ of objects, represented as $a^j \in \mathbb{R}^n$, $j \in J$. The problem is to find k cluster centers (centroids) $c^i \in \mathbb{R}^n$, $i \in I = \{1, \ldots, k\}$ such that the total sum of squared Euclidean distances between objects and their closest centers is minimized:

$$\min_{C \subset \mathbb{R}^n} \left\{ \sum_{j=1}^m \min_{c \in C} \|a^j - c\|^2, \ |C| = k \right\}, \tag{1}$$

where $\| \cdot \|$ is the Euclidean distance.

The problem is NP-hard even in the plane for arbitrary number of clusters k [27]. It is also NP-hard in general dimension even for $k = 2$ [1] and NP-hard when the dimension n is a part of the input, whereas the number of clusters k is not [10].

Being one of the most popular clustering models, MSSC has been attracting much attention from different communities of researchers. Together with the

k-medoids clustering problem, MSSC was first formulated as an optimization problem in [39]. The first exact algorithm, branch and bound, was proposed in [12] and then refined in [9]. The most effective exact algorithm was developed in [11] where the authors proposed an extended formulation of the problem with an exponential number of columns that correspond to all possible clusters. They devised a branch and price method where an auxiliary fractional quadratic binary problem of determining the entering column is solved by a combination of Dinkelbach's algorithm and a variable neighborhood search. This approach was further improved much later in [2] where the authors proposed a geometric approach to solving the auxiliary problem being the main bottleneck of the algorithm from [11].

As exact methods are usually not tractable in practical clustering settings, many research efforts have been focused on devising various heuristic techniques, most of which are modifications of the conventional k-means algorithm. The rise of this research strand is also caused by some drawbacks of Lloyd's algorithm, especially sensitivity to initial solutions and low efficiency for large-scale datasets. For more than 50 years of research nearly all metaheuristics have been applied to MSSC (for a survey see [23,28]). The most prominent heuristics are the global k-means [5,24,31], harmonic clustering [7], j-means [19] etc. For example, the global k-means is an incremental algorithm that, starting from the mean of the initial dataset, finds the next centroid as the data item minimizing the objective value the most. The algorithm halts when all k cluster centers are determined.

There is also a relatively small number of solution methods based on representation of MSSC as an optimization problem with DC (difference of convex) functions. For example, in [8] the authors considered a DC representation of the non-smooth problem (1) and developed a truncated codifferential descent method. In [6] the authors proposed an incremental algorithm for MSSC based on a procedure from [31] for finding starting solutions and a DC approach rested upon a non-smooth formulation of MSSC. Another approach was considered in [21], where a DCA algorithm was developed. It is based on a mixed programming formulation of MSSC that is reformulated into an unconstrained DC minimization problem by replacing binary variables and applying a penalty method.

In this paper, we devise and implement a solution approach to MSSC based on the special global search theory and the global optimality conditions. It is built upon the conventional k-means algorithm that we use to find local optimal solutions. To escape a local optimum found by k-means, we develop a special procedure based on the so-called global optimality conditions. We tested such an improved implementation of k-means in a series of computation experiments on a well-known test library of medium-size datasets and compared it with the k-means algorithm and its modifications.

2   K-Means Algorithm

As was noted above, well-known Lloyd's algorithm remains one of the most popular clustering algorithms. It is an alternate heuristic similar to Cooper's

---

Algorithm 1. k-means (Lloyd's) algorithm

1: Choose initial cluster centers $c^i \in \mathbb{R}^n$, $i \in I$ (often $c^i \in \{a^1 \ldots, a^m\}$).

2: **while** clusters C_i change **do**

3:    Assign data items a^j, $j \in J$, to the closest cluster center i^* by computing $i^* \leftarrow \underset{i \in I}{\operatorname{argmin}} \|a^j - c^i\|^2$. Form clusters C_i of data items assigned to the center c^i.

4:    For each cluster C_i, $i \in I$, compute c^i as the center of gravity (mean) of all the data items from C_i: $c^i \leftarrow \frac{1}{|C_i|} \sum\limits_{s:a^s \in C_i} a^s$.

5: **end while**

---

algorithm proposed for the multi-source Weber problem. Its simplicity and relatively high efficiency for small size datasets made it popular in diverse applications. The idea of the algorithm is to sequentially assign data items to the current closest cluster centers (according to the squared Euclidean distance) and then recompute the centers as means of objects assigned to the same clusters. The algorithm halts when there are no changes in cluster assignments of a sufficiently large number of data items (see Algorithm 1).

One can observe that an iteration of Algorithm 1 requires $\mathcal{O}(kmn)$ time. The k-means algorithm is guaranteed to converge only to local optimal solutions. Moreover, as was noted in [4], it may provide arbitrary bad partitions and has superpolynomial running time, e.g. for some datasets and specified initial cluster centers it requires $2^{\Omega(\sqrt{m})}$ iterations. Moreover, the running time remains superpolynomial with high probability even if the initial centroids are chosen uniformly at random [3]. An upper bound on the running time for the case of the real line is $\mathcal{O}(m\Delta^2)$ where Δ is the ratio between the largest and the smallest distances between data items.

Besides Lloyd's algorithm, there are some other local search heuristics for MSSC that are also often referred to as k-means in the literature. For example, MacQueen's algorithm [26] follows a little bit different flow of operations: i.e. it starts with k arbitrary chosen data items as cluster centers and assigns each new data item one by one to the closest center. After a data item is assigned to a cluster, its center is recomputed [26]. Note that MacQueen's approach can be viewed as an online version of k-means. Another variant of local search for MSSC implemented in many machine learning packages as a default k-means method is Hartigan and Wong's algorithm [20]. It first iteration is identical to Lloyd's algorithm. However, it then swaps data items between clusters in order to improve the objective value. If such a improving swap exists, it is performed and the next data item is considered.

Sensitivity of Lloyd's algorithm to the choice of initial solution has resulted in multiple attempts to develop seeding procedures that guarantee finding quality solutions. One of the most well-known such a modification of Lloyd's algorithm is k-means++ [4] that picks initial centers according to the procedure originally proposed in [32]. Its main idea is to choose a first cluster center uniformly at random and the next ones with probability proportional to the ratio between distance from the new center to the closest already chosen center and the sum of distances from all data items to closest already chosen centers. Note that such a simple seeding technique guarantees that Lloyd's algorithm obtains $\Theta(\log k)$-approximate solution for MSSC [4].

3 Minimum-Sum-of-Squares Clustering Problem

As was observed in the introduction, MSSC can be formulated as a mathematical programming problem in several ways. For example, it can quite naturally be cast as an integer program. Let us remind that given data items a^j, $j \in J$, which are supposed to be partitioned into k clusters. MSSC is to determine cluster centers c^i, $i \in I$, such that the total sum of squared Euclidean distances between data items and their closest centers is minimized.

Let us introduce the following binary variables

$$x_{ij} = \begin{cases} 1, & \text{if data item j is assigned to cluster i,} \\ 0, & \text{otherwise,} \end{cases} \quad i = 1, \ldots, k, \ j = 1, \ldots, m.$$

which are often referred to as assignment variables.

We also introduce the variables $y^i \in \mathbb{R}^n$ that define locations of cluster centers $i = \{1, \ldots, k\}$. With these notations, the problem can be written as the following mixed integer program:

$$\sum_{i=1}^{k} \sum_{j=1}^{m} x_{ij} \|y^i - a^j\|^2 \downarrow \min_{(x,y)}, \tag{2}$$

$$\sum_{i=1}^{k} x_{ij} = 1 \qquad \forall j = 1, \ldots, m; \tag{3}$$

$$x_{ij} \in \{0, 1\} \qquad \forall i = 1, \ldots, k; \ \forall j = 1, \ldots, m. \tag{4}$$

The objective function (2) minimizes the sum of squared Euclidean distances, whereas constraints (3) ensure that each data item is assigned to exactly one cluster center. The integrality of the assignment variables is guaranteed by constraints (4). However, as in many facility location problems aimed at minimizing service costs, the integrality of x_{ij} can be relaxed to $0 \leq x_{ij} \leq 1$, $i = 1, \ldots, k$; $j = 1, \ldots, m$. Indeed, if some variables x_{ij} take not integer values (data item j is assigned to several equidistant centers), then an equivalent integer x_{ij} can be obtained by fixing one of them to one and the rest to zero. Thus, the problem (2)–(4) can be reduced to a nonconvex continuous optimization problem. In general, it has plenty of local solutions that are not globally optimal ones.

In the following we demonstrate how to leverage the nonconvex continuous formulation to obtain a procedure of escaping from local optimal solutions which is based on the special global optimality conditions.

4   Global Search Theory for the DC Minimization Problem

First, let us consider the main ingredients of the global search theory for DC minimization problems. Let us consider the following optimization problem

$$f(x) = g(x) - h(x) \downarrow \min, \quad x \in D, \qquad (\mathcal{DC})$$

where $g(\cdot)$ and $h(\cdot)$ are convex functions, whereas D is a convex set, $D \subset \mathbf{R}^n$.

The general global search method includes two main components:

1. Special local search;
2. Procedures of escaping from a point provided by a local search method.

The idea of the special local search method is simple and consists in solving a series of the following (partially) linearized (at a current point $x^s \in D$) problems [22, 34]

$$\Psi_s(x) := g(x) - \langle \nabla h(x^s), x \rangle \downarrow \min_x, \quad x \in D, \qquad (\mathcal{DCL}_s)$$

which are convex. Hence, such linearazied problems can be solved with any conventional convex optimization methods.

The second part, a procedure of escaping from a solution found by the local search, is the most important step of the approach. The procedure is built upon a theoretical basis guaranteed by the Global Optimality Conditions, which takes the following form for Problem $(\mathcal{DC})$.

Theorem 1 [35, 36]. *Suppose that $\exists \, q \in \mathbb{R}^n : f(q) > f(z) =: \zeta$.*

Then, a point z is a global solution to Problem $(\mathcal{DC})$ if and only if

$$\left. \begin{array}{l} \forall (y, \beta) \in \mathbb{R}^n \times \mathbb{R}: \quad h(y) = \beta - \zeta, \\ g(x) - \beta \geq \langle \nabla h(y), x - y \rangle \ \forall x \in D. \end{array} \right\} \qquad (\mathcal{E})$$

Now we try to violate the so-called principal inequality of $(\mathcal{E})$ by selecting the "perturbation parameters" (y, β) and solving the linearized problem (cf. $(\mathcal{DCL}_s)$)

$$\Psi_y(x) := g(x) - \langle \nabla h(y), x \rangle \downarrow \min_x, \quad x \in D.$$

In this case, we obtain a set of starting points $x(y, \beta)$ to initialize the local search and find a feasible solution with a better objective value.

Note that on each level $\zeta_k = f(z^k)$, it is not necessary to test all the pairs (y, β) satisfying the equation $\zeta_k = \beta - h(y)$, but it is enough to discover a pair $(\hat{y}, \hat{\beta})$ that violates the principal inequality of $(\mathcal{E})$. We denote by $\hat{x}$ the solution found by the local search that starts from the point $(\hat{y}, \hat{\beta})$.

After that, we proceed to the next iteration of the global search: $z^{k+1} := \hat{x}$, $\zeta_{k+1} := f(z^{k+1})$, and start the global search procedure from the very beginning.

Hence, due to Theorem 1, the basic stages of the global search scheme can be described as follows (see Algorithm 2) [35, 36].

Algorithm 2. Global Search Scheme

1: Find a local solution z.

2: Choose a number $\beta \in [\beta_-, \beta_+]$, where the numbers $\beta_- = \inf(g, D)$, $\beta_+ = \sup(g, D)$ can be approximated by rather rough estimates.

3: Construct an approximation

$$R(\beta) = \{y^1, \ldots, y^N \mid h(y^i) = \beta - \zeta, \ i = 1, \ldots, N = N(\beta)\}$$

of the level surface $\{y \in I\!\!R^n \mid h(y) = \beta - \zeta\}$ of the function $h(\cdot)$.

4: Starting at each point y^i of the approximation $R(\beta)$, find a feasible point u^i by means of the local search method.

5: Verify the principal inequality from the Global Optimality Condition $(\mathcal{E})$

$$g(u^i) - \beta \geq \langle \nabla h(w^i), u^i - w^i \rangle \quad \forall i = 1, \ldots, N, \tag{5}$$

where w^i can be found as the projection of the point u^i onto the convex set

$$\mathcal{L}(h, \beta - \zeta) = \{x \in I\!\!R^n \mid h(x) \leq \beta - \zeta\}$$

6: If $\exists j \in \{1, \ldots, N\}$ such that (5) is violated, then set $z \leftarrow u^j$, $\zeta \leftarrow f(u^j)$ and return to step 3. Otherwise, return to step 2.

Note that this approach was successfully applied to solving various practical DC optimization problems [13–15,17]. In the following we consider how the aforementioned principals can be adapted to solve MSSC formulated as the problem (2)–(4).

5   Implementation for K-Means Clustering

Recall that we consider the problem (2)–(4) where the integrality of assignment variables $x_{ij} \in \{0,1\}$ is relaxed to $x_{ij} \in [0,1]$. The resultant problem is to minimize a nonconvex function over a convex feasible set:

$$f(x,y) = \sum_{i=1}^{k} \sum_{j=1}^{m} x_{ij} \|y^i - a^j\|^2 \downarrow \min_{(x,y)}, \ x \in S, \ y \in I\!\!R^{k \times n}, \tag{6}$$

where $S = \{x_{ij} \in [0,1] : \ \sum_{i=1}^{k} x_{ij} = 1, \ j = 1, \ldots, m\}$.

The first step in applying the Global Search Theory for MSSC is to determine an explicit DC representation of the nonconvex objective function.

As the DC representation is known to be not unique, we propose to represent the objective function of the problem (6) as the following difference of two convex functions

$$f(x,y) = g(x,y) - h(x,y), \tag{7}$$

where

$$g(x,y) = \sum_{i=1}^{k} \sum_{j=1}^{m} \left[d_1 \parallel y^i - a^j \parallel^2 + d_2 x_{ij}^2 \right],$$

$$h(x,y) = \sum_{i=1}^{k} \sum_{j=1}^{m} \left[d_1 \parallel y^i - a^j \parallel^2 + d_2 x_{ij}^2 - x_{ij} \parallel y^i - a^j \parallel^2 \right],$$

with some constants $d_1, d_2 > 0$.

It is obvious that the function $g(\cdot)$ is convex. The convexity of the function $h(\cdot)$ with $d_1 > 1$, $d_2 > \dfrac{1}{2d_1} \max\limits_{i,j} \parallel y^i - a^j \parallel^2$, was proved in [16].

The key component of the procedure of escaping from a local optimum is an approximation of the level surface of the convex function $h(\cdot)$. The latter generates the basic nonconvexity in Problem ($\mathcal{P}$) (see step 3 of Algorithm 2). The way the approximation is constructed is crucial for finding high-quality solutions by a local search algorithm. There are actually many ways and techniques to define the approximation. To take into account the particularities of MSSC, we construct the approximation by varying only variables y^i, $i \in I$ (cluster centers) of the function $h(\cdot)$. Note that in [16], we tested another type of approximation, i.e. we used the unit vectors e^l from the Euclidean basis of $\mathbb{R}^n$ to construct the approximation $R(\beta)$ of the level surface $\{h(\cdot) = \beta - \zeta\}$ by the following rule: $w^{il} = z^i + \mu_{il} e^l$, $i = 1, \ldots, k$, $l = 1, \ldots, n$.

Here we employ another approach which is based on conjugated vectors instead of the unit vectors from the Euclidean basis. In this case, we may vary not one but up to n coordinates of each cluster center choosing one at a time:

$$w = (z^1, z^2, \ldots, z^{i-1}, z^i + \nu_i p^i, z^{i+1}, \ldots, z^k) \qquad (8)$$

where $z = (z^1, \ldots, z^k)$ is the current local solution, i is the varied cluster center and p^i, $i = 1, \ldots, k$, are the vectors conjugated with respect to the ($n \times n$) matrices A^s, $s = 1, 2, \ldots$, the rows of which are n data items chosen at random ($n \leq k$). The parameters ν_i may be found analytically, as it requires solving the following quadratic equation of one variable ν_i for the quadratic (with fixed variables x_{ij}) function $h(\hat{x}, y)$:

$$\nu_i^2 \parallel p^i \parallel^2 \sum_{j=1}^{m} (d_1 - \hat{x}_{ij}) - 2\nu_i \sum_{j=1}^{m} (d_1 - \hat{x}_{ij}) \langle z^i - a^j, p^i \rangle + \gamma = 0,$$

where $\gamma = h(\hat{x}, z) - \beta + \zeta$. If for some index $\hat{i}$ the discriminant turns out to be negative, then the variation of cluster center $\hat{i}$ is not performed.

Thus, we construct the set $P = \{p^1, \ldots, p^n, p^{n+1}, \ldots, p^{2n}, \ldots, p^k\}$ of vectors satisfying the following condition:

$$\langle p^i, A^s p^t \rangle = 0 \ \forall i \neq t, \ i, t = 1, \ldots, n, \ s = 1, 2, \ldots. \qquad (9)$$

Algorithm 3. Conjugated vectors construction

1: $i \leftarrow 1$, $p^i \leftarrow (1, 1, \ldots, 1)^\top$, $r^i \leftarrow -p^i$.

2: $\alpha_i^1 \leftarrow \dfrac{\langle r^i, r^i \rangle}{\langle p^i, Ap^i \rangle}$.

3: $r^{i+1} \leftarrow r^i + \alpha_i^1 \cdot Ap^i$.

4: $\alpha_{i+1}^2 \leftarrow \dfrac{\langle r^{i+1}, r^{i+1} \rangle}{\langle r^i, r^i \rangle}$.

5: $p^{i+1} \leftarrow -r^{i+1} + \alpha_{i+1}^2 p^i$.

6: $i \leftarrow i + 1$. If $i < n$, then loop to step 2, else STOP.

In other words, the vectors p^i are conjugated one to another with respect to the matrix $A := A^{\hat{s}}$. Such vectors can be constructed according to Algorithm 3 (e.g. see [29]).

Thus, we can enhance Lloyd's algorithm by employing the aforementioned global optimality conditions for DC minimization problems and approximation of the level surface. Thus, when Lloyd's algorithm halts and returns a local optimal solution, we apply the procedure described above to escape it and try to find a better local solution. Note that our approach can also be considered as the global search algorithm presented in Sect. 4, where the k-means algorithm is employed as a local search method.

6 Computational Experiments

To test our modification of the k-means algorithm enhanced by the DC minimization optimality conditions, we carried out a series of computational experiments. In particular, we compared it with the conventional widespread Lloyd's algorithm and its most popular modification—k-means++. We implemented the algorithms using C++ and compiled with Microsoft Visual C++ 16.2.0 compiler. We ran them on a PC with Intel Core i7-4790K CPU 4.0 GHz. Only one processing core was utilized.

We compared the algorithms on the so-called BIRCH test data collection widely used to test solution algorithms for clustering and facility location problems [18, 37, 38]. The BIRCH dataset includes test instances of two types (Types I and III). Each instance consists of two-dimensional points distributed according to a Gaussian mixture model. The cluster means are either located on a grid (Type I) or chosen uniformly at random (Type III). Note that the instances of Type I are easier to solve than ones of Type III. The test problems are relatively large and contains from $10,000$ to $20,000$ points, whereas the number of clusters k varies from 25 to 100.

For all the competing algorithms we chose random initial solutions. As k-means and k-means++ are converge in general to only local optimal solutions, we restarted them 30 times. For k-means and k-means++ we report maximal, minimal, and average objective values found over all reruns. The stopping criteria for k-means and k-means++ are the number of iterations (500) and the fraction of points that changed their assignment (0.001).

First, we report the computational results on our enhanced k -means algorithm (EKA) in Tables 1, 2, where the following denotations are employed:

- *Name*—problem's name;
- *m*—number of data items;
- *k*—number of clusters;
- *Start Obj.Val.*—the objective value at starting point;
- *Best Obj.Val.*—the objective value at the solution found by our algorithm;
- *St*—number of local solutions passed by our algorithm (note that *St* indicates the number of local optima where the objective value was improved);
- *Time*—CPU time in seconds.

Table 1. The computational results of the developed enhanced k-means on BIRCH problems of Type I

Name	m	k	Start Obj. Val.	Best Obj. Val.	St	Time
ds1x1	10000	100	5093172.87	790.16	123	16.84
ds1x2	15000	100	7758020.84	1254.45	198	56.96
ds1x3	20000	100	10821594.33	1574.27	91	95.94
ds1x4	9600	64	3509505.16	1006.46	188	16.79
ds1x5	12800	64	4702545.87	1307.28	97	21.84
ds1x6	16000	64	5881998.58	1694.58	240	31.98
ds1x7	19200	64	6065932.30	2015.26	118	44.74
ds1x8	10000	25	1106641.41	1591.18	67	4.51
ds1x9	12500	25	1716771.50	1966.56	28	5.12
ds1x10	15000	25	1869527.89	2394.94	80	7.26
ds1xA	17500	25	1903583.29	2893.65	122	7.38
ds1xB	20000	25	1965303.63	3402.85	11	18.09

Table 2. The computational results of the developed enhanced k-means on BIRCH problems of Type III

Name	m	k	Start Obj. Val.	Best Obj. Val.	St	Time
ds3x1	10000	100	14584810.00	1586.35	15	34.31
ds3x2	15000	100	28114197.90	2348.98	189	63.35
ds3x3	20000	100	43819743.68	3064.72	64	102.48
ds3x4	9600	64	6650014.66	1427.52	88	22.78
ds3x5	12800	64	6719172.05	2231.62	153	25.74
ds3x6	16000	64	11040283.26	2382.12	100	41.87
ds3x7	19200	64	12414395.87	2721.93	71	42.72
ds3x8	10000	25	861147.58	1498.45	34	4.70
ds3x9	12500	25	637772.19	1746.02	91	7.59
ds3x10	15000	25	1151099.92	2430.12	36	8.65
ds3xA	17500	25	1655681.98	2744.12	34	10.39
ds3xB	20000	25	1866104.81	2646.70	58	12.97

One can observe that the "complexity" of test instances (with respect to the number of "escaped" local optima), in general, does not depend on the problem size. Indeed, we can see that our algorithm identified and left (with an improvement of the objective value) 240 local solutions for $ds1x6$ ($m = 16,000$ and $k = 64$), whereas it passed only 91 and 11 local solutions for problems $ds1x3$ and $ds1xB$, respectively. Note that the latter problems involves larger number of points and the number of clusters is 100 and 25, respectively. This behavior can be explained by the choice of initial solutions as well as the special structure of some specific data instances. Similar results are obtained for Type III instances. Again, we can see some problems that force our approach to find a particularly large number of sequentially better local optima (e.g. see problem $ds3x2$ in Table 2). We can see that the run time may vary even for problem instances of the same size. It happens since the algorithm may pass a large number of local solutions that do not improve the current best objective value.

In Tables 3, 4 we report a computational comparison of our algorithm with the conventional k-means and k-means++ tested on the same set of instances. Note that we used the same starting points for all the competing algorithms. We follow the same notations in the tables. The columns k-means and k-means++ contains the minimum, maximum, and average values of the objective function over 30 reruns. Note that we ran our algorithm (EKA) only one time.

Table 3. A computational comparison of the developed algorithm with k-means and k-means++ on BIRCH instances of Type I

Name	m	k	k-means			k-means++			EKA
			min	max	av	min	max	av	
ds1x1	10000	100	818.12	902.62	862.13	803.77	844.17	825.19	**790.16**
ds1x2	15000	100	1207.59	1356.93	1281.34	**1199.49**	1265.38	1231.47	1254.45
ds1x3	20000	100	1638.97	1845.56	1723.98	1589.51	1674.54	1621.32	**1574.27**
ds1x4	9600	64	996.39	1131.78	1047.54	**969.63**	1004.36	988.40	1006.46
ds1x5	12800	64	1330.30	1508.91	1392.35	**1302.51**	1356.00	1324.60	1307.28
ds1x6	16000	64	1671.98	1842.77	1768.30	**1657.11**	1691.79	1675.75	1694.58
ds1x7	19200	64	2034.20	2167.71	2088.88	**1964.77**	2131.66	2016.23	2015.26
ds1x8	10000	25	1603.36	1770.01	1681.10	**1589.35**	1880.99	1657.17	1591.18
ds1x9	12500	25	2003.10	2695.17	2156.42	1979.51	2224.23	2075.02	**1966.56**
ds1x10	15000	25	2441.11	2725.58	2544.07	2436.76	2592.44	2492.98	**2394.94**
ds1xA	17500	25	2762.81	3268.65	3003.41	**2767.87**	2993.84	2846.67	2893.65
ds1xB	20000	25	3201.55	3769.73	3463.38	**3189.64**	3473.32	3330.53	3402.85

We can observe that our approach provides very competitive results: it outperforms k-means in most of the cases. Moreover, it even finds the best solutions for 4 problems of Type I and for 3 problems of Type III. In general, we can observe that EKA performs in average better than k-means that may demonstrate highly unstable results (see $ds3x3$ in Table 4). At the same time, k-means++ turns out to be more effective than k-means and converges to the local optimal solutions relatively similar to those found by EKA.

Table 4. A computational comparison of the developed algorithm with k-means and k-means++ on BIRCH instances of Type III

Name	m	k	k-means			k-means++			EKA
			min	max	av	min	max	av	
ds3x1	10000	100	1587.29	2048.38	1781.85	**1408.93**	1592.24	1470.34	1586.35
ds3x2	15000	100	2373.46	2788.77	2586.53	**2309.06**	2465.19	2374.93	2348.98
ds3x3	20000	100	3174.60	10703.25	4865.63	**2952.78**	3162.77	3059.66	3064.72
ds3x4	9600	64	1407.40	4488.79	1894.66	**1363.60**	1462.85	1399.58	1427.52
ds3x5	12800	64	1897.60	4797.09	2660.50	**1849.69**	1973.51	1911.26	2231.62
ds3x6	16000	64	2490.46	5875.91	4294.89	**2228.66**	2341.86	2308.67	2382.12
ds3x7	19200	64	2740.70	5572.09	3392.88	**2638.77**	2958.72	2715.73	2721.93
ds3x8	10000	25	1559.70	2100.35	1916.96	1500.41	1947.88	1647.11	**1498.45**
ds3x9	12500	25	2123.51	2468.39	2314.77	1757.23	2232.98	1957.40	**1746.02**
ds3x10	15000	25	2459.91	2627.91	2535.81	**2114.44**	2476.05	2246.28	2430.12
ds3xA	17500	25	2477.85	2950.68	2820.16	**2467.23**	2807.57	2571.58	2744.12
ds3xB	20000	25	2997.86	3440.25	3173.35	2690.94	3060.20	2865.75	**2646.70**

7   Conclusion

In this paper, we developed an approach to enhancing the well-known k-means algorithm by leveraging the so-called global optimality conditions for DC minimization problems. We developed a procedure for escaping local optimal solutions found by k-means which is based on a DC representation of the minimum-sum-of-squares clustering problem. We tested our approach in a series of computational experiments and compared it with the conventional k-means clustering algorithms. The obtained results demonstrated the effectiveness of the proposed algorithm in the case of medium-size problem instances.

Our future research may be focused on improving the procedure of escaping local optimal solutions, which can substantially decrease the run time of our enhanced k-means algorithm, e.g. by reducing the number of local search iterations that do not result in finding better local optimal solutions.

References

1. Aloise, D., Deshpande, A., Hansen, P., Popat, P.: NP-hardness of Euclidean sum-of-squares clustering. Mach. Learn. **75**, 245–248 (2009). https://doi.org/10.1007/s10994-009-5103-0

2. Aloise, D., Hansen, P., Liberti, L.: An improved column generation algorithm for minimum sum-of-squares clustering. Math. Program. **131**(1–2), 195–220 (2012). https://doi.org/10.1007/s10107-010-0349-7
3. Arthur, D., Vassilvitskii, S.: How slow is the k-means method? In: Amenta, N., Cheong, O. (eds.) Proceedings of the Twenty-Second Annual Symposium on Computational Geometry, SCG 2006, pp. 144–153. ACM, New York (2006). https://doi.org/10.1145/1137856.1137880
4. Arthur, D., Vassilvitskii, S.: K-means++: the advantages of careful seeding. In: Proceedings of the Eighteenth Annual ACM-SIAM Symposium on Discrete Algorithms, SODA 2007, pp. 1027–1035. SIAM, Philadelphia (2007)
5. Bagirov, A.M.: Modified global k-means algorithm for minimum sum-of-squares clustering problems. Pattern Recogn. **41**(10), 3192–3199 (2008). https://doi.org/10.1016/j.patcog.2008.04.004
6. Bagirov, A.M., Taheri, S., Ugon, J.: Nonsmooth DC programming approach to the minimum sum-of-squares clustering problems. Pattern Recogn. **53**, 12–24 (2016). https://doi.org/10.1016/j.patcog.2015.11.011
7. Carrizosa, E., Alguwaizani, A., Hansen, P., Mladenović, N.: New heuristic for harmonic means clustering. J. Glob. Optim. **63**(3), 427–443 (2014). https://doi.org/10.1007/s10898-014-0175-1
8. Demyanov, V., Bagirov, A., Rubinov, A.: A method of truncated codifferential with application to some problems of cluster analysis. J. Glob. Optim. **23**, 63–80 (2002). https://doi.org/10.1023/A:1014075113874
9. Diehr, G.: Evaluation of a branch and bound algorithm for clustering. SIAM J. Sci. Stat. Comput. **6**(2), 268–284 (1985). https://doi.org/10.1137/0906020
10. Dolgushev, A.V., Kel'manov, A.V.: On the algorithmic complexity of a problem in cluster analysis. J. Appl. Ind. Math. **5**(2), 191–194 (2011). https://doi.org/10.1134/S1990478911020050
11. du Merle, O., Hansen, P., Jaumard, B., Mladenovic, N.: An interior point algorithm for minimum sum-of-squares clustering. SIAM J. Sci. Comput. **21**(4), 1485–1505 (1999). https://doi.org/10.1137/S1064827597328327
12. Fukunaga, K., Narendra, P., Koontz, W.: A branch and bound clustering algorithm. IEEE Trans. Comput. **24**(09), 908–915 (1975). https://doi.org/10.1109/T-C.1975.224336
13. Gaudioso, M., Gruzdeva, T.V., Strekalovsky, A.S.: On numerical solving the spherical separability problem. J. Glob. Optim. **66**(1), 21–34 (2015). https://doi.org/10.1007/s10898-015-0319-y
14. Gruzdeva, T.V.: On a continuous approach for the maximum weighted clique problem. J. Glob. Optim. **56**(3), 971–981 (2013). https://doi.org/10.1007/s10898-012-9885-4
15. Gruzdeva, T.V., Strekalovsky, A.S.: On solving the sum-of-ratios problem. Appl. Math. Comput. **318**, 260–269 (2018). https://doi.org/10.1016/j.amc.2017.07.074
16. Gruzdeva, T.V., Ushakov, A.V.: K-means clustering via a nonconvex optimization approach. In: Pardalos, P., Khachay, M., Kazakov, A. (eds.) MOTOR 2021. LNCS, vol. 12755, pp. 462–476. Springer, Cham (2021). https://doi.org/10.1007/978-3-030-77876-7_31
17. Gruzdeva, T.V., Ushakov, A.V., Enkhbat, R.: A biobjective DC programming approach to optimization of rougher flotation process. Comput. Chem. Eng. **108**, 349–359 (2018). https://doi.org/10.1016/j.compchemeng.2017.10.001

18. Hansen, P., Brimberg, J., Urosević, D., Mladenović, N.: Solving large p-median clustering problems by primal-dual variable neighborhood search. Data Min. Knowl. Discov. **19**(3), 351–375 (2009). https://doi.org/10.1007/s10618-009-0135-4
19. Hansen, P., Mladenović, N.: J-means: a new local search heuristic for minimum sum of squares clustering. Pattern Recogn. **34**(2), 405–413 (2001). https://doi.org/10.1016/S0031-3203(99)00216-2
20. Hartigan, J.A., Wong, M.A.: Algorithm AS 136: A k-means clustering algorithm. J. R. Stat. Soc. Ser. C **28**(1), 100–108 (1979). https://doi.org/10.2307/2346830
21. Hoai An, L.T., Hoai Minh, L., Tao, P.D.: New and efficient DCA based algorithms for minimum sum-of-squares clustering. Pattern Recogn. **47**(1), 388–401 (2014). https://doi.org/10.1016/j.patcog.2013.07.012
22. Hoai An, L.T., Tao, P.D.: The DC (difference of convex functions) programming and DCA revisited with DC models of real world nonconvex optimization problems. Ann. Oper. Res. **133**, 23–46 (2005). https://doi.org/10.1007/s10479-004-5022-1
23. José-García, A., Gómez-Flores, W.: Automatic clustering using nature-inspired metaheuristics: a survey. Appl. Soft Comput. **41**, 192–213 (2016). https://doi.org/10.1016/j.asoc.2015.12.001
24. Likas, A., Vlassis, N., Verbeek, J.J.: The global k-means clustering algorithm. Pattern Recogn. **36**(2), 451–461 (2003). https://doi.org/10.1016/S0031-3203(02)00060-2
25. Lloyd, S.: Least squares quantization in PCM. IEEE Trans. Inf. Theory **28**(2), 129–137 (1982). https://doi.org/10.1109/TIT.1982.1056489
26. MacQueen, J.: Some methods for classification and analysis of multivariate observations. In: Cam, L.M.L., Neyman, J. (eds.) Proceedings of the Fifth Berkeley Symposium on Mathematical Statistics and Probability, vol. 1, pp. 281–297. University of California Press, Berkeley (1967)
27. Mahajan, M., Nimbhorkar, P., Varadarajan, K.: The planar k-means problem is NP-hard. Theor. Comput. Sci. **442**, 13–21 (2012). Special Issue on the Workshop on Algorithms and Computation (WALCOM 2009). https://doi.org/10.1016/j.tcs.2010.05.034
28. Mansueto, P., Schoen, F.: Memetic differential evolution methods for clustering problems. Pattern Recogn. **114**, 107849 (2021). https://doi.org/10.1016/j.patcog.2021.107849
29. Nocedal, J., Wright, S.J.: Numerical Optimization. Operations Research and Financial Engineering, 2nd edn. Springer, New York (2006). https://doi.org/10.1007/978-0-387-40065-5
30. Okafor, A., Pardalos, P.: K-Means Clustering Using Entropy Minimization. Series on Computers and Operations Research, vol. 4, pp. 339–351. World Scientific Publishing, Singapore (2004). https://doi.org/10.1142/9789812796592_0015
31. Ordin, B., Bagirov, A.M.: A heuristic algorithm for solving the minimum sum-of-squares clustering problems. J. Glob. Optim. **61**, 341–361 (2015). https://doi.org/10.1007/s10898-014-0171-5
32. Ostrovsky, R., Rabani, Y., Schulman, L.J., Swamy, C.: The effectiveness of Lloyd-type methods for the k-means problem. J. ACM **59**(6) (2013). https://doi.org/10.1145/2395116.2395117
33. Shao, G., Li, D., Zhang, J., Yang, J., Shangguan, Y.: Automatic microarray image segmentation with clustering-based algorithms. PLoS ONE **14**(1), e0210075 (2019). https://doi.org/10.1371/journal.pone.0210075
34. Strekalovsky, A.S.: On local search in D.C. optimization problems. Appl. Math. Comput. **255**, 73–83 (2015)

35. Strekalovsky, A.: On the minimization of the difference of convex functions on a feasible set. Comput. Math. Math. Phys. **43**, 380–390 (2003)
36. Strekalovsky, A.S.: On solving optimization problems with hidden nonconvex structures. In: Rassias, T.M., Floudas, C.A., Butenko, S. (eds.) Optimization in Science and Engineering, pp. 465–502. Springer, New York (2014). https://doi.org/10.1007/978-1-4939-0808-0_23
37. Ushakov, A.V., Vasilyev, I.: Near-optimal large-scale k-medoids clustering. Inf. Sci. **545**, 344–362 (2021). https://doi.org/10.1016/j.ins.2020.08.121
38. Ushakov, A.V., Vasilyev, I.L., Gruzdeva, T.V.: A computational comparison of the p-median clustering and k-means. Int. J. Artif. Intell. **13**(1), 229–242 (2015)
39. Vinod, H.D.: Integer programming and the theory of grouping. J. Am. Stat. Assoc. **64**(326), 506–519 (1969). https://doi.org/10.2307/2283635

Computational Study of Local Search Methods for a D.C. Optimization Problem with Inequality Constraints

M. V. Barkova[✉][iD] and A. S. Strekalovskiy[iD]

Matrosov Institute for System Dynamics and Control Theory of Siberian Branch of Russian Academy of Sciences, Irkutsk 664033, Russia
{mbarkova,strekal}@icc.ru

Abstract. This paper addresses a nonconvex optimization problem where the cost function and inequality constraints are d.c. functions. Two special local search methods based on the idea of the consecutive solution of partially linearized problems are developed. The latter problems turn out to be convex and therefore solvable with the help of software packages for convex optimization. The first method linearizes both the objective function of the problem and all constraints functions. The second approach is based on the reduction of the original problem to a penalized problem without constraints via the exact penalization theory. The methods developed were computationally tested on some well-known test examples and specially generated problems with known local and global solutions.

Keywords: Nonconvex quadratic problem · D.C. functions · Local search methods · Exact penalty · Linearized problem

1 Introduction

As well-known, almost all real-life optimization problems are explicitly or implicitly nonconvex. Such problems may have a lot (often a large number) of stationary vectors and local pits [12,13,17] that may be rather far from global solutions. As a consequence, classical convex optimization methods (conjugate gradients, Newton's and quasi-Newton's methods, TRM, SQP, IPM, etc. [2,17]) turn out to be inoperative as to find a global solution, in general, and often fail when directly applied to nonconvex problems (providing only the KKT points and sometimes, only feasible points). At the same time, some methods of Global Optimization (B&B, cut's methods, etc.) [8,13,24] suffer the so-called "curse of dimensionality" when the exponential growth of computational efforts corresponds to an increase in dimension of the problem in question.

The research was funded by the Ministry of Education and Science of the Russian Federation within the framework of the project "Theoretical foundations, methods and high-performance algorithms for continuous and discrete optimization to support interdisciplinary research" (No. of state registration: 121041300065-9).

© Springer Nature Switzerland AG 2021
N. N. Olenev et al. (Eds.): OPTIMA 2021, LNCS 13078, pp. 94–109, 2021.
https://doi.org/10.1007/978-3-030-91059-4_7

Because of this, the development of new algorithms especially for special classes of nonconvex problems can be viewed as one of the promising directions in nonconvex optimization. Over the last two decades, researchers pay more and more attention to special local search methods for nonconvex optimization problems. In these fields, one can find several different and very interesting propositions. In particular, it was developed several local search methods that can find stationary points in such nonconvex problems as d.c. minimization, reverse convex, and convex maximization problems, etc. (see [2, 8, 9, 13, 20–23]).

In this paper, we investigate two local search methods for solving the nonconvex optimization problem with d.c. inequality constrains. The first one, proposed in [21], operates with the original problem and is based on linearization with respect to the basic nonconvexities. The second combines the exact penalty methodology with the principal ideas of special local search methods. It deals with the penalized problem and dynamically changes penalty parameter during the computational process [20–22].

The paper is organized as follows. After the statement of Problem ($\mathcal{P}$) in Sect. 2, we recall the principal ideas for constructing a special local search method for this problem. Further, in Sect. 4, the original problem is reduced to a problem without nonconvex constraints by the exact penalization theory, so that the reduced (penalized) problem is also a d.c. minimization problem. After that, we introduce a special local search method for the penalized problem ($\mathcal{P}_\sigma$), where the penalty parameter may be changed during computational solving of linearized problems. Furthermore, in Sect. 5, we present the results of computational testings of two local searches on some low-dimensional test problems from the literature. Finally, we give the results of a comparison of the two special local search methods with well-known computational software.

2 Problem Statement

Consider a d.c. minimization problem with d.c. inequality constraints:

$$(\mathcal{P}): \qquad \left. \begin{array}{l} f_0(x) = g_0(x) - h_0(x) \downarrow \min\limits_{x}, \ x \in S, \\ f_i(x) = g_i(x) - h_i(x) \le 0, \quad i \in \mathcal{I} = \{1, \dots, m\}, \end{array} \right\} \qquad (1)$$

where functions $g_i, h_i, \ i \in \mathcal{I} \cup \{0\}$ are convex, $S \in \mathbb{R}^n$ is a closed convex set.

Below, assume that the feasible set of Problem ($\mathcal{P}$) is nonempty: $\mathcal{F} := \{x \in S \mid f_i(x) \le 0, \ i \in \mathcal{I}\} \ne 0$, and the optimal value of Problem ($\mathcal{P}$) is finite: $\mathcal{V}(\mathcal{P}) = \inf\limits_{x}\{f_0(x) \mid x \in \mathcal{F}\} > -\infty$.

3 Special Local Search Method I

The main idea of this local search method (LSMI), first described in [21], consists in the consecutive solution of partially linearized problems. We linearize not only the objective function of Problem ($\mathcal{P}$) but all nonconvex constraints.

Let us given a feasible starting point $x^0 \in \mathcal{F}$ and a current iterate $x^s \in \mathcal{F}$. Then we find the next iterate x^{s+1} as a solution to the following problem:

$$(\mathcal{P}L_s): \qquad \left.\begin{array}{l} \Phi_{0s}(x) := g_0(x) - \langle \nabla h_0(x^s), x \rangle \downarrow \min_x, \quad x \in S, \\ \Phi_{is}(x) := g_i(x) - \langle \nabla h_i(x^s), x - x^s \rangle - h_i(x^s) \le 0, \quad i \in \mathcal{I}. \end{array}\right\} \qquad (2)$$

Since the objective function and the feasible set $\mathcal{F}_s$ of Problem $(\mathcal{P}L_s)$ are convex,

$$\mathcal{F}_s := \mathcal{F}L_s = \{x \in S \mid g_i(x) - \langle \nabla h_i(x^s), x - x^s \rangle - h_i(x^s) \le 0, \ i \in \mathcal{I}\}, \qquad (3)$$

then, Problem $(\mathcal{P}L_s)$ turns out to be also convex and can be solved by suitable classical optimization methods. Therefore, the next point x^{s+1} is produced to satisfy the following inequality

$$\Phi_{0s}(x^{s+1}) := g_0(x^{s+1}) - \langle \nabla h_0(x^s), x^{s+1} \rangle \le V(\mathcal{P}L_s) + \delta_s, \qquad (4)$$

and $V(\mathcal{P}L_s) := \inf_x \{g_0(x) - \langle \nabla h_0(x^s), x \rangle \mid x \in \mathcal{F}_s\}$ is the optimal value of Problem $(\mathcal{P}L_s)$, where the sequence δ_s is chosen according the next conditions

$$\delta_s \ge 0, \ s = 1, 2, \ldots, \sum_{s}^{\infty} \delta_s < \infty. \qquad (5)$$

Furthermore, every new iterate $x^{s+1} \in Sol(\mathcal{P}L_s)$ is feasible not only in the linearized Problem $(\mathcal{P}L_s)$ but also in the original Problem $(\mathcal{P})$: $x^{s+1} \in \mathcal{F}$. Indeed, due to the convexity of the functions $h_i(\cdot)$, $i \in \mathcal{I}$, we have:

$$\begin{aligned} 0 &\ge g_i(x^{s+1}) - \langle \nabla h_i(x^s), x^{s+1} - x^s \rangle - h_i(x^s) = \Phi_{is}(x^{s+1}) \\ &\ge g_i(x^{s+1}) - [h_i(x^{s+1}) - h_i(x^s)] - h_i(x^s) \\ &= g_i(x^{s+1}) - h_i(x^{s+1}) = f_i(x^{s+1}), \ i \in \mathcal{I}. \end{aligned} \qquad (6)$$

Thus, $f_i(x^{s+1}) \le 0 \ \forall i \in \mathcal{I}$, and the sequence $\{x^s\}$, generated by the rule (4)–(5), turns out to be feasible in the original Problem $(\mathcal{P})$.

Definition 1. *A point x_* is called to be a critical point in Problem $(\mathcal{P})$ with respect to the LSM (4)–(5) if it is a solution to the Problem $(\mathcal{P}L_*)$, i.e.*

$$g_0(x_*) - \langle \nabla h_0(x_*), x_* \rangle = \inf_x \{g_0(x) - \langle \nabla h_0(x_*), x \rangle \mid x \in \mathcal{F}L_*\}. \qquad (7)$$

where

$$(\mathcal{P}L_*): \qquad \left.\begin{array}{l} g_0(x) - \langle \nabla h_0(x_*), x \rangle \downarrow \min_x, \quad x \in S, \\ g_i(x) - \langle \nabla h_i(x_*), x - x_* \rangle - h_i(x_*) \le 0, \quad i \in \mathcal{I}. \end{array}\right\} \qquad (8)$$

Using theoretical results on the convergence of LSMI proposed in [21], we can use one of the following inequalities as a stopping criterion:

$$\begin{cases} \text{a)} \ f_0(x^s) - f_0(x^{s+1}) \le \dfrac{\tau}{2}, \\ \text{b)} \ \Phi_{0s}(x^s) - \Phi_{0s}(x^{s+1}) \overset{\Delta}{=} g_0(x^s) - g_0(x^{s+1}) - \langle \nabla h_0(x^s), x^s - x^{s+1} \rangle \le \dfrac{\tau}{2}, \end{cases} \qquad (9)$$

where τ is an accuracy. If one of the inequalities (9) is satisfied and $\delta_s \leq \tau/2$, then the point x^s is a τ-critical point in Problem $(\mathcal{PL}_s)$, which means that it is a τ-solution to the Problem $(\mathcal{PL}_s)$, or equivalent the next inequality holds:

$$g_0(x^s) - \langle \nabla h_0(x^s), x^s \rangle \leq \inf_x \{ g_0(x) - \langle \nabla h_0(x^s), x \rangle \mid x \in \mathcal{FL}_s \} + \tau. \qquad (10)$$

Let us be given a starting point $x_0 \in \mathcal{F}$ and a sequence $\{\delta_s\}$ satisfying (5).

Local Search Scheme I

Step 0. Set $s := 0$, $x^s := x_0 \in \mathcal{F}$.
Step 1. Find x^{s+1} as a δ_s-solution to the linearized problem $(\mathcal{PL}_s)$.
Step 2. (Stopping criteria) If the inequality (9 b) holds and $\delta_s \leq \dfrac{\tau}{2}$ then STOP: x^{s+1} is a τ-critical point relatively to LSMI.
Step 3. Set $s := s + 1$, and return to Step 1.

It is worth noting that the starting point x_0 should be feasible in $(\mathcal{P})$, $x_0 \in \mathcal{F}$ ensuring that the set $\mathcal{FL}_s$ is nonempty. Therefore, the sequence $\{x^s\}$ produced by the rule (4)–(5) is a sequence of feasible points in Problem $(\mathcal{P})$ as shown in (6).

4   The Penalization and Local Search Method II

Let us introduce the penalized problem [11,12,17]

$$(\mathcal{P}_\sigma): \qquad\qquad \Theta_\sigma(x) := f_0(x) + \sigma W(x) \downarrow \min_x, \ x \in S, \qquad (11)$$

where $\sigma \geq 0$ is a penalty parameter, and the penalty function $W(x)$ is defined as follows:

$$W(x) := \max\{0, f_1(x), \ldots, f_m(x)\}. \qquad (12)$$

It means that Problem $(\mathcal{P})$ is reduced to a penalized problem $(\mathcal{P}_\sigma)$, which is a d.c. minimization problem without nonconvex constraints. Let us verify that the objective function $\Theta_\sigma(x)$ of the penalized problem $(\mathcal{P}_\sigma)$ is a really d.c. function. Indeed, the function $\Theta_\sigma(\cdot)$ can be represented as follows:

$$\Theta_\sigma(x) = f_0(x) + \sigma \max\{0, f_1(x), \ldots, f_m(x)\} = G_\sigma(x) - H_\sigma(x), \qquad (13)$$

where the functions

$$G_\sigma(x) = g_0(x) + \sigma \max \left\{ \sum_{p \in \mathcal{I}} h_p(x); \ \left[g_i(x) + \sum_{p \in \mathcal{I}}^{p \neq i} h_p(x) \right], i \in \mathcal{I} \right\},$$

$$H_\sigma(x) = h_0(x) + \sigma \sum_{p \in \mathcal{I}} h_i(x)$$

are convex, thanks to the properties of this class of functions (see [12]).

Let us describe a local search method (LSMII), producing a sequence $\{x^s\}$ every member of which is an approximate solution to the next linearized problem

$$(\mathcal{P}_s L_s): \qquad \Phi_s(x) := G_s(x) - \langle \nabla H_s(x^s), x \rangle \downarrow \min_x, \ x \in S, \qquad (14)$$

where $G_s := G_{\sigma_s}, H_s := H_{\sigma_s}, \nabla H_s(x^s) = \nabla h_0(x^s) + \sigma_s \sum_{i \in \mathcal{I}} \nabla h_i(x^s)$. It can be readily seen [11,12,17,20] that Problem $(\mathcal{P}_s L_s)$ is a convex problem, where the function $H_s(\cdot)$, that accumulates all nonconvexities of the Problems $(\mathcal{P})$ and $(\mathcal{P}_\sigma)$, was linearized. It means that a point x^{s+1} is produced to satisfy the following inequality:

$$\Phi_s(x^{s+1}) \stackrel{\triangle}{=} G_s(x^{s+1}) - \langle \nabla H_s(x^{s+1}), x^{s+1} \rangle \leq \mathcal{V}(\mathcal{P}_s L_s) + \delta_s, \qquad (15)$$

where $\mathcal{V}(\mathcal{P}_s L_s) := \inf_x \{G_s(x) - \langle \nabla H_s(x^{s+1}), x \rangle \mid x \in S\}$ is an optimal value of Problem $(\mathcal{P}_s L_s)$ and the sequence $\{\delta_s\}$ is defined as in (5).

Moreover, let us introduce the following auxiliary problem:

$$(\mathcal{AP}_W L_s): \qquad \Phi_W(x) := G_W(x) - \langle \nabla H_W(x^s), x \rangle \downarrow \min_x, \ x \in S, \qquad (16)$$

where $G_W(x) = \sigma_s^{-1}[G_s(x) - g_0(x)] = \max\{\sum_{p \in \mathcal{I}} h_p(x); [g_i(x) + \sum_{p \in \mathcal{I}}^{p \neq i} h_p(x)], i \in \mathcal{I}\},$
$\nabla H_W(x^s) = \sigma_s^{-1}[\nabla H_s(x^s) - \nabla h_0(x)] = \sum_{i \in \mathcal{I}} \nabla h_i(x^s).$

It is easy to verify that Problem $(\mathcal{AP}_W L_s)$ is a linearized problem of the penalty function minimization:

$$(\mathcal{P}_W): \qquad W(x) \stackrel{\triangle}{=} G_W(x) - H_W(x) \downarrow \min_x, \ x \in S, \qquad (17)$$

where $G_W(x) = \max\{\sum_{p \in \mathcal{I}} h_p(x); [g_i(x) + \sum_{p \in \mathcal{I}}^{p \neq i} h_p(x)], i \in \mathcal{I}\}, H_W(x) = \sum_{i \in \mathcal{I}} h_i(x).$

Let us given a starting point $x^0 \in S$, the initial value of the penalty parameter $\sigma_0 > 0$, and the method parameters $\eta_1, \eta_2 \in]0, 1[$.

Local Search Scheme II

Step 0. Set $s := 0$, $x^s := x_0$, $\sigma_s := \sigma^0$ (say, $\sigma^0 = 1$).
Step 1. Find a δ_s-solution $x(\sigma_s)$ to the linearized problem $(\mathcal{P}_s L_s)$.
Step 2. If $W(x(\sigma_s)) = 0$, then set $\sigma_+ := \sigma_s$, $x(\sigma_+) := x(\sigma_s)$, and go to Step 7.
Step 3. If $W(x(\sigma_s)) > 0$, find a solution $x_W^s \in Sol(\mathcal{AP}_W L_s)$ to the auxiliary problem $(\mathcal{AP}_W L_s)$.
Step 4. If $W(x_W^s) = 0$, then, starting at x_W^s, (by increasing, if necessary, the value $\sigma > \sigma_s$) solve consecutively the problems $(\mathcal{P}_s L_s)$, and find $\sigma_+ > \sigma_s$ such that $W(x(\sigma_+)) = 0$, $x(\sigma_+) \in Sol(\mathcal{P}_{\sigma_+} L_s)$, and go to Step 7.

Step 5. If $W(x_W^s) > 0$, and a feasible penalty parameter $\sigma_+ > \sigma_s$, such that $W(x(\sigma_+)) = 0$, is not found (during Step 4), find $\sigma_+ > \sigma_s$ which satisfies the following inequality:

$$W(x^s) - W(x(\sigma_+)) \geq \eta_1[W(x^s) - W(x_W^s)]. \tag{18}$$

Step 6. Increase σ_+, (if necessary), to fulfill the inequality

$$\Phi_s(x^s) - \Phi_{\sigma_+}(x(\sigma_+)) \geq \eta_2\sigma_+[W(x^s) - W(x(\sigma_+))]. \tag{19}$$

Step 7. Set $s := s + 1$, $\sigma_{s+1} := \sigma_+$, $x^{s+1} := x(\sigma_+)$, and return to Step 1.

The inequalities (18) and (19) estimate that the change in the penalty function is a good measure of progress made by linearized and penalty functions [2]. The theoretical results of the convergence of the LSMII were proposed and discussed in detail in [22]. According to it, we use the following inequalities as a stopping criterion:

$$\begin{cases} \text{a) } W(x(\sigma_+)) \leq \nu, \\ \text{b) } \Phi_{s+1}(x^s) - \Phi_{s+1}(x(\sigma_+)) \leq \dfrac{\tau}{2}, \end{cases} \tag{20}$$

where ν, τ are given accuracies. It is worth noting that inequality (20 a) ensures that the point $x(\sigma_+)$ is a feasible point of Problem $(\mathcal{P}_\sigma)$ (meanwhile LSMII can start at unfeasible points). Moreover, if the inequality (20 b) is satisfied with $\delta_s \leq \tau/2$, then according to Remark 7.6 [22] the point x^s generated by Local search scheme II is τ-critical point and τ-solution to the linearized problem $(\mathcal{P}_sL_s)$ (linearized at the point x^s with the penalty parameter $\sigma_+ > 0$). Hence the stopping criterion (20) can be used in a computational simulation.

5   Numerical Testing

In this section, we perform a comparative numerical testing of the two local search methods described above. The testing was carried out on one of the most frequent subclasses of Problem $(\mathcal{P})$, where $f_i(\cdot)$ $i \in \mathcal{I} \cup \{0\}$ are nonconvex quadratic functions. First, we tested local search methods on the small-scale problems from the open-access data sets to verify the algorithm performance. After that, we used specially generated test problems to observe algorithms' behavior on the multiextremal problems and to demonstrate their efficiency during the numerical solution of these high-dimension problems.

5.1   Low-Dimensional Instances

For the first stage of numerical testing, we chose 4 low-dimensional test examples from the literature [5,6,8,9] with known global solutions. For solving, we used a computer with the Intel Core i5-4670K 3.40 GHz processor. The auxiliary linearized problems (convex problems) were solved by the built-in fmincon procedure (MATLAB R2011b [14]), in which we employed tree algorithms: sqp method, active-set approach, and interior point method.

The first test instance "Hesse" is a nonconvex minimization problem with reverse-convex inequality constraints. The feasible set of the problem contains two nonconvex inequalities, linear and box constraints. Hence the nonconvexity of the problem is generated by the cost function and two inequality constraints. The problem has 18 local extrema and a unique global solution $z = (5, 1, 5, 0, 5, 10)$.

Example 1. "Hesse" [8,9] $(n = 6, m = 2, \mathcal{V}(\mathcal{P}) = -310)$

$$f_0(x) = -25(x_1-2)^2 - (x_2-2)^2 - (x_3-1)^2 - (x_4-4)^2 - (x_5-1)^2 - (x_6-4)^2 \downarrow \min_x,$$
$$f_1(x) = -x_3^2 + 6x_3 - x_4 \leq 5, \quad f_2(x) = -x_5^2 + 6x_5 - x_6 \leq 5,$$
$$S = \{x \in \mathbb{R}^6 : x_1 - 3x_2 \leq 2, \ -x_1 + x_2 \leq 2, \ x_1 + x_2 \leq 6, \ x_1 + x_2 \geq 2,$$
$$x_1 \geq 0, \ x_2 \geq 0, \ 1 \leq x_3 \leq 5, \ 0 \leq x_4 \leq 6, \ 1 \leq x_5 \leq 5, \ 0 \leq x_6 \leq 10.\}$$

Table 1 presents results illustrating the performance of two local search methods. We use the following denotations: the value of the objective function at the starting point is $f_0(x_0)$, this one at the obtained point is $f_0(x_*)$, PL stands for the number of solved linearized problems. The second local search method table presents the value of the final penalty parameter σ_* and the number of iterations

Table 1. Results of solving the problem "Hesse"

#	$f_0(x_0)$	Solver	LSMI			LSMII				
			$f_0(x_*)$	PL	$Time$	$f_0(x_*)$	σ_*	PL	Itr	$Time$
1	−58	sqp	−132	8	0.34	−132	0.1	2	1	0.04
		as	−132	8	0.15	−132	0.1	2	1	0.07
		ip	−131.99	7	0.25	−131.99	0.1	12	6	0.28
2	−80.5	sqp	−298	3	0.09	−298	0.1	2	1	0.14
		as	−298	3	0.05	−298	0.1	2	1	0.47
		ip	−297.99	4	0.17	−298	0.1	2	1	0.11
3	−44	sqp	−132	8	0.09	−132	0.1	2	1	0.07
		as	−132	8	0.09	−132	0.1	2	1	0.04
		ip	−131.99	7	0.16	−131.99	0.1	12	6	0.29
4	−181.34	sqp	−257.99	11	0.35	−258	0.1	4	2	0.07
		as	−36	13	0.67	−258	0.1	4	2	0.09
		ip	−35.99	15	0.57	−258	0.1	10	5	0.49
5	−497.67	sqp	−132	17	0.42	−132	0.1	4	2	0.12
		as	−132	13	0.44	−132	0.1	4	2	0.16
		ip	−131.99	11	0.39	−131.99	0.1	18	9	0.53
6	−64.35	sqp	−132	9	0.32	−148	0.1	10	5	0.21
		as	−132	10	0.54	−148	0.1	10	5	0.19
		ip	−131.99	8	0.38	−163.99	0.1	24	12	0.82

of the local search Itr. The LSMI and the LSMII were launched from six starting points using the built-in matlab algorithm denoted in the column $Solver$. The cases where any special local search method found a global solution (with the accuracy $\varepsilon = 10^{-4}$) are highlighted in bold.

Note that starting from all points both local search methods found the feasible points. Further, LSMI, is starting at a feasible point, produces the sequence of feasible vectors $\{x^s\}$ according to the inequalities (6), meanwhile, the LSMII (which can start at unfeasible points) generates the sequence $\{x^s\}$ such that $\lim\limits_{s \to \infty} W(x^s) = 0$.

In addition, Table 1 shows that both methods provide a rather good improvement of the values of the cost function in less than a second. Indeed, one can see that starting from point #2, the LSMII needs only 2 iterations to achieve the point close to the global solution, according to the value of the objective function. However, the interior-point algorithm showed results worse than other methods. It required twice as many iterations to reach the same points.

Note that the values of the objective function obtained by the LSMII launched from the starting points #4, #6 turn out to be better than those of LSMI. Although in the LSMII, the penalty parameter value starting from $\sigma_0 = 0.1$ remains the same for all points. On the other hand, it is clear that in this test example, both local search methods failed to find global solutions $\mathcal{V}(\mathcal{P}) = -310$ starting from all initial points.

The next two test instances "Hs108" and "Mistake" have a similar structure. The objective function of examples contains bilinear elements highlighting the d.c. structure of the function. They have four and five nonconvex constraints, respectively, which are also bilinear. The best known solution to the problem "Hs108" is $z = (0.9238, 0.3828, 0.1304, 0.9915, 0.9238, 0.3828, 0.1304, 0.9915, 0)$. For the problem "Mistake", the point $z = (0.7575, 0.7958, 0.3433, 0.8463, 0.9961, 0.0888, 0.6688, 1.7918, 0.1430)$ provides the best known solution to the problem. The starting points for the two examples are the same, moreover, three of them are infeasible.

Example 2. "Hs108" [6] $(n = 9, m = 4, \mathcal{V}(\mathcal{P}) = -0.866025)$

$$f_0(x) = -0.5(x_1x_4 - x_3x_2 + x_3x_9 - x_5x_9 + x_8x_5 - x_6x_7) \downarrow \min_x,$$

$$f_1(x) = x_2x_3 - x_1x_4 \leq 0, \quad f_2(x) = -x_3x_9 \leq 0,$$

$$f_3(x) = x_5x_9 \leq 0, \quad f_4(x) = x_6x_7 - x_5x_8 \leq 0,$$

$$S = \{x \in \mathbb{R}^9 : x_3^2 + x_4^2 \leq 1, \quad (x_1 - x_5)^2 + (x_2 + x_6)^2 \leq 1, \quad x_5^2 + x_6^2 \leq 1,$$

$$(x_1 - x_7)^2 + (x_2 + x_8)^2 \leq 1, \quad x_9^2 \leq 1, \quad x_9 \geq 0, \quad x_1^2 + (x_2 + x_9)^2 \leq 1,$$

$$(x_3 - x_7)^2 + (x_4 + x_8)^2 \leq 1, \quad (x_3 - x_5)^2 + (x_4 + x_6)^2 \leq 1.\}$$

The results of solving the problem "Hs108" in Table 2 demonstrate that the LSMI launched from the starting point #2 found the best known solution. The biggest number of solved linearized problems for this problem shows the interior point method, by which the LSMI found the global solution by solving 97 linearized problems. And, except for the time which interior point method takes for solving

Table 2. Results of solving the problem "Hs108"

#	$f_0(x_0)$	Solver	LSMI			LSMII				
			$f_0(x_*)$	PL	Time	$f_0(x_*)$	σ_*	PL	Itr	Time
1	−0.31	sqp	−0.67	64	3.08	−0.8659	0.1	50	25	2.67
		as	−0.67	75	2.31	−0.8659	0.1	50	25	2.54
		ip	−0.67	64	7.92	−0.8657	0.1	50	25	7.88
2	−0.34	sqp	**−0.87**	37	1.34	−0.8653	0.3	102	50	1.98
		as	**−0.87**	27	0.78	−0.8655	0.3	104	51	1.99
		ip	**−0.87**	97	12.11	−0.8652	0.4	99	48	8.19
3	−0.42	sqp	−0.67	45	2.25	−0.6726	0.1	44	22	1.74
		as	−0.67	46	1.54	−0.6725	0.1	42	21	1.42
		ip	−0.67	60	6.28	−0.6726	0.1	44	22	8.02
4	−1.47	sqp	−0.53	7	1.99	−0.8657	0.1	22	11	0.93
		as	−0.40	10	3.43	−0.8658	0.1	24	12	1.13
		ip	−0.51	14	6.63	**−0.866**	0.2	44	16	7.15
5	−32.73	sqp	−0.67	9	2.84	−0.8658	0.1	86	43	2.98
		as	−0.85	9	3.08	−0.8659	0.1	86	43	2.80
		ip	−0.63	13	11.44	−0.8659	0.3	112	49	15.77
6	−11.43	sqp	−0.61	7	2.53	−0.8658	0.1	18	9	0.67
		as	−0.41	8	2.96	−0.8659	0.1	20	10	0.76
		ip	−0.27	22	14.36	−0.8659	0.1	20	10	3.02

the test example we can conclude that the LSMI solves the problem "Hs108" in less than four seconds.

Moreover, in this case, for the starting points #2, #4, and #5, we can see that the LSMII doubled (or even tripled) the penalty parameter during the computational process. Even though LSMII found a global solution (with accuracy $\varepsilon = 10^{-4}$) only at the starting point #4 (using the interior-point method), all other produced points, except #3, are too close to the optimal one.

Example 3. "Mistake" [6] ($n = 9, m = 5, \mathcal{V}(\mathcal{P}) = -1.0$)

$$f_0(x) = -0.5(x_1 x_4 - x_3 x_2 + x_3 x_9 - x_5 x_9 + x_8 x_5 - x_6 x_7) \downarrow \min_x,$$
$$f_1(x) = x_2 x_3 - x_1 x_4 \leq 0, \quad f_2(x) = -x_8 x_9 \leq 0, \quad f_3(x) = -x_5 x_9 \leq 0,$$
$$f_4(x) = x_6 x_7 - x_5 x_8 \leq 0, \quad f_5(x) = x_7^2 + x_9 x_8 \leq 1,$$
$$S = \{x \in \mathbb{R}^9 : (x_1 - x_7)^2 + (x_2 + x_8)^2 \leq 1, \quad (x_3 - x_7)^2 + (x_4 + x_8)^2 \leq 1,$$
$$x_9^2 \leq 1, \quad x_3^2 + x_4^2 \leq 1, \quad x_1^2 + (x_2 + x_9)^2 \leq 1, \quad (x_1 - x_5)^2 + (x_2 + x_6)^2 \leq 1,$$
$$x_5^2 + x_6^2 \leq 1, \quad (x_3 - x_5)^2 + (x_4 + x_6)^2 \leq 1.\}$$

Table 3 demonstrates computational results for the problem "Mistake". The LSMI starting from the points #4, #5, and #6, found the best known solution

Table 3. Results of solving the problem "Mistake"

#	$f_0(x_0)$	Solver	LSMI			LSMII				
			$f_0(x_*)$	PL	Time	$f_0(x_*)$	σ_*	PL	Itr	Time
1	−0.31	sqp	−0.49	23	1.57	−0.932	0.8	209	102	6.28
		as	−0.49	20	1.19	−0.932	0.8	209	102	5.38
		ip	−0.93	156	34.67	−0.932	0.8	229	112	29.65
2	−0.34	sqp	−0.86	15	0.67	−0.931	0.8	179	87	4.43
		as	−0.86	15	0.53	−0.931	0.8	181	88	3.81
		ip	−0.86	15	1.73	−0.931	0.8	185	90	18.74
3	−0.42	sqp	−0.49	33	1.74	−0.998	0.3	348	173	8.41
		as	−0.66	87	2.92	−0.998	0.3	348	173	7.21
		ip	−0.49	44	20.93	−0.997	0.4	313	155	37.22
4	−1.47	sqp	**−1**	123	6.82	−0.998	0.2	239	119	6.27
		as	−0.34	13	4.97	−0.998	0.3	238	118	5.26
		ip	**−0.99**	371	92.34	−0.997	0.4	239	118	26.13
5	−32.73	sqp	**−1**	36	2.25	**−0.999**	0.1	24	12	1.07
		as	**−1**	25	2.82	**−0.999**	0.1	24	12	0.85
		ip	**−0.99**	247	35.77	**−0.999**	0.1	24	12	4.46
6	−11.43	sqp	**−1**	29	2.16	**−0.999**	0.1	62	31	2.45
		as	**−0.99**	32	2.37	**−0.999**	0.1	62	31	2.11
		ip	**−0.99**	297	71.78	**−0.999**	0.1	62	31	12.12

while the LSMII only from the points #5 and #6. Similar to the previous example, for almost all starting points the LSMII found the critical ones, which are too close to the optimal solution. For all points, the running times of the interior point algorithm in both local search methods confirm once more that we should not use it in further testing.

The final problem "Colville" is particularly interesting at this stage of numerical testing. It contains a nonconvex quadratic objective function subject to six nonconvex inequality constraints. The data of this test example are specially modeled to be unscaled, i.e. the left-hand side of the constraints is considerably smaller than the right-hand side. In addition, these constraints have a d.c. structure. The global optimum is $z = (78, 33, 29.99, 45, 36.77)$.

Example 4. "Colville" [5,8] $(n = 5, m = 6, \mathcal{V}(\mathcal{P}) = -30665.54)$

$f_0(x) = 5.3578547x_3^2 + 0.8356891x_1x_5 + 37.293239x_1 - 40792.141 \downarrow \min_x,$

$f_1(x) = 0.0056858x_2x_5 - 0.0022053x_3x_5 + 0.0006262x_1x_4 \leq 6.665593,$

$f_2(x) = 0.0022053x_3x_5 - 0.0056858x_2x_5 - 0.0006262x_1x_4 \leq 85.334407,$

$f_3(x) = 0.0071317x_2x_5 + 0.0021813x_3^2 + 0.0029955x_1x_2 \leq 29.48751,$

$f_4(x) = -0.0071317x_2x_5 - 0.0021813x_3^2 - 0.0029955x_1x_2 \leq -9.48751,$

$f_5(x) = 0.0047026x_3x_5 + 0.0019085x_3x_4 + 0.0012547x_1x_3 \leq 15.599039,$

$f_6(x) = -0.0047026x_3x_5 - 0.0019085x_3x_4 - 0.0012547x_1x_3 \leq -10.699039,$

$S = \{x \in I\!R^5 : 78 \leq x_1 \leq 102, \ 33 \leq x_2 \leq 45, \ 27 \leq x_3 \leq 45,$
$$27 \leq x_4 \leq 45, \ 27 \leq x_5 \leq 45.\}$$

Table 4. Results of solving the problem "Colville"

#	$f_0(x_0)$	Solver	LSMI			LSMII				
			$f_0(x_*)$	PL	$Time$	$f_0(x_*)$	σ_*	PL	Itr	$Time$
1	−28716.82	sqp	−**30665.54**	1125	79.12	−	−	−	−	−
		as	−**30665.54**	1233	101.82	−**30665.54**	1297	32702	16345	433.89
		ip	−**30665.54**	1235	85.11	−30349.28	1946	6605	3296	118.86
2	−24803.19	sqp	−**30665.54**	1130	78.19	−**30665.54**	1297	10694	5341	179.94
		as	−**30665.54**	1057	91.36	−**30665.54**	1297	34158	17073	810.71
		ip	−**30665.54**	1056	73.71	−30499.75	1297	12156	6072	217.40
3	−29459.92	sqp	−**30665.54**	548	36.11	−**30665.54**	1297	12452	6220	214.05
		as	−**30665.54**	548	50.17	−**30665.54**	1297	34960	17474	669.58
		ip	−**30665.54**	548	36.5	−30549.08	1946	7155	3571	130.31
4	22776.82	sqp	−**30665.54**	1364	76.85	−	−	−	−	−
		as	−**30665.54**	1212	95.69	−**30665.54**	1297	35542	17765	430.41
		ip	−**30665.54**	1396	100.3	−30342.82	1297	4736	2362	91.69
5	−8156.53	sqp	−**30665.54**	263	14.71	−**30665.54**	1946	537977	268982	2639.45
		as	−**30665.54**	267	18.65	−**30665.54**	1297	27320	13654	335.40
		ip	−**30665.54**	263	15.74	−30531.86	1297	13642	6815	230.15
6	5788.81	sqp	−**30665.54**	1511	105.07	−	−	−	−	−
		as	−**30665.54**	573	99.49	−**30665.54**	1297	5770	17879	424.72
		ip	−**30665.54**	1321	104.89	−30152.50	1946	8783	4385	154.89

The results of Table 4 show that the running times increased significantly for both local search methods compared to the previous test examples. Partly, this is due to the increased number of solved linearized problems. It is also interesting that already at this (local!) stage the LSMI found the global solution for all starting points. Let us note that the LSMII, using sqp algorithm for solving auxiliary convex problems $(\mathcal{P}_\sigma L_s)$, failed to find any results starting from 3 initial points.

Analyzing the results of numerical testing of two local search methods on low-dimensional problems, one can see that both methods have shown their efficiency in solving non-convex quadratic problems. In problems "Hs108", "Mistake" and "Colville", from some starting points it was possible to find a global solution (with an accuracy $\varepsilon = 10^{-4}$). The best and the most stable results have been achieved by using the active-set algorithm as a built-in Matlab algorithm for solving auxiliary linearized problems. Therefore, it will be used as the principal one at the next stage of the numerical experiment.

5.2   Generated Examples

However, one can conclude that the existing range of test problems turns out to be insufficient for the further comparison of the two algorithms. The necessity to have a test problem collection for general nonconvex problems of high dimension and the lack of such data in free libraries motivates us to generate a new field of test instances with known local and global solutions. To this end, we developed a technique for generating nonconvex test problems with quadratic data [1]. Let us remind the main idea of this method of generation and its basic stages.

Based on the idea of P. Calamai and L. Vicente [3], the proposed method of generation consists of three stages. First of all, we construct low-dimensional ($n = 2$) kernel problems and find all local and global solutions to these problems. Then, by uniting a finite number of kernel problems with different properties we obtain a separable problem of the required dimension. Finally, the separable problem is transformed to eliminate the separability of the constructed problem. In this manner, one can construct test instances of any dimension with the required properties and known local and global solutions [1,3].

At this stage of numerical testing, we compare two local search methods between them. To this end, using the technique proposed, we generated a series of 8 examples in the following form:

$$\begin{cases} f_0(x) := \langle x, Q^0 x \rangle + \langle b^0, x \rangle + d_0 \downarrow \min_{x}, \ x \in S, \\ f_i(x) := \langle x, Q^i x \rangle + \langle b^i, x \rangle + d_i \leq 0, \ i \in \mathcal{I} = \{1, ..., m\}, \\ S = \{x \in \mathbb{R}^n : f_j(x) := \langle x, Q^j x \rangle + \langle b^j, x \rangle + d_j \leq 0, \ j \in \mathcal{J} = \{m+1, ..., 2m\}\}. \end{cases}$$

where matrices $Q^i \in \mathbb{R}^{n \times n}$, $i \in \mathcal{I} \cup \{0\}$ are indefinite, matrices $Q^j \in \mathbb{R}^{n \times n}$, $j \in \mathcal{J}$ are positive-definite, therefore $S \subset \mathbb{R}^n$ is a convex, closed set.

Table 5 shows the parameters of constructed examples, such as the problem's dimension (n), a number of nonconvex inequalities (m), numbers of stationary points (st), and global solutions (gl) in the problem. The column "density" shows the density of constructed matrices $Q^i \in \mathbb{R}^{n \times n}$, $i \in \mathcal{I} \cup \{0\}$. Note that the complexity of the test problem depends on the number of kernel problems of a particular class (r_i, $i = 1, 2, 3$).

Table 5. Parameters of generated examples

#	n	m	r_1	r_2	r_3	Density	st	gl
Q10K1-2-2	10	5	1	2	2	0.7	3^5	2
Q10K2-1-2	10	5	2	1	2	0.7	3^5	4
Q20K0-5-5	20	10	0	5	5	0.75	3^{10}	1
Q20K1-4-5	20	10	1	4	5	0.75	3^{10}	2
Q30K0-8-7	30	15	0	8	7	0.5	3^{15}	1
Q30K1-7-7	30	15	1	7	7	0.7	3^{15}	2
Q40K2-9-9	40	20	2	9	9	0.5	3^{20}	4
Q50K2-12-11	50	25	2	12	11	0.5	3^{25}	4

The results of the numerical experiment on specially generated examples are given in Table 6. They demonstrate a considerable difference between the effectiveness of the two developed methods of local search. We can conclude that the LSMII tends to find a point close to the global solution by solving a large number of linearized problems. Moreover, it chooses a penalty parameter to achieve a global solution (already at the local stage). At the same time, the LSMI quickly passes through feasible points and stops at a first critical point by solving a small number of linearized problems.

Table 6. Comparison of two local search methods

x_0	$f_0(x_0)$	LSMI			LSMII				
		$f_0(x_*)$	PL	$Time$	$f_0(x_*)$	σ_*	PL	Itr	$Time$
Q10K1-2-2									
1	3.75	−8.6	9	0.29	**−9.6**	1.2	346	170	3.84
2	12.84	−8.8	19	0.52	**−9.6**	1.2	234	111	3.24
3	11.46	**−9.6**	17	0.57	**−9.6**	1.2	354	174	4.25
4	17.3054	−7.8	24	0.79	**−9.6**	1.2	358	176	4.68
Q10K2-1-2									
1	3.575	−8.0375	20	0.77	**−9.0375**	1.7	278	132	5.09
2	1.775	−7.2375	55	1.63	**−9.0375**	1.7	306	146	6.82
3	18.19	−8.6375	71	2.33	**−9.0375**	1.7	286	136	5.72
4	−185.69	−8.0375	63	2.66	**−9.0375**	1.7	272	129	4.71
Q20K0-5-5									
1	3.74	−17.5374	110	9.04	**−19.9375**	2.6	1362	673	48.76
2	7.46	−18.3375	183	15.03	−19.5375	3.8	1046	514	43.64
3	−0.65	−16.9375	209	17.46	**−19.9375**	2.6	1396	690	54.96
4	−170.31	−15.4731	117	13.26	**−19.9375**	2.6	1424	704	52.06
Q20K1-4-5									
1	5.64	−17.5749	51	6.82	**−19.375**	3.8	1792	882	57.03
2	3.18	−17.7749	128	15.0	**−19.375**	3.8	2118	1049	77.25
3	5.19	−16.7749	178	16.55	**−19.375**	3.8	1800	891	60.54
4	−766.13	−16.175	301	37.15	−18.975	3.8	1216	597	62.39
Q30K0-8-7									
1	8.63	−25.1	12	8.58	**−30.1**	3.8	714	347	52.82
2	34.28	−27.7	18	7.97	**−30.1**	3.8	784	382	54.13
3	42.13	−26.3	21	8.28	**−30.1**	3.8	810	395	59.04
4	−387.89	−28.1	21	12.61	**−30.1**	3.8	766	373	62.74
Q30K1-7-7									
1	13.05	−26.7375	444	72.42	**−29.5375**	5.8	1698	839	270.98
2	14.95	−26.3375	505	88.79	**−29.5375**	5.8	1748	863	293.50
3	−2693.00	−26.7374	753	155.34	**−29.5375**	5.8	1836	908	317.07
4	−3050.35	−27.1375	528	125.32	**−29.5375**	5.8	1758	869	277.00
Q40K2-9-9									
1	6.15	−32.1375	135	56.37	−33.08	5.8	3160	1569	1553.65
2	5.75	−35.5375	695	194.33	−31.04	5.8	1878	928	773.16
3	−2690.59	−33.1373	1109	373.31	**−39.14**	5.8	4794	2385	1691.55
4	−2438.47	−34.1368	1579	508.84	**−39.14**	5.8	4978	2477	1792.09
Q50K2-12-11									
1	11.6139	−43.8447	17	34.85	**−49.3**	7.6	4345	2170	2327.53
2	30.4595	−43.597	14	23.32	**−49.3**	7.6	4307	2151	2115.11
3	−3073.65	−43.0446	68	156.48	**−49.3**	7.6	4507	2251	2923.64
4	17.9451	−44.3731	16	27.62	**−49.3**	7.6	4599	2297	2251.93

5.3 Comparison with Modern Software Packages

Nowadays, most modeling systems and solvers include tools for dealing with non-convex optimization problems. New software packages offer the possibility to solve increasingly difficult problems with a large number of variables and constraints. However, almost all of them use the well-known approaches based on branch-and-bound and whose direct implementation for high dimensional problems may lead to huge search trees. Nevertheless, these methods are widely employed for finding a global solution to applied problems. Among those are ANTIGONE [16], BARON [18], COUENNE [7], LINDOGlobal [15] that implement a spatial branch-and-bound algorithm, Conopt [4] and SCIP [19], which, in addition to branch-and-bound approaches, use various primal heuristics.

Table 7. Results of solving generated examples via software packages

#	CONOPT		COUENNE		LINDOGL		MINOS		SCIP	
	$fval$	$Time$	$fval$	$Time$	$fval$	$Time$	$fval$	$Time$	$fval$	$Time$
Q10K1-2-2	−7.6	0.2	−7.6	>3600	−7.6	0.7	−7.6	0.2	**−9.6**	>3600
Q10K2-1-2	−8.038	0.5	−6.88	>3600	−8.04	0.7	−8.038	18.2	**−9.038**	86.2
Q20K0-5-5	−14.94	0.3	−17.74	>3600	−14.94	0.3	−14.94	0.2	−15.54	>3600
Q20K1-4-5	−15.38	0.3	−15.38	>3600	−15.38	0.3	−18.98	0.2	−16.58	96.3
Q30K0-8-7	−22.1	0.2	**−30.1**	0.4	−22.1	0.3	−22.1	0.1	−28.7	87.3
Q30K1-7-7	−22.54	1.2	−26.69	>3600	−22.54	1.1	−22.54	0.3	−26.69	129.8
Q40K2-9-9	−30.14	1.3	**−39.14**	>3600	−30.14	0.8	−30.14	0.31	−32.79	>3600
Q50K2-12-11	−37.3	5.9	**−49.3**	1.14	−37.29	2.7	−37.3	0.9	−43.76	93.74

To verify the competitiveness of developed local search methods with respect to modern software packages, we test some of them numerically. As an interface for these packages, we use the modeling system GAMS 25.1.1 [10].

The results in Table 7 show that the generated instances have indeed proved to be complex. The table uses the following denotations: $fval$ is an obtained value, all time values $Time$ are expressed in seconds. Only two modern solvers (Couenne, Scip) managed to find (in some cases) the global solution.

Hence, comparing Tables 6 and 7, it can be readily seen that on the specially generated examples of dimensions $n = 10, 20, 30, 40, 50$, the LSMII demonstrated considerable advantages with respect to all applied software packages.

6 Conclusion

In this paper, we considered a nonconvex optimization problem with inequality constraints given by d.c. functions. With the help of exact penalization techniques, the original problem was reduced to a penalized problem, the goal function of which was presented as d.c. function. Two local search methods based on linearization of the objective function and constraints of the original and penalized problems were studied. Finally, preliminary computational testing of two LSMs was carried out.

It is worth noting that the obtained numerical results have demonstrated a considerable difference between the effectiveness of the two methods of local search. On the other hand, the results of comparison of these methods with popular approaches on specially generated instances have shown that in some cases the proposed local search methods considerably outperform some state-of-the-art techniques for solving nonconvex problems. Thus, the results obtained have demonstrated the promising effectiveness of the LSMs and the possibilities of their use in future studies within global search algorithms in large-scale nonconvex problems.

References

1. Barkova, M.V.: On generating nonconvex optimization test problems. In: Khachay, M., Kochetov, Y., Pardalos, P. (eds.) MOTOR 2019. LNCS, vol. 11548, pp. 21–33. Springer, Cham (2019). https://doi.org/10.1007/978-3-030-22629-9_2
2. Byrd, R.H., Nocedal, J., Waltz, R.A.: Steering exact penalty methods for nonlinear programming. Optim. Methods Softw. **23**(2), 197–213 (2008)
3. Calamai, P.H., Vicente, L.N., Judice, J.J.: A new technique for generating quadratic programming test problems. Math. Program. **61**, 215–231 (1993). https://doi.org/10.1007/BF01582148
4. Calasan, M., Nikitović, L., Mujovic, S.: CONOPT solver embedded in GAMS for optimal power flow. J. Renew. Sustain. Energy **11**, 046301 (2019)
5. Colville, A.R.: A comparative study of nonlinear programming codes. In: Kuhn, H.W. (ed.) Princeton Symposium on Mathematical Programming, pp. 487–501. Princeton University Press, Princeton (1970)
6. The COCONUT Benchmark. https://www.mat.univie.ac.at/~neum/glopt/coconut/Benchmark/Benchmark.html. Accessed 1 June 2021
7. Convex Over and Under ENvelopes for Nonlinear Estimation, Couenne, https://www.coin-or.org/Couenne. Accessed 1 June 2021
8. Floudas, C.A., Pardalos, P.M.: Handbook of Test Problems in Local and Global Optimization. Kluwer Academic Publishers, Dordrecht (1999)
9. Hesse, R.: A heuristic search procedure for estimating a global solution of nonconvex programming problems. Oper. Res. **21**(6), 1177–1326, 1267 (1973)
10. The General Algebraic Modeling System, GAMS. https://www.gams.com/. Accessed 1 June 2021
11. Eremin, I.I.: The penalty method in convex programming. Soviet Math. Dokl. **8**, 459–462 (1966)
12. Hiriart-Urruty, J.-B., Lemarechal, C.: Convex Analysis and Minimization Algorithms I. Springer, Berlin (1993). https://doi.org/10.1007/978-3-662-02796-7
13. Horst, R., Tuy, H.: Global Optimization: Deterministic Approaches. Springer, Berlin (1996). https://doi.org/10.1007/978-3-662-03199-5
14. MathWorks, MATLAB. https://www.mathworks.com. Accessed 1 June 2021
15. LINDOGlobal. https://www.lindo.com/. Accessed 1 June 2021
16. Misener, R., Floudas, C.A.: ANTIGONE: algorithms for coNTinuous/integer global optimization of nonlinear equations. J. Glob. Optim. **59**(2), 503–526 (2014). https://doi.org/10.1007/s10898-014-0166-2

17. Nocedal, J., Wright, S.: Numerical Optimization. Springer, New York (2006). https://doi.org/10.1007/978-0-387-40065-5
18. Sahinidis, N.V.: BARON: a general purpose global optimization software package. J. Glob. Optim. 8(2), 201–205 (1996). https://doi.org/10.1007/BF00138693
19. SCIP: Solving Constraint Integer Programs. https://www.scipopt.org/. Accessed 1 June 2021
20. Strekalovsky, A.S.: Elements of Nonconvex Optimization. Nauka, Novosibirsk (2003).(in Russian)
21. Strekalovsky, A.S.: On local search in d.c. optimization problems. Appl. Math. Comput. 255, 73–83 (2015)
22. Strekalovsky, A.S.: Local search for nonsmooth DC optimization with DC equality and inequality constraints. In: Bagirov, A.M., Gaudioso, M., Karmitsa, N., Mäkelä, M.M., Taheri, S. (eds.) Numerical Nonsmooth Optimization, pp. 229–261. Springer, Cham (2020). https://doi.org/10.1007/978-3-030-34910-3_7
23. Strekalovsky, A.S., Minarchenko, I.M.: A local search method for optimization problem with d.c. inequality constraints. Appl. Math. Model. 58, 229–244 (2018)
24. Tuy, H.: D.C. optimization: theory, methods and algorithms. In: Horst, R., Pardalos, P.M. (eds.) Handbook of Global optimization, pp. 149–216. Kluwer Academic Publisher, Dordrecht (1995)

On Search for All Roots of a System of Quadratic Equations

Tatiana V. Gruzdeva[1]([⊠]) [ID] and Oleg V. Khamisov[2]

[1] Matrosov Institute for System Dynamics and Control Theory of SB RAS,
134 Lermontov Street, 664033 Irkutsk, Russia
gruzdeva@icc.ru
[2] Melentiev Energy Systems Institute of SB RAS, 130 Lermontov Street,
664033 Irkutsk, Russia

Abstract. We propose an approach to finding the roots of systems of quadratic equations in a box. This approach is based on a reduction to an auxiliary optimization problem. The auxiliary problem turns out to be, in general, a nonconvex optimization problem, with the objective function and inequality constraints given by d.c. functions. We use the linearization technique with respect to the basic nonconvexity and box partition procedure to try to find all solutions of the system or proof that there are no solutions in the box. The results of the computational simulation are given.

Keywords: System of quadratic equations · Finding roots of system of nonlinear equations · DC programming · Quadratic programming · Linearized problem · Local search

1 Introduction

Finding solutions of nonlinear equations is an important problem which is widely encountered in science and engineering. One of the important problems arising in energy systems and requiring an effective solution is the problem of solving nonlinear equations of steady-state modes of electric power systems (see, for instance, [17]). Nonlinear equations of nodal voltages describe the steady-state of the electrical system.

The mathematical and computational theory for solutions of systems of algebraic equations developed well when solving linear systems. The situation is much more complicated when the equations in the system given by nonlinear (quadratic, nonconvex) functions.

It is worth mentioning that in this case there is a multitude of critical points (generated by Newton's methods) that are rather far from the root set. As a consequence, the classical optimization methods for the nonconvex problem and numerous variants of Newton's schemes turn out to be, in general, inoperative and ineffective when it comes to finding a solution to the equation system because they fail to escape a local pit (and therefore find all roots of the system) in the case of an arbitrary starting point (see [7,14,19]).

© Springer Nature Switzerland AG 2021
N. N. Olenev et al. (Eds.): OPTIMA 2021, LNCS 13078, pp. 110–120, 2021.
https://doi.org/10.1007/978-3-030-91059-4_8

A recent review of iterative methods of solving systems of nonlinear equations is present by [2]. Among the modern effective methods, the LP-Newton method can be mentioned (see [9,13] and references therein). At the same time, the problem of finding all the roots of a system of nonlinear equations remains relevant enough. For the system of multivariate polynomial equations there exists quite comprehensive theoretical approach based on numerical polynomial algebra [6,8,26]. Within this approach it is possible to describe roots varieties of polynomial systems [3]. However, practical implementation of the developed algorithms demands calculation of extremely large number of auxiliary polynomials. Another approach uses interval arithmetics techniques [22,23,31]. Global optimization methods in combination with multistart and other heuristics are suggested in [10,12,20,29,30]. A partition technique with Newton's method similar to the branch and bound methodology in global optimization is described in [24]. A sequential determining roots method by means of cutting planes is presented in [5].

In this paper, we propose to apply an optimization approach for solving systems of quadratic equations, that lead us to an optimization problem which is a nonconvex one. The paper is structured exactly as follows. In Sect. 2, we investigate the problem statement and reduce the original problem to nonconvex optimization problem. In Sect. 3 we describe the local search method, which is often called DCA. Section 4 concerns the question about construction low and upper bounds on objective function over box constraint. In Sect. 5 we developed a scheme for finding all roots of the system of quadratic equations and in Sect. 6 we demonstrate the result of computational simulations on two test systems from literature.

2   Problem Statement

Consider the following system of quadratic equations

$$f_i(x) = x^\top Q_i x + c_i^\top x + r_i = 0, \quad i \in \mathcal{I} = \{1, \ldots, n\}, \tag{1}$$

where Q_i are $n \times n$ matrices, $c_i \in \mathbb{R}^n$ and $r_i \in \mathbb{R}$, $i \in \mathcal{I}$, and try to find all roots of system (1) on the following box:

$$\Pi = \{x \in \mathbb{R}^n \mid a_i \leq x_i \leq b_i, \ i \in \mathcal{I}\}. \tag{2}$$

To find all roots of system of quadratic equations (1) in a box (2) we propose optimization approach based on solving the auxiliary nonconvex quadratically constrained quadratic problem, as follows

$$\left. \begin{array}{c} F(x) \triangleq \sum\limits_{i=1}^{n} f_i(x) \to \min, \quad x \in \Pi, \\[2mm] f_i(x) \geq 0, \quad i \in \mathcal{I}. \end{array} \right\} \tag{$\mathcal{P}$}$$

Denote $D = \{x \in \Pi \mid f_i(x) \geq 0, \ i \in \mathcal{I}\}$. Reduction ($\mathcal{P}$) was earlier suggested in [4]. Previous reductions are described by the two following auxiliary problems

$$\Psi_1(x) = \sum_{i=1}^{n} f_i^2(x) \to \min, \quad x \in \Pi \tag{3}$$

and

$$\Psi_2(x) = \sum_{i=1}^{n} |f_i(x)| \to \min, \quad x \in \Pi. \tag{4}$$

Problems (3) and (4) have very nice property, namely, if $\Psi_1(\hat{x}) = 0$ or $\Psi_2(\hat{x}) = 0$ for some feasible $\hat{x}$, then $\hat{x}$ is a root of system (1). Moreover, if lower bounds $\underline{\Psi}_1 : \Psi_1(x) \geq \underline{\Psi}_1 \ \forall x \in \Pi$ and $\underline{\Psi}_2 : \Psi_2(x) \geq \underline{\Psi}_2 \ \forall x \in \Pi$ such that $\underline{\Phi}_1 > 0$ or $\underline{\Psi}_2 > 0$ are available, then system (1) is inconsistent over Π. These two properties can by considered as a root-certificate and system-inconsistency properties.

The relations between the problem of finding roots of system (1) in a box and optimization Problem ($\mathcal{P}$) are obviously stated by following result.

Proposition 1. *Any solution x_* to Problem ($\mathcal{P}$) such that $F(x_*) = 0$ is the root of the system (1).*

Proof. Suppose that $x_* \in Sol(\mathcal{P})$, $F(x_*) = 0$, but $\exists j \in \{1, \ldots, n\} : f_j(x_*) \neq 0$. Then two following options are possible
1) $f_j(x_*) < 0$. But in this case x_* turns out to be infeasible to Problem ($\mathcal{P}$), i.e. $x_* \notin D$.
2) $f_j(x_*) > 0$. Then $\sum_{i \neq j} f_i(x_*) + f_j(x_*) > 0$ which contradicts the fact that $F(x_*) = 0$.

Hence, problem ($\mathcal{P}$) also has root-certificate and system-inconsistency properties. In what follows we use smooth optimization methods, so problem (4) will not be considered any more since its objective function is not differentiable. As for the problem (3) we see that the objective is a sum of squared quadratic functions. In contrast to that, in our investigation we would like to remain in the field of quadratic optimization. The difference between (3) and ($\mathcal{P}$) consists in the following: in problem (3) feasible (starting) points are readily available. In problem ($\mathcal{P}$) feasible domain may be nonconvex and disconnected and direct search for a feasible point can be a nontrivial task. However, the following trick was suggested in [15]. Take a point $\hat{x} \in \Pi$. Define functions $\tilde{f}_i(\cdot)$, $i = 1, \ldots, n$ in the following way. If $f_i(\hat{x}) \geq 0$ then $\tilde{f}_i(x) = f_i(x)$, if $f_i(\hat{x}) < 0$ then $\tilde{f}_i(x) = -f_i(x)$. Obviously, point $\hat{x}$ is feasible for system $\tilde{f}_i(x) \geq 0$, $i = 1, \ldots, n$ and systems $f_i(x) = 0$, $i = 1, \ldots, n$ and $\tilde{f}_i(x) = 0$, $i = 1, \ldots, n$ have the same roots.

According to Proposition 1 the application of any method for solving the continuous optimization Problem ($\mathcal{P}$) can produce an approximate solution to the system (1) if the objective function is sufficiently small. Therefore, we are able to avoid the direct solution of the system (1) and address the search for points which are feasible to Problem ($\mathcal{P}$), and the value of the objective function at these points is equal to zero. Hence, we propose to combine a solution of Problem $(\mathcal{P})$ with a box partition procedure and [16] relaxation of a quadratic terms.

3   Local Search

In order to find the feasible (to Problem $(\mathcal{P})$) point $z :\ F(z) = 0$, we apply the special local search method for the general DC optimization problem (see [11, 27, 28]). Its main idea consists in the linearization of the function, which defines "the basic non-convexity" of Problem $(\mathcal{P})$, at a current point with the subsequent minimization of the convex approximation of the objective function over the convex set obtained by replacing nonconvex constraints with their linearizations. Observe that the algorithm designed in that way provides critical points by employing only tools and methods of convex programming. However, we have to construct the DC representation of the objective function and constraint functions.

It is known that any symmetric quadratic matrix Q may be represented as the difference of two symmetric positive definite matrices Q^1 and Q^2. Thus, we can get the following representation of the quadratic part of function $f_i(x)$:

$$x^\top Q_i x = x^\top Q_i^1 x - x^\top Q_i^2 x, \ i \in \mathcal{I}.$$

Therefore Problem $(\mathcal{P})$ can be present as its DC form, as follows

$$\left.\begin{aligned} F(x) = G(x) - H(x) \to \min, \quad x \in \Pi, \\ -f_i(x) = g_i(x) - h_i(x) \le 0, \quad i \in \mathcal{I}, \end{aligned}\right\} \tag{5}$$

where

$$G(x) = \sum_{i=1}^{n} \left[x^\top Q_i^1 x + c_i^\top x + r_i \right], \ \ H(x) = \sum_{i=1}^{n} x^\top Q_i^2 x; \tag{6}$$

$$g_i(x) = x^\top Q_i^2 x - c_i^\top x - r_i, \ \ h_i(x) = x^\top Q_i^1 x, \ \ i \in \mathcal{I},$$

are strongly convex functions (since $Q_i^1, Q_i^2, i \in \mathcal{I}$, are positive definite matrices). Note, that the representations (6) are not unique.

Assume further that a feasible starting point $x^0 \in \Pi$ is given and, furthermore, that after several successive iterations we find a current point $x^s \in \Pi$, $s \in \{1, 2, \dots\}$, so the linearized (at x^s) problem can be written as follows:

$$\left.\begin{aligned} \Phi_s(x) = G(x) - [\nabla H(x^s)]^\top x \to \min_x, \ x \in \Pi, \\ g_i(x) - [\nabla h_i(x^s)]^\top (x - x^s) - h_i(x^s) \le 0, \ \ i \in \mathcal{I}. \end{aligned}\right\} \quad (\mathcal{PL}_s)$$

Note that Problem $(\mathcal{PL}_s)$ is convex, since both its objective function and feasible set

$$D_s = \{ x \in \Pi \mid g_i(x) - [\nabla h_i(x^s)]^\top (x - x^s) - h_i(x^s) \le 0, \ i \in \mathcal{I} \}$$

are convex, meanwhile Problem $(\mathcal{P})$ was a nonconvex one. Hence, Problem $(\mathcal{PL}_s)$ can be solved with a suitable convex optimization method (for example, [18]) at any given precision. Let us compute a new iteration x^{s+1} as an approximate solution to the linearized problem $(\mathcal{PL}_s)$, so that x^{s+1} is feasible, i.e. $x^{s+1} \in D_s$, and satisfies the following inequality:

$$\Phi_s(x^{s+1}) = G(x^{s+1}) - [\nabla H(x^s)]^\top x^{s+1} \le V_s + \Delta_s, \qquad (7)$$

where V_s is the optimal value of Problem $(\mathcal{PL}_s)$, i.e. $V_s = \inf_x\{\Phi_s(x) \mid x \in D_s\}$, while a given sequence $\{\Delta_s\}$ is such that

$$\Delta_s \ge 0, \; s = 0, 1, 2, \ldots; \sum_{s=0}^{\infty} \Delta_s < \infty.$$

One can see that $D_s \subset D$. Hence x^{s+1} is feasible not only for the linearized problem $(\mathcal{PL}_s)$, but also for the original Problem $(\mathcal{P})$, since, due to convexity of $h_i(\cdot)$, we have

$$0 \ge g_i(x^{s+1}) - [\nabla h_i(x^s)]^\top (x^{s+1} - x^s) - h_i(x^s)$$

$$\ge g_i(x^{s+1}) - h_i(x^{s+1}) = -f_i(x^{s+1}).$$

As was proposed by [27], we can use one of the following inequalities as the stopping criterion of the local search:

$$F(x^{s+1}) - F(x^s) \le \frac{\tau}{2}, \quad \Delta_s \le \frac{\tau}{2}, \qquad (8)$$

or

$$\Phi_s(x^s) - \Phi_s(x^{s+1}) \le G(x^s) - G(x^{s+1}) - [\nabla H(x^s)]^\top (x^s - x^{s+1}) \le \frac{\tau}{2},$$

$$\Delta_s \le \frac{\tau}{2}, \qquad (9)$$

Thus, if one of the inequalities (8)–(9) holds, the point x^s turns out to be a critical point for Problem $(\mathcal{PL}_s)$ with the accuracy τ under the assumption that $\Delta_s \le \frac{\tau}{2}$. Indeed, (8)–(9) and the inequality (7) imply that

$$G(x^s) - [\nabla H(x^s)]^\top x^s \le \frac{\tau}{2} + G(x^{s+1}) - [\nabla H(x^s)]^\top x^{s+1} \le V_s + \frac{\tau}{2} + \Delta_s.$$

Therefore, if $\Delta_s \le \frac{\tau}{2}$, the point x^s is a τ-solution to Problem $(\mathcal{PL}_s)$.

We intend to use the local search method described in this section, on boxes $\Pi_k \subset \Pi$, $k = 1, 2, \ldots$, (see the following section) where we cannot determine, according low and upper bounds, if the root of the system is in the k-box or not.

4 Low and Upper Bounds

In order to determine the areas with roots of the system of quadratic equations, we use the relaxation [16] of a bi-linear term $x^{\top}y$ with $y = Qx + c$ for finding low and upper bounds on objective function $F(x)$ over box constraint Π.

Therefore we relaxed the sets

$$S_j = \left\{ (x_j, y_j, v_j) \in \left[x_j^L, x_j^U\right] \times \left[y_j^L, y_j^U\right] \times \mathbb{R} \mid v_j = x_j y_j \right\},\ j = 1, \dots, n,$$

with the following four inequalities $(j = 1, \dots, n)$:

$$v_j \geq x_j^L y_j + y_j^L x_j - x_j^L y_j^L \overset{\triangle}{=} l_j^1,$$

$$v_j \geq x_j^U y_j + y_j^U x_j - x_j^U y_j^U \overset{\triangle}{=} l_j^2; \tag{10}$$

$$v_j \leq x_j^L y_j + y_j^U x_j - x_j^L y_j^U \overset{\triangle}{=} l_j^3,$$

$$v_j \leq x_j^U y_j + y_j^L x_j - x_j^U y_j^L \overset{\triangle}{=} l_j^4. \tag{11}$$

For a single bi-linear term $v_j = x_j y_j$, the relaxations (10)–(11) describe the convex hull of set S_j, $j = 1, \dots, n$ (see [1]).

Using (10) and (11) we easy get

$$\alpha_j \overset{\triangle}{=} \max\left\{l_j^1; l_j^2\right\} \leq v_j \leq \min\left\{l_j^3; l_j^4\right\} \overset{\triangle}{=} \beta_j,\ j = 1, \dots, n. \tag{12}$$

Therefore $L_i^k = \sum\limits_{j=1}^{n} \alpha_j^k + r_i$ and $U_i^k = \sum\limits_{j=1}^{n} \beta_j^k + r_i$ turn out to be low and upper bounds, respectively, on box Π_k, $k = 1, 2, \dots$, for quadratic functions $f_i(\cdot)$, $i \in \mathcal{I}$, from the system (1).

Note that boundaries y_j^L and y_j^U on auxiliary variables y_j can easily be calculated using boundaries x_j^L and x_j^U on variables x_j, $j = 1, \dots, n$, equality $y = Qx + c$ and interval arithmetic (see, for example, [25]).

The following statement is quite obvious.

Proposition 2. *Let L_i^k and U_i^k be low and upper bounds, respectively, for functions $f_i(\cdot)$, $i \in \mathcal{I}$, on some box Π_k. If there exists number $p \in \{1, \dots, n\}$ such that $L_p^k > 0$ or $U_p^k < 0$, then there are no roots of the system (1) in box Π_k.*

5 Search Scheme for Quadratic System Roots

In order to find the roots of the system (1), we combined a box partition procedure, function value estimations, and method for finding local solutions to the problem (5) into an algorithm, applying the Search Scheme for System Roots (Algorithm 1).

Let π be the set of boxes $\Pi_l \subset \Pi$, $|\pi| \leq M$, ε be the accuracy by the objective function of the reducing optimization Problem $(\mathcal{P})$.

The cardinality M of the set π, the minimum length of the box edge resulting from partition, and the accuracy ε of equality to zero the objective function $F(\cdot)$ are the algorithm parameters that can be varied.

Algorithm 1. Search Scheme for System Roots

$\pi := \{\Pi\}$.

repeat

Select the box Π_k with the maximum edge from the set of boxes π. Remove Π_k from π.

Split the box Π_k into 2 parts by bisection the maximum edge at midpoint: $\Pi_k = \Pi_k^1 \bigcup \Pi_k^2$.

    for $l = 1, 2; \ i \in \mathcal{I}$ **do**

        Find low L_{li}^k and upper U_{li}^k bounds for function $f_i(\cdot)$ on box Π_k^l.

    end for

    if there are no j such that $L_{lj}^k > 0$ or $U_{lj}^k < 0$ **then**

        add Π_k^l in π.

    end if

until $|\pi| \leq M$

for $l = 1, 2, ..., M$ **do**

Applying local search method from Sect. 3, find solution z_l to Problem $(\mathcal{P})$ with additional constraint $x \in \Pi_l, \ \Pi_l \in \pi$.

    if $|F(z_l)| \leq \varepsilon$ **then**

        z_l is the $\frac{\varepsilon}{n}$-root of the system (1).

    end if

end for

---

6 Computational Simulations

The algorithm was coded in C++ language and applied to search all roots of some systems of quadratic equations from [21] test collection on box $\Pi = [-10n; 10n]^n$. These test systems are with sparse matrices, which are consistent with the properties of problems arising in energy systems.

    System 1 (Test 222, [21])

$$f_i = (3 - 2x_i)x_i - x_{i-1} - 2x_{i+1} + 1 = 0, \ i = 1, \ldots, n$$

$$x_0 = x_{n+1} = 0.$$

System 2 (Test 217, [21])

$$f_i = 3x_i(x_{i+1} - 2x_i + x_{i-1}) + \frac{(x_{i+1} - x_{i-1})^2}{4} = 0,$$

$$x_0 = 0, \ x_{n+1} = 20, \ i = 1, \ldots, n.$$

Note that System 2 is more complicated because the functions $f_i(\cdot)$, $i \in \mathcal{I}$, are nonconvex.

For $n = 2$ all roots of Systems 1:

$$z_1 = (1.6456; 0.260407), \quad z_2 = (-0.453289; -0.385405);$$

and of System 2:

$$z_1 = (0.104444; -1.43569), \quad z_2 = (-0.81506; 11.2038), \quad z_3 = (8.33828; 14.5584);$$

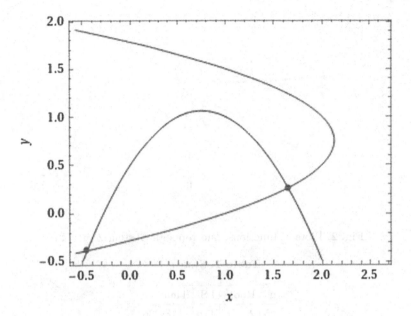

Fig. 1. Plots of functions, and roots for System 1 ($n = 2$)

are displayed on Fig. 1 and Fig. 2, respectively.

The computational experiments were performed on the Intel Core i3-10110U CPU 2.10 GHz. All auxiliary convex (linearized) problems arising during the implementation of the local search method were solved by the software package IBM ILOG CPLEX 12.6.2.

Tables 1 and 2 show the results of computational testing of Algorithm 1 for finding all roots of Systems 1 and 2, respectively, and employ the following denotations:

- n is the number of equations;
- $roots$ stands for the number of roots finding by the algorithm;
- LS is the number of local search method startups;
- $Time$ stands for the CPU time in seconds.

We have chosen the following parameters of the algorithm: $M = 100n$, $\varepsilon = 10^{-5}$. The minimum length of the box edge was equal to 0.1.

Most of the CPU time is spent on local search in Problems ($\mathcal{P}$) with additional constraint $x \in \Pi_l$, $l = 1, \ldots, M$, therefore faster method for finding roots of the system of quadratic equations on the small box will significantly reduce the algorithm's operation time.

The results of the computational experiment showed that the number of found roots of System 2 directly depends on the accuracy of the equalities $f_i(\cdot) = 0$, $i \in \mathcal{I}$, and System 2 (as the system with equations given by non-convex quadratic functions) is rather interesting for further research to improve the algorithm.

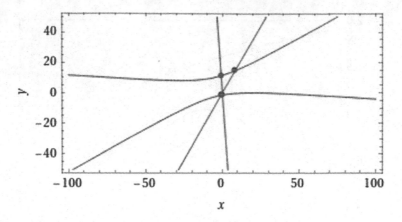

Fig. 2. Plots of functions, and roots for System 2 ($n = 2$)

Table 1. System 1

n	Roots	LS	Time
5	2	115	18.236
8	2	191	151.873
10	2	237	359.098
12	2	291	658.444
15	2	365	879.154
18	2	453	1058.134
20	2	478	1134.475
25	2	602	3195.599

Table 2. System 2

n	Roots	LS	Time
3	4	309	114.520
4	6	455	174.846
5	7	478	199.728
6	8	622	330.085
7	9	699	367.306
8	10	825	735.348
9	6	938	1133.535
10	4	1538	2140.265

Acknowledgments. The authors would like to thank the reviewers for their work and valuable comments.

The research was carried out under State Assignment Projects (no. FWEU-2021-0006, FWEW-2021-0003) of the Fundamental Research Program of Russian Federation 2021–2030.

References

1. Al-Khayyal, F., Falk, J.: Jointly constrained biconvex programming. Math. Oper. Res. 8(2), 273–286 (1983)
2. Amat, S., Busquier, S. (eds.): Advances in Iterative Methods for Nonlinear Equations. SEMA SIMAI, vol. 10. Springer, Cham (2016). https://doi.org/10.1007/978-3-319-39228-8
3. Bates, D., Hauenstein, J., Peterson, C., Sommese, A.: A numerical local dimension test for points on the solution set of a system of polynomial equations. SIAM J. Numer. Anal. 47(5), 3608–3623 (2009)
4. Bulatov, V.: Methods for solving multi-extremal problems (global search). Ann. Oper. Res. 25, 253–277 (1990)
5. Bulatov, V.: Numerical method of funding all real roots of systems of nonlinear equations. Comput. Math. Math. Phys. 40, 331–338 (2000)
6. Cox, D.A., Little, J., O'Shea, D.: Ideals, Varieties, and Algorithms. An Introduction to Computational Algebraic Geometry and Commutative Algebra. UTM, Springer, Cham (2015). https://doi.org/10.1007/978-3-319-16721-3
7. Dennis, J.E., Schnabel, R.: Numerical Methods for Unconstrained Optimization and Nonlinear Equation. SIAM, Philadelphia (1996)
8. Dickenstein, A., Emiris, I. (eds.): Solving Polynomial Equations. Foundations, Algorithms, and Applications. AACIM, vol. 14. Springer-Verlag, Heidelberg (2005). https://doi.org/10.1007/b138957
9. Facchinei, F., Fischer, A., Herrich, M.: A family of newton methods for nonsmooth constrained systems with nonisolated solutions. Math. Meth. Oper. Res. 77, 433–443 (2013)
10. Gong, W., Wang, Y., Cai, Z., Wang, L.: Finding multiple roots of nonlinear equation systems via a repulsion-based adaptive differential evolution. IEEE Trans. Syst. Man Cybern. Syst. 50(4), 1499–1513 (2020)
11. Gruzdeva, T.V., Strekalovsky, A.S.: Local search in problems with nonconvex constraints. Comput. Math. Math. Phys. 47(3), 381–396 (2007)
12. Hirsch, M., Pardalos, P., Resende, M.: Solving systems of nonlinear equations with continuous grasp. Nonlinear Anal. Real World Appl. 10(4), 2000–2006 (2009)
13. Izmailov, A., Kurennoy, A., Solodov, M.: Critical solutions of nonlinear equations: local attraction for newton type-methods. Math. Program. 167, 355–379 (2018)
14. Kelley, C.: Iterative Methods for Linear and Nonlinear Equations. SIAM, Philadelphia (1995)
15. Khamisov, O., Kolosnitsyn, A.: An optimization approach to finding roots of systems of nonlinear equations. IOP Conf. Ser. Mater. Sci. Eng. 537, 042007 (2019)
16. McCormick, G.P.: Computability of global solutions to factorable nonconvex programs: Part i-convex underestimating problems. Math. Program. 10(1), 147–175 (1976)
17. Milano, F.: Power System Modelling and Scripting. POWSYS, Springer-Verlag, Heidelberg (2010). https://doi.org/10.1007/978-3-642-13669-6

18. Nocedal, J., Wright, S.J.: Numerical Optimization. Springer, New York (2006). https://doi.org/10.1007/978-3-540-35447-5
19. Ortega, J., Rheinboldt, W.: Iterative Solution of Nonlinear Equations in Several Variables. Academic Press, New York (1970)
20. Ramadas, G., Fernandes, E., Rocha, A.: Finding multiple roots of systems of nonlinear equations by a hybrid harmony search-based multistart method. Appl. Math. Inf. Sci. **12**(1), 21–32 (2018)
21. Roose, A., Kulla, V., Lomp, M., Meressoo, T.: Test Examples of Systems of Nonlinear Equations. Estonian Software and Computer Service Company, Tallin (1990)
22. Schichl, H., Neumaier, A.: Exclusion regions for systems of equations. SIAM J. Numer. Anal. **42**(1), 383–408 (2004)
23. Semenov, V.Y.: A method to find all the roots of the system of nonlinear algebraic equations based on the krawczyk operator. Cybern. Syst. Anal. **51**, 819–825 (2015)
24. Semenov, V.: The method of determining all real nonmultiple roots of systems of nonlinear equations. Comput. Math. Math. Phys. **47**, 1428–1434 (2007)
25. Shary, S.: Finite-dimensional interval analysis. XYZ, Novosibirsk (2020)
26. Stetter, H.: Numerical Polynomial Algebra. SIAM, New York (2004)
27. Strekalovsky, A.S.: On local search in d.c. optimization problems. Appl. Math. Comput. **255**, 73–83 (2015)
28. Tao, P.D., An, L.T.H.: The DC (difference of convex functions) programming and DCA revisited with DC models of real world nonconvex optimization problems. Ann. Oper. Res. **133**(2), 23–46 (2005)
29. Tsoulos, I., Stavrakoudis, A.: On locating all roots of systems of nonlinear equations inside bounded domain using global optimization methods. Nonlinear Anal. Real World Appl. **11**(4), 2465–2471 (2010)
30. Wang, W.-X., Shang, Y.-L., Wang, G.S., Zhang, Y.: Finding the roots of system of nonlinear equations by a novel filled function method. Abstract Appl. Anal. **2011**. Article ID 209083 (2011)
31. Yamamura, K., Fujioka, T.: Finding all solutions of nonlinear equations using the dual simplex method. J. Comput. Appl. Math. **152**(1–2), 587–595 (2004)

Discrete and Combinatorial Optimization

MIP-Based Heuristics for a Robust Transfer Lines Balancing Problem

Pavel Borisovsky[1]($\boxtimes$) (ID) and Olga Battaïa[2] (ID)

[1] Sobolev Institute of Mathematics SB RAS, Novosibirsk, Russia
pborisovsky@ofim.oscsbras.ru
[2] KEDGE Business School, Bordeaux, France
olga.battaia@kedgebs.com

Abstract. We consider a problem of optimal allocation of processing tools to a conveyor belt with parallel execution of tasks, which is known as a Transfer Lines Balancing Problem. It requires to form a set of blocks of tasks and assign them to machines so that the cycle time constraint is satisfied. The execution times of some tasks are supposed to be uncertain. Since this uncertainty has an impact on the feasibility of the solution because of existing constraint on cycle time, it is required to find a solution with the maximum stability radius, i.e. the maximum deviation from the initial data for which the feasibility of the solution can assured. In our case, it also can be viewed as an application of the threshold robustness approach. The relations between tasks are defined by a precedence graph. In addition to the earlier formulation from the literature we extend the problem with exclusion and inclusion constraints that play an important role in the area of machining lines balancing. We propose MIP-based greedy and local search algorithms, in which a given solution is iteratively built or improved by formulating subproblems of smaller size and solving them with a MIP solver. The numerical experiments showed that our algorithms outperform the straightforward application of a MIP solver on large-scale problems.

Keywords: Transfer line · Balancing · Stability radius · Threshold robustness · Uncertainty · Robust optimization · MIP · Heuristics

1 Introduction

The classic lines balancing problems usually deal with assignment of a set of tasks $V = \{1, \dots, n\}$ to a sequence of machines $W = \{1, \dots, m\}$. For each task $j \in V$, its processing time t_j is known, and the limit on the total workload time of each machine (*cycle time*) is given and denoted by T.

The research of the first author was supported by the Russian Science Foundation grant RSF-ANR 21-41-09017. The research of the second author was supported by the French Agency for Research grant ANR-20-CE40-0021.

© Springer Nature Switzerland AG 2021
N. N. Olenev et al. (Eds.): OPTIMA 2021, LNCS 13078, pp. 123–135, 2021.
https://doi.org/10.1007/978-3-030-91059-4_9

Differences in industrial environment generated an important number of versions of production line balancing problem formulations. In this paper, in particular, we consider the machining environment, in which all the tasks are automated and realised by a sequence of machines connected by an automated conveyor belt. In the literature, this type of line is known as *transfer* line.

In a transfer line, the tasks may be grouped into blocks, the tasks of the same block are performed simultaneously. The processing time of a block is equal to the longest task included in it. Each machine can perform several blocks in a sequential order. As a consequence, the workload time of a machine is a sum of processing times of its blocks. The number of tasks in a block is limited by the given parameter r_{max}.

A partial order on the set of tasks is given in a form of directed acyclic graph $G = (V, A)$, where $(i, j) \in A$ means that the block containing task j must be placed after the block containing task i either on the same or on some subsequent machine. The block exclusion constraints are defined by the family E^b of sets of tasks, such that the tasks from $e \in E^b$ cannot be assigned to the same block. Similarly, the families E^n and I^m define the machine exclusion and machine inclusion constraints, which mean that the tasks from $e \in E^m$ cannot be assigned to one machine, and tasks from $e \in I^m$ must be assigned to the same machine.

A mathematical model for the deterministic version of this problem has been introduced in [6]. The problem has been shown to be NP-hard. A number of heuristic methods have been therefore developed to deal with large-size problem instances [4] including decomposition methods [5,9] and matheuristic approaches [10]. A comparative study between exact and heuristic methods has been conducted in [8] which provided useful guidelines for decision makers for choosing the most efficient solution methods depending on the structure of the problem instance and its size.

Further, uncertainty in task processing times has been introduced in [7] and the notion of stability measure has been developed. It defines the maximum possible deviation in task processing time which does not impact the feasibility of the solution or optimality of the solution. The uncertainty of task processing time is important to take into account at the design stage in order to not face the infeasible situation at the operational level when the transfer line is implemented.

The first version of balancing problem with the objective to maximize the stability radius has been introduced in [13] for a simple line balancing problem. The first robust version maximizing the stability radius for a transfer line balancing problem has been proposed in [11]. In this paper, we extend this formulation and develop a new solution method that outperforms the straightforward application of a MIP solver in numerical experiments on large-scale problems.

The rest of the paper is organised in the following way. Section 2 introduces the considered objective function and presents the corresponding mathematical model. Section 3 develops a new solution method. Section 4 presents the results obtained in numerical experiment and Sect. 5 concludes this study.

2 Problem Formulation and MIP Model

Usually, in the line balancing problems it is required to minimize the cycle time given the limited number of machines or vice versa. Other criteria like some particular cost or cumulative functions can be also applied. In this paper, we consider the robust formulation of the problem as in [11], in which a set of *uncertain tasks* $\widetilde{V} \subseteq V$ is given. For such tasks the processing times may deviate from their nominal values and their probability distribution is not supposed to be known.

In this robust version of the problem, the objective is to find a solution S with the maximum *stability radius*, i.e. the maximum value of deviation supported by the solution which keeps its feasibility:

$$\rho(S,t) = \max\{\varepsilon \mid \forall \xi \in B(\varepsilon) \text{ solution } S \text{ stays feasible if } t \text{ is replaced by } t + \xi\},$$

where $t = (t_1, ..., t_n)$ is the vector of execution times and

$$B(\varepsilon) = \{\xi \in R^n \mid \xi_j > 0 \text{ for } j \in \widetilde{V}, \xi_j = 0 \text{ for } j \notin \widetilde{V}, \text{ and } \|\xi\| \le \varepsilon\}$$

is a space of all possible execution time deviations of the uncertain tasks, bounded by ε. In this study, we consider the case of l_1 norm: $\|\xi\|_1 = \sum_j \xi_j$.

Note that it can be also viewed as an application of the *threshold robustness* approach [3], in which the robust version of the optimisation problem is obtained by bounding the objective function by a certain threshold and considering the robustness as the smallest deviation of the uncertain parameters that causes violation of this threshold limit. In our case, suppose that an original optimisation problem (without uncertainity) asks to minimize the cycle time with the fixed number of machines. Define by $T(t)$ the optimal value of the cycle time for some given vector of processing times t. Then in the uncertain version, we bound the objective function by a fixed threshold T and define the robustness

$$\rho(S,t) = \inf\{\|\xi\| \; : \; \xi \in R^n, \xi_j > 0 \text{ for } j \in \widetilde{V}, T(t + \xi) > T\}.$$

It is easy to see that both definitions of $\rho(S,t)$ are equivalent, because the inequality $T(t + \xi) > T$ means that for $t + \xi$ there are no feasible solutions and vice versa.

A first MIP model for the robust transfer line balancing problem has been proposed in [11].

To evaluate the stability radius, the following approach is used. Suppose a solution of the considered transfer line balancing problem is given. For any uncertain block k (a block containing at least one uncertain task) of machine p define a *save time* as the difference between the block working time τ_k and the processing time of its longest uncertain task, i.e., $\Delta_k^{(p)} := \tau_k - \max_{j \in \widetilde{V}_k} t_j$. For machine p a *minimal save time* $\Delta_{\min}^{(p)}$ is defined as $\Delta_{\min}^{(p)} := \min_{k \in \widetilde{U}(p)} \Delta_k^{(p)}$, where $\widetilde{U}(p)$ is the set of uncertain blocks on machine p. For blocks and machines without uncertain tasks these values do not need to be defined.

It was proven in [11] that the stability radius ρ corresponding to l_1-norm for a given feasible solution can be calculated as follows

$$\rho = \min_{p \in \widetilde{W}} \rho_p, \quad \text{where} \quad \rho_p = T - \sum_{k \in U(p)} \tau_k + \Delta_{\min}^{(p)}, \tag{1}$$

$\widetilde{W}$ is the set of machines containing uncertain tasks and $U(p)$ is the set of all blocks installed on machine p. This expression helps to evaluate and optimize ρ by means of mixed linear programming technique.

To formulate a MIP model it is assumed that the set of all blocks in the line is $U = \{1, 2, ..., m \cdot b_{\max}\}$ where $b_{\max}$ is the maximal number of blocks on one machine. If it is not given in the input data it is estimated as $b_{\max} = \max\{k | \sum_{i=1}^{k} t_{\pi_j} \leq T\}$, where $(\pi_1, ..., \pi_n)$ is a sequence of tasks sorted by decreasing order of their processing times. The subset of blocks corresponding to machine p is $U(p) = \{(p-1)b_{\max} + 1, ..., pb_{\max}\}$ for all $p = 1, ..., m$. The decision variables of the MIP formulation are:

$x_{jk} \in \{0, 1\}$ is equal to 1 if and only if task j is allocated to block k;
$y_k \in \{0, 1\}$ is equal to 1 if block k is not empty;
$\tau_k \geq 0$ is the working time of block k;
$\Delta_{\min}^{(p)} \geq 0$ is the minimal save time of machine p;
$a_p \in \{0, 1\}$ is equal to 1 if and only if machine p contains at least one uncertain task (it can be shown that variables a_p can be regarded as real-valued in interval $[0, 1]$);
$z_k \in \{0, 1\}$ is equal to 1 if an uncertain task is allocated to block k.

For the sake of convenience when modeling the inclusion constraints in (16) we will assume that each inclusion set $e \in I^m$ is formally represented as an ordered sequence of tasks $e = (j_1^e, ..., j_{|e|}^e)$, in which the particular order does not matter and can be arbitrary.

The MIP model is as follows:

$$\text{Maximise } \rho; \tag{2}$$

$$\sum_{k \in U} x_{jk} = 1, \quad \forall j \in V; \tag{3}$$

$$\sum_{j \in V} x_{jk} \leq r_{\max}, \quad \forall k \in U; \tag{4}$$

$$x_{jk} \leq y_k, \quad \forall k \in U, \, j \in V; \tag{5}$$

$$y_k \leq \sum_{j \in V} x_{jk}, \quad \forall k \in U; \tag{6}$$

$$x_{jk} \leq z_k, \quad \forall k \in U, \, \forall j \in \widetilde{V}; \tag{7}$$

$$x_{jk} \leq a_p, \quad \forall p \in W, \ k \in U(p), \ j \in \tilde{V}; \tag{8}$$

$$t_j \cdot x_{j,k} \leq \tau_k, \quad \forall k \in U, j \in V; \tag{9}$$

$$\sum_{k \in U(p)} \tau_k \leq T, \quad \forall p \in W; \tag{10}$$

$$\sum_{k \in U} k x_{ik} \leq \sum_{k \in U} k x_{jk} - 1, \quad \forall (i,j) \in A; \tag{11}$$

$$\Delta_{\min}^{(p)} \leq T \cdot (2 - y_k - z_k) + \tau_k - t_j \cdot x_{jk}, \quad \forall p \in W, \ k \in U(p), \ j \in \tilde{V}; \tag{12}$$

$$\rho \leq T \cdot (2 - a_p) - \sum_{k \in U(p)} \tau_k + \Delta_{\min}^{(p)}, \quad \forall p \in W; \tag{13}$$

$$\sum_{j \in e} x_{jk} \leq |e| - 1, \quad \forall k \in U, \ e \in E^b; \tag{14}$$

$$\sum_{j \in e, k \in U(p)} x_{jk} \leq |e| - 1, \quad \forall p \in W, \ e \in E^m; \tag{15}$$

$$\sum_{k \in U(p)} x_{j_h,k} = \sum_{k \in U(p)} x_{j_{h+1},k}, \quad \forall p \in W, \ e \in I^m, \ j_h, j_{h+1} \in e; \tag{16}$$

$$\rho, \ \Delta_{\min}^{(p)}, \ a_p, \ \tau_k \geq 0, \quad \forall p \in W, \ k \in U; \tag{17}$$

$$x_{jk}, \ z_k, \ y_k \in \{0,1\}, \quad \forall j \in V, \ k \in U. \tag{18}$$

Equation (3) means that each task must be assigned to one block. Constraint (4) limits the number of tasks in each block by $r_{\max}$. Constraints (5)–(8) are required to support the definitions of variables y_k, z_k, and a_p. Constraints (9) and (10) estimate the blocks working times and ensure the cycle time constraints. Constraints (11) reflect the precedence requirements. Constraints (12) and (13) estimate the minimal save time for each machine and evaluate the objective function. Constraints (14)–(16) correspond to block exclusion, machine exclusion, and machine inclusion constraints.

To solve this problem, we develop a new matheuristic approach which is presented in the next section.

3   Solution Approach

To solve the described problem we propose two algorithms of matheuristic type. Both of them are based on the idea of creating subproblems of smaller size and applying a MIP solver for their solution. This is also known as the "relax-and-fix" approach, which is amply presented in [12]. The first algorithm is a constructive greedy heuristic that processes one machine at a time solving a problem by balancing a given set of tasks. The second one is a local search method that tries to withdraw some tasks from the most loaded machine and reallocate them to some other machines. This basic scheme it quite generic and can be applied for different problems, formulated in a MIP form. For example, it was used for a transfer line balancing problem (without uncertainity) in [10], and for an industrial scheduling problem in [2]. The main difficulty in this approach consists in formulating the subproblems of rather small size to be easily solved but at the same time large enough to provide a good overall solution quality.

3.1   Greedy Algorithm

The basic idea of the greedy algorithm is similar to the one of the multi-start heuristic [11], in which a certain lower bound on the objective function ρ is fixed and tasks are added to the current machine until its stability radius falls below this bound. When this happens, a new machine is added to the line and the process continues. In our approach, we also construct machines one by one, but we do this with solving an appropriate MIP model. This model is basically the same as the one formulated above, but with the following modifications. New variables $v_j \in \{0, 1\}$ are introduced, so that $v_j = 1$ if and only if task j is chosen to be assigned to the current machine. Instead of objective function (2), we consider ρ as a fixed parameter and maximize the total duration of chosen tasks:

$$\text{Maximise} \sum_j t_j v_j \tag{19}$$

Equation (3) is rewritten as follows:

$$\sum_{k \in U} x_{jk} = v_j, \quad \forall j \in V \tag{20}$$

The precedence constraint for any $(i, j) \in A$ is not used, if the second task of the arc is not chosen, so Eq. (11) turn to:

$$\sum_{k \in U} k x_{ik} \leq \sum_{k \in U} k x_{jk} - 1 + n(1 - v_j), \quad \forall (i, j) \in A \tag{21}$$

Besides, if the second task of the arc is chosen, then the first task must be chosen as well, this requires to add new equations to the model:

$$v_i \geq v_j, \quad \forall (i, j) \in A. \tag{22}$$

Thus, the model used in the greedy algorithm has objective function (19), constraints (20), (4)–(10), (21)–(22), (12)–(18), the set of machines W consists of one machine, and the value of ρ in (13) is fixed. This model is labeled as M^{grd}.

To keep the problem small, only a reduced subset of tasks is fed into the model. To do this, a random sequence of tasks consistent with the precedence relations is generated. To generate it, firstly, a random permutation π is built and modified as follows: iteratively choose the most-left task from π without predecessors in π and move this task to the end of new sequence S. This step is repeated until all tasks are moved from π to S. As a result, a new random sequence S is obtained, in which the tasks are sorted according to the given partial order.

At each iteration, the greedy algorithm extracts the first h elements from the sequence for passing them to model M^{grd}. The number h is a tunable parameter. If a taken subset contains at least one task from an inclusion set $e \in I^m$, then the whole set e is also added. If for taken tasks i and l, there exist tasks i, j, l such that $(i, j) \in A, (j, l) \in A$, then j must be taken too. Model M^{grd} is then solved for the taken tasks, the obtained solution is saved, the used tasks are removed from the sequence π, and the process continues. If the model cannot be solved, the whole run of the algorithm is considered as failed with the given tasks sequence and parameter ρ, and it should be restarted with different parameters. If the number of stations reaches m, but there are unallocated tasks, this run is also considered as failed. The described algorithm is referred to as procedure Greedy, and its formal outline is as follows.

Algorithm 1. Procedure Greedy(target radius ρ, sequence of tasks seq):

1: Let S be a copy of seq and $p := 1$.
2: **while** S is not empty and $p \leq m$ **do**
3:     Let F be the first h elements from S. For all $e \in I^m : e \cap F \neq \emptyset$ add e to F. For all tasks i, j, l such that $(i, j) \in A, (j, l) \in A$ and $i, l \in F$ add j to F.
4:     Solve the model M^{grd} for the machine p, the set of tasks F, and radius ρ.
5:     **if** a feasible solution is found **then**
6:         Save the obtained assignment of the tasks to the blocks and remove the tasks j with $v_j = 1$ from S.
7:     **else**
8:         Return "solution is not found".
9:     **end if**
10: **end while**
11: **if** S is empty **then**
12:     Return the constructed solution.
13: **else**
14:     Return "solution is not found".
15: **end if**

To find a solution of the initial problem, it is necessary to find maximal ρ for which the procedure Greedy returns a feasible solution. The most evident way to do this is to apply the bisection search starting from some lower and upper bounds on ρ (denoted by LB and UB). Initially, LB may be set to 0, the way to estimate UB will be described below.

Algorithm 2. Bisection Algorithm

1: Build a random sequence seq of tasks consistent with the precedence relations.
2: **while** $UB - LB > \epsilon$ **do**
3:      Let $\rho := (LB + UB)/2$.
4:      Call procedure Greedy(ρ, seq).
5:      **if** a feasible solution is obtained **then**
6:          Set $LB := \rho$.
7:      **else**
8:          Set $UB := \rho$.
9:      **end if**
10: **end while**
11: Return the best found solution.

Upper Bound Estimation. To start the bisection algorithm an initial upper bound on ρ is needed. The value of the line cycle time T may serve for it, but two simple procedures provide better bounds. The first one identifies the worst case when some uncertain task is placed alone on a machine, which gives

$$UB_1 = \max_{j \in \widetilde{V}}(T - t_j). \tag{23}$$

The second procedure takes into account the machine inclusion constraints. Within family I^m, for all tasks of each set $e \in I^m$ the reduced one-machine problem is solved. Since such a problem is quite small-sized, the MIP model can be used. The lowest objective value over all $e \in I^m$ gives a valid upper bound UB_2. In addition, this procedure may prove infeasibility in case of a conflict between inclusion constraints, cycle time, and $r_{\max}$ limit, precisely, when the tasks from an inclusion set have large execution times and/or $r_{\max}$ is too small to run them in parallel.

3.2   Local Search Algorithm

The proposed local search algorithm is based on the generic scheme described in [1]. Starting from an initial solution of the MIP problem, an attempt is made to move to a better solution. Each time a subset of binary variables is chosen to be considered as "free" and other binary variables are "fixed" to their current values. This problem with some fixed variables is solved by a MIP solver and the solution is updated. The way of choosing the subset of free variables is quite a challenging task which has a major influence on the performance of the algorithm. For the considered problem, the difficulties arise because the successful move from one feasible solution to another may involve changes in many machines or even in all of them. This is illustrated in Fig. 1, where task 3 moves to machine 2 and task 5 moves to machine 3. The direct jump of task 3 to machine 3 could be impossible due to precedence or exclusion constraints.

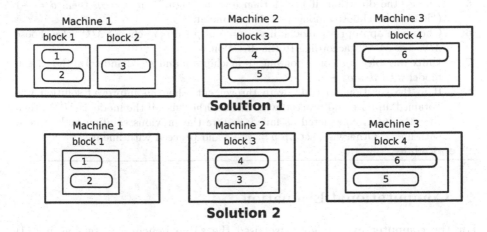

Fig. 1. An example of a move from one feasible solution to another.

In this study, a different approach is proposed, in which infeasible solutions are allowed in some intermediate steps. In the current solution, the worst machine (i.e. the one with the smallest value of ρ) is chosen and a MIP model is solved in order to increase ρ even if some tasks should be thrown out. These tasks are supposed to be assigned to some other machines at the next steps. Clearly, due to precedence constraints a task can be shifted either "to the left", so that the processed machine may contain its successors but not predecessors, or, on the contrary, "to the right". Accordingly, we formulate three models: M^{right} is the same as the one used in the greedy algorithm; model M^{left}, in which constraints (21)–(22) are changed to

$$\sum_{k \in U} kx_{ik} \leq \sum_{k \in U} kx_{jk} - 1 + n(1 - v_i), \quad \forall (i,j) \in A \qquad (24)$$

$$v_i \leq v_j, \quad \forall (i,j) \in A \qquad (25)$$

The third model, M^{fin}, is the original model (2)–(18) and it is used when there are no more non-viewed machines in the chosen direction and the tasks cannot be thrown out. Unlike in the greedy algorithm, the models are applied to two consecutive machines rather than one, which given an additional possibility to reallocate tasks within two machines.

Algorithm 3. Local Search Algorithm

1: Start from a given initial solution.
2: **while** the given time limit is not exceeded **do**
3:    Fix all variables x_{jk} to the values of the current solution.
4:    Compute $\rho_1, ..., \rho_m$ and find machine p with the minimal ρ_p.
5:    Choose the direction: if $p = 1$ then $d = 1$ ("right"), if $p = m$ them $d = -1$
      ("left"), else choose among $\{-1, 1\}$ at random.
6:    Choose an appropriate model: if $p + d = 1$ or $p + d = m$ choose M^{fin} else choose
      M^{left} or M^{right} according to the direction.
7:    Unfix variables x_{jk} corresponding to machines p and $p + d$ and solve the chosen
      model with $\rho = \rho_p + \delta$.
8:    If in the solution no tasks were thrown out, then an improved solution was
      obtained, update and proceed with line 2. Otherwise, if the model is M^{last}, then
      the attempt is considered as failed; restore the previous solution and proceed
      with line 2. Otherwise, set up $p := p + d$ and proceed with line 6.
9: **end while**

4   Computational Evaluation

For the computer experiments, we used the same benchmark sets as in [11]
but extended with the data for inclusion/exclusion constraints. Originally, these
instances were generated in [8] and [10], where all the data were assumed to
be certain[1]. In [11], they were simplified by removing the inclusion/exclusion
data, and different limits for the maximal number of machines were set. In
addition, each instance contains a random permutation of tasks that is used
to define the set of uncertain tasks, namely a parameter $ratio \in [0, 1]$ must
be chosen, then the first $\lceil ratio * n \rceil$ tasks of the indicated permutation are
considered as uncertain[2]. Varying parameters $ratio$ and $r_{\max}$ different problems
can be obtained. In this study, we took the last mentioned dataset and restored
the inclusion/exclusion data from the original benchmarks[3]. In our experiments,
we consider $ratio \in \{0.5, 0.75, 1\}$ and $r_{\max} \in \{2, 3\}$.

The algorithms were coded in Java and run on the server with AMD EPYC
7502 CPU, OS Ubuntu 20.04 and OpenJDK 14. As a MILP solver, Gurobi 9.0.3
was used.

In the first part of the experiments, we analyse the performance of the algorithms and compare them with the straightforward use of Gurobi on the small-
sized instances of series S1. These series consist of 50 instances with 25 tasks
and 5 machines. The results are summarized in Table 1. Column "Gurobi # opt"
shows the number of optimally solved instances when running Gurobi with the
30 min time limit and allowing up to 8 threads. For each instance, the greedy

---

[1] The original TLBP instances are available at
   http://www.math.nsc.ru/AP/benchmarks/english.html.
[2] The TLBP instances with uncertainity are available at
   http://pagesperso.ls2n.fr/~gurevsky-e/data/R-TLBP.zip.
[3] The TLBP instances with uncertainity and inclusion/exclusion data are available at
   https://github.com/pborisovsky/TLBP/tree/main/instances.

Table 1. Results for small-sized instances of series S1.

r_{max}	ratio	Gurobi # opt	Greedy			LS		
			avgGAP, %	# 1%GAP	avgTime	avgGAP, %	# 1%GAP	avgTime
2	0.5	18	8.9	8	5.2	3.8	25	68
2	0.75	16	7	9	5.6	2.3	26	75
2	1	17	7.4	10	6.5	2.7	25	39.8
3	0.5	48	6.6	17	3.1	0.9	42	14.2
3	0.75	46	5.9	20	3.1	0.9	41	4.4
3	1	48	5	18	3.4	0.4	44	4.3

algorithm was run five times and the best solution was returned. In order to reduce the number of iterations in the bisection algorithm, the lower bound was updated after each run for using it in the next computations. The parameter h of the greedy algorithm was chosen as $h = 2n/m$ which is twice as much as the average number of tasks on one machine.

The columns in the table have the following meaning: "avgGAP" is a gap (relative error) between the greedy and Gurobi solutions, i.e. ($\rho^{gurobi} - \rho^{greedy})/\rho^{gurobi} \cdot 100\%$, taken on average over all instances; the next column, "# 1%GAP", is the number of cases, in which this gap is less than 1%; the last column is an average running time in seconds. The local search was given ten iterations starting from the best greedy solution. The three columns in the table have the same meaning as before. In both greedy and LS, the MIP subproblems were solved by Gurobi. The execution time of each individual call of the solver was limited by 20 s. The parameter δ of the LS was set to 0.1.

Table 2. Results for large-sized instances of series S7.

r_{max}	ratio	# feas	Greedy		LS		
			# opt	avgTime	# opt	Impr, %	avgTime
2	0.5	8	1	853	4	3.4	793
2	0.75	8	2	808	3	2.7	1013
2	1	8	2	717	4	4.5	766
3	0.5	17	2	538	8	2	510
3	0.75	17	3	505	5	7.8	625
3	1	17	2	635	4	4	634

As it can be seen, the problems with $r_{max} = 2$ are rather hard for both the MIP solver and heuristics. The greedy solutions have large gaps but are obtained in a short time. The local search provides significant improvements in reasonable time, in many cases the solutions are close to the best solutions obtained by Gurobi. The problems with $r_{max} = 3$ are much easier, almost all instances are solved optimally, and the gaps and the running times are smaller for both heuristics. The solutions obtained by the LS are very close to optima.

The second experiment was carried out for the large-sized problems of series S7. It consists of 20 instances having from 87 to 125 tasks and from 19 to 36 machines. An attempt to solve them straightforwardly applying Gurobi to the model (13)–(18) setting the five hours time limit for each instance was made, but no feasible solutions were found. The results obtained by the developed algorithms are summarized in Table 2. The settings of the tunable parameters were the same as before. The upper bound estimation procedure described in Sect. 3.1 proved infeasibility of some instances with certain values of r_{max}. In all other cases, feasible solutions were obtained by the greedy algorithm. The number of feasible instances is shown in column "# feas". Column "#opt" gives the number of instances for which the value of ρ of the best found solution coincides with the upper bound. Column "avgTime" provides the average solution time in seconds. For the local search, the relative improvement of the solution comparing to the initial one provided by the greedy algorithm is shown in column "Impr". It is computed as $(\rho^{ls} - \rho^{greedy})/\rho^{greedy} \cdot 100\%$ and is averaged over all instances, for which greedy solutions are not optimal (because otherwise the improvement is always equal to zero).

It can be seen again that with $r_{max} = 3$ problems are easier to solve than with $r_{max} = 2$. Unfortunately, the optimal solutions are not known, and the available upper bounds are too rough for an adequate estimation of the algorithms performance, but we may conclude that the heuristics show good ability to find feasible solutions in a reasonable time, and the local search increases the number of optima and provides a notable improvement of the objective function. Note that some instances were solved optimally even by the greedy algorithm, which happens when the number of machines is large enough and the problems are easy to solve. The complete results for all the considered benchmarks can be found at https://github.com/pborisovsky/TLBP/tree/main/results.

5   Conclusions

In this paper, a robust formulation of the transfer line balancing problem is extended with exclusion and inclusion relations. Two matheuristic algorithms are developed to solve large-size problem instances which are intractable by conventional solvers. Experimentally, for the small-sized benchmark problems the algorithms showed good approximation with respect to the straightforward application of the MIP solver. For the larger problems, which were not solved by the MIP solver within quite a large time limit, the proposed algorithms were able to find feasible solutions in rather short time. Since the basic schemes of the algorithms are quite generic, it would be worthwhile to implement them for other line balancing problems and investigate their performance. Another promising direction for the further research could be using this approach within the metaheuristic frameworks such as tabu search or genetic algorithm.

Acknowledgement. An AMD EPYC[TM] based server of Sobolev Institute of Mathematics, Omsk Branch is used for computing.

References

1. Blum, C., Puchinger, J., Raidl, G., Roli, A.: Hybrid metaheuristics in combinatorial optimization: a survey. Appl. Soft Comput. **11**(6), 4135–4151 (2011)
2. Borisovsky, P., Eremeev, A., Kallrath, J.: Multi-product continuous plant scheduling: combination of decomposition, genetic algorithm, and constructive heuristic. Int. J. Prod. Res. **58**(9), 2677–2695 (2019)
3. Carrizosa, E., Nickel, S.: Robust facility location. Math. Methods Oper. Res. **58**, 331–349 (2003)
4. Dolgui, A., Finel, B., Guschinsky, N., Levin, G., Vernadat, F.: A heuristic approach for transfer lines balancing. J. Intell. Manuf. **16**(2), 159–171 (2005)
5. Dolgui, A., Finel, B., Guschinskaya, O., Guschinsky, N., Levin, G., Vernadat, F.: Balancing large-scale machining lines with multi-spindle heads using decomposition. Int. J. Prod. Res. **44**(18–19), 4105–4120 (2006)
6. Dolgui, A., Finel, B., Guschinsky, N., Levin, G., Vernadat, F.: MIP approach to balancing transfer lines with blocks of parallel operations. IIE Trans. **38**(10), 869–882 (2006)
7. Gurevsky, E., Battaïa, O., Dolgui, A.: Stability measure for a generalized assembly line balancing problem. Discrete Appl. Math. **161**(3), 377–394 (2013)
8. Guschinskaya, O., Dolgui, A.: Comparison of exact and heuristic methods for a transfer line balancing problem. Int. J. Prod. Econ. **120**(2), 276–286 (2009)
9. Guschinskaya, O., Dolgui, A., Guschinsky, N., Levin, G.: A heuristic multi-start decomposition approach for optimal design of serial machining lines. Eur. J. Oper. Res. **189**(3), 902–913 (2008)
10. Guschinskaya, O., Gurevsky, E., Dolgui, A., Eremeev, A.: Metaheuristic approaches for the design of machining lines. The Int. J. Adv. Manuf. Technol. **55**(1–4), 11–22 (2011)
11. Pirogov, A., Gurevsky, E., Rossi, A., Dolgui, A.: Robust balancing of transfer lines with blocks of uncertain parallel tasks under fixed cycle time and space restrictions. Eur. J. Oper. Res. **290**(3), 946–955 (2021)
12. Pochet, Y., Wolsey, L.A.: Production Planning by Mixed Integer Programming. ORFE, Springer, New York (2006). https://doi.org/10.1007/0-387-33477-7
13. Rossi, A., Gurevsky, E., Battaïa, O., Dolgui, A.: Maximizing the robustness for simple assembly lines with fixed cycle time and limited number of workstations. Discrete Appl. Math. **208**, 123–136 (2016)

Problem-Specific Branch-and-Bound Algorithms for the Precedence Constrained Generalized Traveling Salesman Problem

Michael Khachay[1,3]([✉]) [iD], Stanislav Ukolov[2] [iD], and Alexander Petunin[1,2] [iD]

[1] Krasovsky Institute of Mathematics and Mechanics, Ekaterinburg, Russia
{mkhachay,a.a.petunin}@imm.uran.ru
[2] Ural Federal University, Ekaterinburg, Russia
s.s.ukolov@urfu.ru
[3] Omsk Technical University, Omsk, Russia

Abstract. The Generalized Traveling Salesman Problem (GTSP) is a well-known combinatorial optimization problem having numerous valuable practical applications in operations research. In the Precedence Constrained GTSP (PCGTSP), any feasible tour is restricted to visit all the clusters according to some given partial order. Unlike the common setting of the GTSP, the PCGTSP appears still weakly studied in terms of algorithmic design and implementation. To the best of our knowledge, all the known algorithmic results for this problem can be exhausted by Salmans's general branching framework, a few MILP models, and the PCGLNS meta-heuristic proposed by the authors recently. In this paper, we present the first problem-specific branch-and-bound algorithm designed with an extension of Salman's approach and exploiting PCGLNS as a powerful primal heuristic. Using the public PCGTSPLIB testbench, we evaluate the performance of the proposed algorithm against the classic Held-Karp dynamic programming scheme with branch-and-bound node fathoming strategy and Gurobi state-of-the-art solver armed by our recently proposed MILP model and PCGLNS-based warm start.

Keywords: Generalized Traveling Salesman Problem · Precedence constraints · Branch-and-bound algorithm

1   Introduction

The Generalized Traveling Salesman Problem (GTSP) is a well-known combinatorial optimization problem introduced in the seminal paper [27] by S. Srivastava et al. and attracted the attention of many researchers (see the survey in [10]).

In the GTSP, for a given weighted digraph $G = (V, E, c)$ and partition $V_1 \cup \ldots \cup V_m$ of the nodeset V into non-empty mutually disjoint clusters, it is required to find a minimum cost closed tour T that visits each cluster V_i exactly once.

In this paper, we consider the Precedence Constrained Generalized Traveling Salesman Problem (PCGTSP), where the clusters should be visited according

© Springer Nature Switzerland AG 2021
N. N. Olenev et al. (Eds.): OPTIMA 2021, LNCS 13078, pp. 136–148, 2021.
https://doi.org/10.1007/978-3-030-91059-4_10

to some given partial order. This extended version of the GTSP has numerous relevant industrial applications including

- toolpath optimization for Computer Numerical Control (CNC) machines [2]
- *air* time minimization in metal sheet cutting [5,20]
- coordinate measuring machinery [24]
- path optimization in multi-hole drilling [6].

Related Work. The GTSP is an extension of the classic Traveling Salesman Problem (TSP). Therefore, any time when the number of clusters m is a part of the input, the problem is strongly NP-hard even on the Euclidean plane [23]. On the other hand, the well-known Held and Karp dynamic programming scheme [11] adapted to the GTSP has running-time bound $O(n^3 m^2 \cdot 2^m)$, i.e. this problem belongs to the class FPT being parameterized by the number of clusters. Furthermore, the GTSP can be solved to optimality in polynomial time, provided $m = O(\log n)$.

As it follows from the literature, algorithmic design for the GTSP developed in several ways.

The first approach is based on the reduction of the initial problem to some corresponding instance of the Asymmetric TSP, after that this auxiliary instance can be solved by an algorithm designed to the ATSP [19,22]. Despite its mathematical elegance, this approach suffers from a couple of shortcomings:

(i) the resulting ATSP instances have a rather unusual shape making their solution hard even for the state-of-the-art MIP solvers like Gurobi and CPLEX
(ii) close-to-optimal solutions of these instances can produce infeasible solutions of the initial problem [13].

Another approach deals with developing problem-specific exact algorithms and approximation algorithms with theoretical performance guarantees. Among them are branch-and-bound and branch-and-cut algorithms (see, e.g. [8,30]) and Polynomial Time Approximation Schemes (PTAS) for several special settings [7,14].

Finally, the third approach is about designing various heuristics and meta-heuristics. Thus, G.Gutin and D.Karapetyan [9] proposed an efficient memetic algorithm, in [12], the famous Lin-Kernighan-Helsgaun heuristic solver was extended to the GTSP, and in [26] the powerful Adaptive Large Neighborhood Search (ALNS) meta-heuristic was developed, which appear to be a best-performer to date.

Unfortunately, for the PCGTSP, algorithmic results still remain quite rare. To the best of our knowledge, the published results are exhausted by

(i) efficient algorithms for the Balas-type special precedence constraints [1,3,4] and the precedence constraints leading to quasi- and pseudo-pyramidal optimal tours [16,17]
(ii) general scheme of a possible branch-and-bound algorithm for this problem [25]

(iii) recent PCGLNS heuristic proposed by the authors [15] as an extension of the results of [26].

In this paper, we try to bridge this gap.

Contribution of this paper is three-fold:

(i) extending the ideas proposed in [25], we design and implement the first problem-specific branch-and-bound algorithm for the PCGTSP
(ii) relying on the classic branching approach [21], we implement the Held and Karp dynamic programming scheme supplemented with an original bounding strategy
(iii) carried out numerical experiments show that the proposed and implemented algorithms are competitive with the state-of-the-art Gurobi solver equipped with the best known MILP model and MIP start solution, both in terms of the running time and accuracy.

2   Problem Statement

We consider the general setting of the Precedence Constrained Generalized Traveling Problem (PCGTSP). An instance of this problem is given by a triplet (G, C, Π), where

- an edge-weighted digraph $G = (V, E, c)$ defines a groundset network supplemented with transportation costs $c(u, v)$ for any arc $(u, v) \in E$
- a partition $C = \{V_1, \ldots, V_m\}$ splits the nodeset V of the graph G into m non-empty pairwise-disjoint *clusters*
- a directed acyclic graph $\Pi = (C, A)$ defines a partial order (*precedence constraints*) on the set of clusters C.

For any node $v \in V$, by $V(v)$ we denote the (only) cluster $V_p \in C$, such that $v \in V_p$. Further, without loss of generality, we assume Π to be *transitively closed* (i.e. $(V_i, V_j) \in A$ and $(V_j, V_k) \in A$ imply $(V_i, V_k) \in A$) and that $(V_1, V_p) \in A$ for any $p \in \{2, \ldots, m\}$.

A closed m-tour $T = v_1, \ldots, v_m$ is called a *feasible* solution of the PCGTSP, if

- it departs for and arrives at some node $v_1 \in V_1$
- it visits each cluster $V_p \in C$ in one node exactly
- the tour T is *consistent* with the partial order Π, i.e. any cluster V_q is visited by the tour T *only after* all the clusters that precede[1] V_q in the order Π.

To any tour T, we assign its cost

$$cost(T) = c(v_m, v_1) + \sum_{i=1}^{m-1} c(v_i, v_{i+1}).$$

The goal is to find a feasible tour T of the minimum cost $cost(T)$.

---

[1] The only evident exception is made for the last arc (v_m, v_1) closing the tour T.

3   Preliminaries

Both algorithms, the branch-and-bound and dynamic programming designed and implemented in this paper exploit the similar main idea.

3.1   Instance Decomposition

At any node of the search tree, before the branching, we decompose the initial problem instance into pair of smaller auxiliary sub-problems as follows:

(i)   consider a subset $C' \subset C$, such that $V_1 \in C'$, fix some cluster $V_l \in C'$ and nodes $v \in V_1$ and $u \in V_l$, respectively
(ii) let $c_{\min}$ be a lower bound for the minimum cost of v-u-paths traversing all the clusters in C' and fulfilling the precedence constraints[2]
(iii) excluding from C' all the inner clusters and connecting V_1 with V_l directly by a zero-cost arc(s), we consider a smaller auxiliary subproblem $\mathcal{P}$, which inherits other transportation costs, clustering, and precedence constraints from the initial instance
(iv) taking

$$\text{LB} = c_{\min} + \text{OPT}(\mathcal{P}_{rel}) \tag{1}$$

as a lower bound, we fathom the current node of search tree each time when LB > UB. Here, $\text{OPT}(\mathcal{P}_{rel})$ is the optimum of some efficiently solvable relaxation of $\mathcal{P}$ and UB is the cost of the best known feasible solution.

3.2   Lower Bounds

In this subsection, we compare the lower bounds obtained by several relaxations of the auxiliary problem $\mathcal{P}$. To relax $\mathcal{P}$, we use the two-stage approach proposed in [25].

At the first stage, we reduce $\mathcal{P}$ to the appropriate ATSP instance by one of the following ways:

(i)   relax the initial precedence constraints by exclusion all the arcs $(v', v'') \in E$, for which $(V(v''), V(v')) \in A$. Then, reduce the obtained instance to ATSP using the classic Noon and Been transformation [22]
(ii) after the same relaxation of the precedence constraints, reduce the relaxed problem to the ATSP instance defined by the auxiliary *cluster* graph $H_1 = (\tilde{C}', A_1, c_1)$, where

$$\tilde{C}' = C \setminus C' \cup \{V_1, V_l\},$$
$$A_1 = \{(V_1, V_l)\} \cup \{(V_i, V_j) \mid i > 2, \{V_i, V_j\} \subset \tilde{C}', \exists (v' \in V_i, v'' \in V_j) \colon (v', v'') \in E\},$$
$$c_1(V_1, V_l) = 0, \ c_1(V_i, V_j) = \min\{c(v', v'') \colon v' \in V_i, v'' \in V_j, (v', v'') \in E\}$$

(iii) reduce the initial problem to the instance of ATSP defined by the digraph $H_2 = (\tilde{C}', A_2, c_2)$, for which

---

[2] In our dynamic programming, this bound is tight.

$$A_2 = \{(V_1, V_i)\} \cup \{(V_i, V_k) \mid i > 2,$$
$$\exists (j > 1) : \{V_i, V_j, V_k\} \subset \tilde{\mathcal{C}}' \wedge (\{(V_j, V_i), (V_k, V_j), (V_k, V_i)\} \cap A = \varnothing)$$
$$\wedge \exists (v' \in V_i, v'' \in V_j, v''' \in V_k) : (\{(v', v''), (v'', v''')\} \subset E)\}$$
$$\cup \{(V_i, V_k) \mid i > 2, (\{V_i, V_k\} \subset \tilde{\mathcal{C}}') \wedge ((V_k, V_i) \notin A)$$
$$\wedge \exists (v' \in V_i, v_1 \in V_1, v'' \in V_k) : \{(v', v_1), (v_1, v'')\} \subset E\},$$

i.e., for any $V_i \in \tilde{\mathcal{C}}' \setminus \{V_1\}$, the ordered pair $(V_i, V_k) \in A_2$, if there exists $V_j \in \tilde{\mathcal{C}}'$ and nodes $v' \in V_i, v'' \in V_j$ and $v''' \in V_k$, such that the path $\pi = v', v'', v'''$ is consistent with the initial precedence constraints. Then,

$$c_2(V_1, V_l) = 0, \ c_2(V_i, V_k) = \min\{c(v', v'') + c(v'', v''') : \pi = v', v'', v''' \text{ is consistent}\}.$$

At the second stage, relaxing the obtained ATSP instance by reduction either to the Minimum Spanning Arborescence Problem (MSAP) or to the Assignment Problem (AP), we compute the appropriate lower bounds by Eq. (1). In addition, to increase the tightness of our lower bounds, we compute optimum values for some ATSP instances obtained by option (ii), using the solver Gurobi. For convenience, we present the designations for all the used lower bounds in Table 1(a). Its columns represent the ways to transform the auxiliary problem $\mathcal{P}$ into ATSP that are Noon and Bean transformation (option (i) at the list above), building cluster graph H_1 (option (ii)), and H_2 (option (iii)) respectively. Its rows are the methods used to solve ATSP instance.

Table 1. Lower bounds: (a) bound names; (b) for each bound we present 95%-confidence interval for its averaged ratio to L_3

	Noon-Bean	H_1	H_2
AP	E_1	L_1	L_2
MSAP	E_2	E_3	E_4
Gurobi	E_5	L_3	E_6

(a)

E_1	$E_2 = E_3$	E_4
0.48 ± 0.03	0.54 ± 0.01	0.60 ± 0.002
L_1	L_2	L_3
0.91 ± 0.02	0.97 ± 0.02	1.00

(b)

Relying on results of the exploratory experiments, we shorten the list of lower bounds employed in the subsequent evaluation (see Table 1(b)). Indeed, the bounds L_1–L_3 appear to be tighter than others, which is statistically significant with a 95% confidence level. Also, we skip bounds E_5 and E_6, whose computation leads to extremely high time consumption. Thus, in Sect. 6, we restrict ourselves to the bounds, defined by the following equation

$$\mathrm{LB}_i = c_{\min} + L_i, \ i \in \{1, 2, 3\}.$$

4   Branch-and-Bound Algorithm

To solve the PCGTSP instance $(G, \mathcal{C}, \Pi)$, we traverse the search tree in Breadth First Search order (see Algorithm 1). Each node of this tree is associated with a prefix $\sigma = (V_{i_1}, V_{i_2}, \ldots V_{i_r})$, where $V_{i_j} \in \mathcal{C}$, $V_{i_1} = V_1$, and $r \in \{1, \ldots m\}$. Clusters V_{i_j} are visited exactly in order specified by σ, while other clusters can be visited in any order (considering partial order Π), thus contributing to the auxiliary problem $\mathcal{P}$, see Subsect. 3.1.

For each node of the search tree we apply the Bounding procedure (Algorithm 2) to perform the following actions:

Algorithm 1. BnB :: Main

Input: the graph G, clusters $\mathcal{C}$, the DAG Π
Output: the tour and cost of optimal solution

1: initialize $Q = $ empty queue
2: start from $Root = V_1$
3: Q.push($Root$)
4: **while** not Q.empty() **do**
5:     get prefix to process: $\sigma = Q$.pop()
6:     $process = Bounding(\sigma)$
7:     **if** not $process$ **then**
8:         prefix is fathomed; **continue**
9:     **end if**
10:     $UpdateLowerBound(\sigma)$
11:     **for each** $child \in Branching(\sigma)$ **do**
12:         queue child prefix Q.push($child$)
13:     **end for**
14: **end while**

Algorithm 2. BnB :: Bounding procedure

Input: the prefix σ
Output: the flag if the prefix survives or is fathomed

1: **global** $D_{ij}^{\mathcal{T}}$
2: **global** $Opt^{\mathcal{T}}$
3: calculate tuple $\mathcal{T} = (V_{i_1}, \{V_{i_1}, V_{i_2}, \ldots V_{i_r}\}, V_{i_r})$
4: $D_{ij} = MinCosts(\sigma)$
5: **if** $D_{ij}^{(\sigma)} \geq D_{ij}^{\mathcal{T}}[\mathcal{T}], \forall i, j$ **then**
6:     **return false**
7: **end if**
8: update best weights $D_{ij}^{\mathcal{T}}[\mathcal{T}] = \min\left(D_{ij}^{\mathcal{T}}[\mathcal{T}], D_{ij}\right), \forall i, j$
9: $c_{min} = \min_{i,j} D_{ij}$
10: **if** $\mathcal{T} \notin Opt^{\mathcal{T}}$ **then**
11:     calculate bounds $Opt^{\mathcal{T}}[\mathcal{T}] = \max\left(L_1(\sigma), L_2(\sigma)\right)$
12: **end if**
13: $LB = c_{min} + Opt^{\mathcal{T}}[\mathcal{T}]$
14: **if** $LB > UB$ **then**
15:     **return false**
16: **end if**
17: **return true**

- for the prefix σ, we assign the tuple

$$T(\sigma) = (V_{i_1}, \{V_{i_1}, V_{i_2}, \ldots V_{i_r}\}, V_{i_r})$$

- at step 4, we compute the matrix $D(\sigma)$ of minimal pairwise costs by the following formula:

$D(\sigma)_{vu} = \min\{cost(P_{v,u}) : v \in V_{i_1}, u \in V_{i_r}, P_{v,u} \text{ is a partial } v\text{-}u \text{ path along } \sigma\}$.

This can be easily calculated incrementally using matrix $D(\sigma')$ of parent tree node
- if, for some σ_1, $T(\sigma) = T(\sigma_1)$ and

$$D(\sigma)_{vu} \geq D(\sigma_1)_{vu}, \quad (v \in V_{i_1}, u \in V_{i_r}),$$

then, prefix σ is dominated by σ_1 and is fathomed
- at step 11, we calculate bounds L_1 and L_2, see Table 1 and assign the global variable Opt^T by the formula

$$Opt^{T(\sigma)} = \max(L_1, L_2)$$

- for current node σ, its lower bound is calculated by the formula

$$\mathrm{LB}(\sigma) = \min_{vu} D(\sigma)_{vu} + Opt^{T(\sigma)}$$

at step 13
- finally, the node σ is fathomed if LB > UB.

Algorithm 3. BnB :: Branching procedure

Input: the prefix σ
Output: the list of children prefixes to process

1: initialize R = empty queue
2: **for each** $V \in \mathcal{C}$ **do**
3:     $valid$ = **true**
4:     **for each** $W \in \sigma$ **do**
5:         **if** $W = V$ or $(V, W) \in \Pi$ **then**
6:             $valid$ = **false**
7:             **break**
8:         **end if**
9:     **end for**
10:    **if** $valid$ **then**
11:        append new prefix R.push($\sigma + V$)
12:    **end if**
13: **end for**
14: **return** R

Nodes that survived are subjected to the *Branching* procedure (Algorithm 3), where we try to enlarge the current prefix σ taking into account the precedence constraint Π.

Algorithm 4. DP :: inductive construction of the lookup table

Input: the graph G, the DAG Π, the layer $\mathcal{L}_k$ of the lookup table, and the current best upper bound UB

Output: the $(k+1)$-th layer $\mathcal{L}_{k+1}$

1: initialize $\mathcal{L}_{k+1} = \varnothing$
2: **for each** $\mathcal{C}' \in \mathfrak{I}_k$ **do**
3:    **for each** cluster $V_l \in \mathcal{C} \setminus \mathcal{C}'$, s.t. $\mathcal{C}' \cup \{V_l\} \in \mathfrak{I}_{k+1}$ **do**
4:       **for each** $v \in V_1$ and $u \in V_l$ **do**
5:          **if** there exists a state $S = (\mathcal{C}', U, v, w) \in \mathcal{L}_k$, s.t. $(w, u) \in E$ **then**
6:            define new state $S' = (\mathcal{C}' \cup \{V_l\}, V_l, v, u)$
7:            $S'[cost] = \min\{S[cost] + c(w, u) : S = (\mathcal{C}', U, v, w) \in \mathcal{L}_k\}$
8:            $S'[pred] = \arg\min\{S[cost] + c(w, u) : S = (\mathcal{C}', U, v, w) \in \mathcal{L}_k\}$
9:            $S'[LB] = S'[cost] + \max\{L_1, L_2, L_3\}$
10:           **if** $S'[LB] \leq$ UB **then**
11:             append S' to $\mathcal{L}_{k+1}$
12:          **end if**
13:         **end if**
14:       **end for**
15:    **end for**
16: **end for**
17: **return** $\mathcal{L}_{k+1}$

---

5   Dynamic Programming

The branch-and-bound algorithm proposed in Sect. 4 appears to be closely related to the classic Dynamic Programming (DP) scheme of Held and Karp [11] adapted to take into account precedence constraints and augmented with one of the bounding strategies introduced in the seminal paper [21].

Therefore, in this paper, we implement the revised version of this scheme to examine numerically the performance of our BnB algorithm. Like to the classic DP, our algorithm consists of two main stages.

(i) at this stage, the lookup table is constructed incrementally, in the forward direction, layer by layer. The optimum of the instance to be solved is computed after the construction of the last m-th layer

(ii) here the optimal tour is reconstructed on the lookup table, in the backward direction.

Each DP state (entry of the lookup table) corresponds to a partial v-u-path and is indexed by a tuple $(\mathcal{C}', V_l, v, u)$, where

(i) $\mathcal{C}' \subset \mathcal{C}$ is an *ideal* of the partially ordered set of clusters $\mathcal{C}$, i.e.

$$\forall (V \in \mathcal{C}', V' \in \mathcal{C})\ (V', V) \in A) \Rightarrow (V' \in \mathcal{C}');$$

obviously, in our setting, V_1 belongs to an arbitrary ideal $\mathcal{C}' \subset \mathcal{C}$

(ii) $V_l \subset \mathcal{C}'$, for which there is no $V \in \mathcal{C}'$, such that $(V_l, V) \in A$
(iii) $v \in V_1$, $u \in V_l$.

Content of each DP entry S consists of the reference $S[pred]$ to the predecessing state, the local lower bound $S[LB]$, and the cost $S[cost]$ of the corresponding partial v-u-path.

Let $\mathfrak{I}_k$ be a subset of ideals of the same size $k \in \{1, \ldots, m\}$. Evidently, $\mathfrak{I}_1 = \{\{V_1\}\}$, therefore, the 1st layer $\mathcal{L}_1$ of the lookup table can be constructed trivially. Inductive construction of other layers is defined in Algorithm 4.

5.1   Remarks

(i) The optimum of the given instance can be found by the classic Bellman's equation

$$OPT = \min_{v \in V_1} \min\{S[cost] + c(u,v) : S = (\mathcal{C}', V_l, v, u) \in \mathcal{L}_m\}$$

(ii) By construction, the size of the lookup table is $O(n^2 m \cdot |\mathfrak{I}|)$. Therefore, the running time of our algorithm is $O(n^3 m^2 \cdot |\mathfrak{I}|)$. In particular, in the case of a partial order of any fixed *width* w, $|\mathfrak{I}| = O(m^w)$ [28]. Therefore, the PCGTSP can be solved to optimality in a polynomial time, even without state fathoming at Steps 10–12.

(iii) After construction of any current layer $\mathcal{L}_k$, we recalculate the global lower bound value, which leads to a decrease in the overall gap.

(iv) In our implementation, to speed up the algorithm, we compute the bound L_3 at Step 9 only for a small number of states, with the smallest lower bounds.

6   Numerical Evaluation

In this section, we report the results of numerical performance evaluation of the proposed branch-and-bound algorithm in comparison with the DP scheme and the Gurobi solver supplemented with our recent MILP model [15].

6.1   Experimental Setup

All the algorithms are tested against the public PCGTSPLIB testbench library [25]. To perform a warm start on each testing instance, all algorithms are supplied by the same feasible solution obtained by the PCGLNS heuristic solver [18]. For the BnB and DP algorithms, all computations are carried out on the same hardware (16-core Intel Xeon 128G RAM) within the same time limit of 10 h. As a stop criteria, we use 5% gap tolerance, where

$$gap = \frac{UB - LB}{LB}. \tag{2}$$

For the baseline, we reproduce numerical experiments of [15] for Gurobi and PCGLNS MIP-start solutions, using exactly aforementioned experimental setup, that is the same hardware, time limit of 10 h and gap tolerance of 5% according to (2).

The source code of our algorithms along with auxiliary scripts is freely available at [29].

Table 2. Experimental results: lines, where BnB or DP algorithms appears to be the best performers are highlighted

Instance				Gurobi			Branch & Bound			DP		
#	ID	n	m	UB_0	Time (sec)	LB	Gap (%)	Time (sec)	LB	Gap (%)	Time (sec)	LB	Gap (%)
1	br17.12	92	17	43	82.00	43	0.00	**11.2**	**43**	**0.00**	27.3	43	0.00
2	ESC07	39	8	1730	0.24	1730	0.00	1.3	1726	0.23	8.37	1730	0.00
3	ESC12	65	13	1390	3.35	1390	0.00	4.3	1385	0.36	14.99	1390	0.00
4	ESC25	133	26	1418	10.61	1383	0.00	32	1383	0.00	60.69	1383	0.00
5	ESC47	244	48	1399	3773	1064	4.93	36000	980	42.76	36000	981	42.61
6	ESC63	349	64	62	25.35	62	0.00	1.3	62	0.00	**0.52**	**62**	**0.00**
7	ESC78	414	79	14872	1278.45	14630	1.66	1.3	14594	1.63	**0.68**	**14594**	**1.63**
8	ft53.1	281	53	6194	36000	5479	13.04	36000	4839	28.27	36000	4839	28.27
9	ft53.2	274	53	6653	36000	5511	20.7	36000	4934	34.84	36000	4940	34.68
10	ft53.3	281	53	8446	36000	6354	32.92	36000	5465	54.55	36000	5465	54.55
11	ft53.4	275	53	11822	20635	11259	5.00	35865	11274	4.86	**2225**	11290	**4.71**
12	ft70.1	346	70	32848	83.70	31521	4.21	36000	31153	5.44	36000	31177	5.36
13	ft70.2	351	70	33486	36000	31787	5.35	36000	31268	7.09	36000	31273	7.08
14	ft70.3	347	70	35309	36000	32775	7.73	36000	32180	9.72	36000	32180	9.72
15	ft70.4	353	70	44497	36000	41160	8.11	36000	38989	14.13	**36000**	**41640**	**6.86**
16	kro124p.1	514	100	33320	36000	29541	12.79	36000	27869	19.56	36000	27943	19.24
17	kro124p.2	524	100	35321	36000	29983	17.80	36000	28155	25.45	36000	28155	25.45
18	kro124p.3	534	100	41340	36000	30669	34.79	36000	28406	45.53	36000	28406	45.53
19	kro124p.4	526	100	62818	36000	46033	36.46	36000	38137	64.72	36000	38511	63.12
20	p43.1	203	43	22545	4691	21677	4.00	36000	738	2954.88	36000	788	2761.04
21	p43.2	198	43	22841	36000	21357	6.94	36000	749	2949.53	36000	877	2504.45
22	p43.3	211	43	23122	36000	15884	45.57	36000	898	2474.83	36000	906	2452.10
23	p43.4	204	43	66857	36000	45198	47.92	4470	66846	0.00	**333.02**	**66846**	**0.00**
24	prob.100	510	99	1474	36000	805	83.10	36000	632	133.23	36000	632	133.23
25	prob.42	208	41	232	13310	196	4.86	36000	149	55.70	36000	153	51.63
26	rbg048a	255	49	282	24.22	282	0.00	0.9	272	3.68	**0.25**	**272**	**3.68**
27	rbg050c	259	51	378	13.83	378	0.00	**0.2**	**372**	**1.61**	0.25	372	1.61
28	rbg109a	573	110	848	6	848	0.00	2407	812	4.43	682	809	4.82
29	rbg150a	871	151	1415	15	1382	2.38	**0.4**	**1353**	**4.58**	0.53	1353	4.58
30	rbg174a	962	175	1644	27	1605	2.43	**0.4**	**1568**	**4.85**	0.67	1568	4.85
31	rbg253a	1389	254	2376	61	2307	2.99	**0.8**	**2269**	**4.72**	1.42	2269	4.72
32	rbg323a	1825	324	2547	416	2490	2.29	**2.0**	**2448**	**4.04**	3.59	2448	4.04
33	rbg341a	1822	342	2101	18470	2033	4.97	36000	1840	14.18	36000	1840	14.18
34	rbg358a	1967	359	2080	17807	1982	4.95	36000	1933	7.60	36000	1933	7.60
35	rbg378a	1973	379	2307	32205	2199	4.91	36000	2032	13.53	36000	2031	13.59
36	ry48p.1	256	48	13135	36000	11965	9.78	36000	10739	22.31	36000	10764	22.03
37	ry48p.2	250	48	13802	36000	12065	14.39	36000	10912	26.48	36000	11000	25.47
38	ry48p.3	254	48	16540	36000	13085	26.40	36000	11732	40.98	36000	11822	39.91
39	ry48p.4	249	48	25977	36000	22084	17.62	18677	25037	3.75	**14001**	**25043**	**3.73**

6.2   Results

The obtained numerical results are reported in Table 2. It is organized as follows: The first column group describes problem instance with its ID, number of nodes (n) and clusters (m), and the weight of the start solution, given by PCGLNS heuristic (UB_0). Then goes three groups of columns for Gurobi solver and two proposed algorithms. Each group reports time to run (in seconds), the best lower bound (LB), and the obtained gap computed using the final UB value (any time, when the gap equals to 0, an optimum value is reported). Each instance, where one of the proposed algorithms outperforms Gurobi, is highlighted in bold.

As it follows from Table 2, for 13 out of 39 instances (33%) one of our algorithms was the best performer. Among them, (sub)optimal solution of the required gap was obtained faster for 12 instances and, for 7 instances approximation ratio appears to be better.

Notice that, the proposed algorithms managed to find an optimal solution for 6 out of 39 instances (although it was not required in this experiment). Further, for 10 (15) out of the remaining tasks including almost the largest instances *rbg323a* and *rbg358a* of 1825 and 1967 nodes respectively, suboptimal solutions were obtained with gap less than 5% (10%) were obtained.

On the other hand, for some instances (e.g. *p43.1, p43.2* and *p43.3*), our algorithms are defeated by Gurobi, we guess the reason is untight lower bounds. Note, however, instances *p43.4* and *ry48p.4*, where, on the contrary, our algorithms significantly outperform Gurobi.

In general, although Gurobi is more likely to win so far, BnB and DP are usually lag slightly behind, with rare exceptions. To be fair, we should note, that Gurobi was provided with very good MIP-start PCGLNS solution, which is rather unusual in such experiments.

7   Conclusion

In this paper, we designed and implemented the first problem-specific branch-and-bound algorithms for the Precedence Constrained GTSP. The algorithms evolve ideas of the classic Held and Karp DP scheme and Salman's bounding framework.

To evaluate performance of the proposed algorithms, we carried out numerical experiments in comparison with Gurobi solver, which show that our algorithms appear to be quite competitive with a state-of-the-art MIP-solver.

To the future work we postpone design of more tight lower bounds. In addition, we believe that further optimization and parallelization can significantly speed up the implementation of our algorithms.

Acknowledgments. The work was performed as a part of research carried out in the Ural Mathematical Center with the financial support of the Ministry of Science and Higher Education of the Russian Federation (Agreement number 075-02-2021-1383).

All the computations were performed on supercomputer 'Uran' at Krasovsky Institute of Mathematics and Mechanics.

References

1. Balas, E., Simonetti, N.: Linear time dynamic-programming algorithms for new classes of restricted TSPs: a computational study. INFORMS J. Comput. **13**(1), 56–75 (2001). https://doi.org/10.1287/ijoc.13.1.56.9748

2. Castelino, K., D'Souza, R., Wright, P.K.: Toolpath optimization for minimizing air-time during machining. J. Manuf. Syst. **22**(3), 173–180 (2003). https://doi.org/10.1016/S0278-6125(03)90018-5. http://www.sciencedirect.com/science/article/pii/S0278612503900185

3. Chentsov, A.G., Khachai, M.Y., Khachai, D.M.: An exact algorithm with linear complexity for a problem of visiting megalopolises. Proc. Steklov Inst. Math. **295**(1), 38–46 (2016). https://doi.org/10.1134/S0081543816090054

4. Chentsov, A., Khachay, M., Khachay, D.: Linear time algorithm for precedence constrained asymmetric generalized traveling salesman problem. IFAC-PapersOnLine **49**(12), 651–655 (2016). 8th IFAC Conference on Manufacturing Modelling, Management and Control MIM 2016. https://doi.org/10.1016/j.ifacol.2016.07.767. http://www.sciencedirect.com/science/article/pii/S2405896316310485

5. Chentsov, A.G., Chentsov, P.A., Petunin, A.A., Sesekin, A.N.: Model of megalopolises in the tool path optimisation for CNC plate cutting machines. Int. J. Prod. Res. **56**(14), 4819–4830 (2018). https://doi.org/10.1080/00207543.2017.1421784

6. Dewil, R., Küçükoğlu, I., Luteyn, C., Cattrysse, D.: A critical review of multi-hole drilling path optimization. Arch. Comput. Methods Eng. **26**(2), 449–459 (2019). https://doi.org/10.1007/s11831-018-9251-x

7. Feremans, C., Grigoriev, A., Sitters, R.: The geometric generalized minimum spanning tree problem with grid clustering. 4OR **4**(4), 319–329 (2006). https://doi.org/10.1007/s10288-006-0012-6

8. Fischetti, M., González, J.J.S., Toth, P.: A branch-and-cut algorithm for the symmetric generalized traveling salesman problem. Oper. Res. **45**(3), 378–394 (1997). https://doi.org/10.1287/opre.45.3.378

9. Gutin, G., Karapetyan, D.: A memetic algorithm for the generalized traveling salesman problem. Nat. Comput. **9**(1), 47–60 (2010). https://doi.org/10.1007/s11047-009-9111-6

10. Gutin, G., Punnen, A.P.: The Traveling Salesman Problem and Its Variations. Springer, Boston (2007). https://doi.org/10.1007/b101971

11. Held, M., Karp, R.M.: A dynamic programming approach to sequencing problems. J. Soc. Ind. Appl. Math. **10**(1), 196–210 (1962). http://www.jstor.org/stable/2098806

12. Helsgaun, K.: Solving the equality generalized traveling salesman problem using the Lin-Kernighan-Helsgaun algorithm. Math. Program. Comput. **7**, 269–287 (2015). https://doi.org/10.1007/s12532-015-0080-8

13. Karapetyan, D., Gutin, G.: Efficient local search algorithms for known and new neighborhoods for the generalized traveling salesman problem. Eur. J. Oper. Res. **219**(2), 234–251 (2012). https://doi.org/10.1016/j.ejor.2012.01.011. https://www.sciencedirect.com/science/article/pii/S0377221712000288

14. Khachai, M.Y., Neznakhina, E.D.: Approximation schemes for the generalized traveling salesman problem. Proc. Steklov Inst. Math. **299**(1), 97–105 (2017). https://doi.org/10.1134/S0081543817090127

15. Khachay, M., Kudriavtsev, A., Petunin, A.: PCGLNS: a heuristic solver for the precedence constrained generalized traveling salesman problem. In: Olenev, N., Evtushenko, Y., Khachay, M., Malkova, V. (eds.) OPTIMA 2020. LNCS, vol. 12422, pp. 196–208. Springer, Cham (2020). https://doi.org/10.1007/978-3-030-62867-3_15

16. Khachay, M., Neznakhina, K.: Towards tractability of the Euclidean generalized traveling salesman problem in grid clusters defined by a grid of bounded height. In: Eremeev, A., Khachay, M., Kochetov, Y., Pardalos, P. (eds.) OPTA 2018. CCIS, vol. 871, pp. 68–77. Springer, Cham (2018). https://doi.org/10.1007/978-3-319-93800-4_6
17. Khachay, M., Neznakhina, K.: Complexity and approximability of the Euclidean generalized traveling salesman problem in grid clusters. Ann. Math. Artif. Intell. **88**(1), 53–69 (2019). https://doi.org/10.1007/s10472-019-09626-w
18. Kudriavtsev, A., Khachay, M.: PCGLNS: adaptive heuristic solver for the Precedence Constrained GTSP (2020). https://github.com/AndreiKud/PCGLNS/
19. Laporte, G., Semet, F.: Computational evaluation of a transformation procedure for the symmetric generalized traveling salesman problem. INFOR: Inf. Syst. Oper. Res. **37**(2), 114–120 (1999). https://doi.org/10.1080/03155986.1999.11732374
20. Makarovskikh, T., Panyukov, A., Savitskiy, E.: Mathematical models and routing algorithms for economical cutting tool paths. Int. J. Prod. Res. **56**(3), 1171–1188 (2018). https://doi.org/10.1080/00207543.2017.1401746
21. Morin, T.L., Marsten, R.E.: Branch-and-bound strategies for dynamic programming. Oper. Res. **24**(4), 611–627 (1976). http://www.jstor.org/stable/169764
22. Noon, C.E., Bean, J.C.: An efficient transformation of the generalized traveling salesman problem. INFOR: Inf. Syst. Oper. Res. **31**(1), 39–44 (1993). https://doi.org/10.1080/03155986.1993.11732212
23. Papadimitriou, C.: Euclidean TSP is NP-complete. Theor. Comput. Sci. **4**, 237–244 (1977)
24. Salman, R., Carlson, J.S., Ekstedt, F., Spensieri, D., Torstensson, J., Söderberg, R.: An industrially validated CMM inspection process with sequence constraints. Procedia CIRP **44**, 138–143 (2016). 6th CIRP Conference on Assembly Technologies and Systems (CATS). https://doi.org/10.1016/j.procir.2016.02.136. http://www.sciencedirect.com/science/article/pii/S2212827116004182
25. Salman, R., Ekstedt, F., Damaschke, P.: Branch-and-bound for the precedence constrained generalized traveling salesman problem. Oper. Res. Lett. **48**(2), 163–166 (2020). https://doi.org/10.1016/j.orl.2020.01.009
26. Smith, S.L., Imeson, F.: GLNS: an effective large neighborhood search heuristic for the generalized traveling salesman problem. Comput. Oper. Res. **87**, 1–19 (2017). https://doi.org/10.1016/j.cor.2017.05.010
27. Srivastava, S., Kumar, S., Garg, R., Sen, P.: Generalized traveling salesman problem through n sets of nodes. CORS J. **7**(2), 97–101 (1969)
28. Steiner, G.: On the complexity of dynamic programming for sequencing problems with precedence constraints. Ann. Oper. Res. **256**, 103–123 (1990). https://doi.org/10.1007/BF02248587
29. Ukolov, S., Khachay, M.: Branch-and-bound algorithm for the Precedence Constrained GTSP (2021). https://github.com/ukoloff/PCGTSP-BnB
30. Yuan, Y., Cattaruzza, D., Ogier, M., Semet, F.: A branch-and-cut algorithm for the generalized traveling salesman problem with time windows. Eur. J. Oper. Res. **286**(3), 849–866 (2020). https://doi.org/10.1016/j.ejor.2020.04.024. https://www.sciencedirect.com/science/article/pii/S0377221720303581

Optimal Control

Optimal Control of Two Linear Programming Problems

Anatoly Antipin[1] and Elena Khoroshilova[2](✉)

[1] FRC, Computer Science and Control, RAS, Vavilov 40, 119333 Moscow, Russia
[2] Lomonosov MSU, CMC Faculty, Leninskiye Gory, 119991 Moscow, Russia

Abstract. On a fixed time interval, a terminal control problem generating a phase trajectory is considered. Three points are selected on the segment: two end points and one intermediate point, they correspond to the values of the trajectory. The left end of the trajectory is fixed. Finite-dimensional linear programming problems are associated with the intermediate and last moments of time, and the corresponding values of the phase trajectory should at the same time be optimal solutions of these problems. It is required to draw a phase trajectory by choosing a control so that, starting from the left end, the trajectory passes through an intermediate point and reaches the right end of the time interval. To solve the problem, a new approach is proposed based on duality theory and Lagrangian formalism. An iterative computational process of the saddle-point type is investigated. The convergence of the process in all components of the solution is proved. It is emphasized that only evidence-based computing technologies transform mathematical models into a tool for making guaranteed solutions.

Keywords: Optimal control · Lagrange function · Duality · Linear programming · Saddle point · Iterative solution methods · Convergence

1 Introduction

A new approach to solving terminal control problems based on saddle-point sufficient optimality conditions is considered. This is the author's approach based on the Lagrangian formalism and the duality theory [3,4]. We study linear controlled dynamics with a phase trajectory loaded at two points of the time interval $[t_0, t_2]$ with linear programming problems. These problems (one is formulated at some intermediate point t_1 of the time interval, the other is formulated at the terminal point t_2) generate solutions in the corresponding finite-dimensional spaces. It is assumed that the phase trajectory $x(t)$ passes through these solutions of the problems.

The computational process is based on the saddle-point gradient method, which simultaneously moves in each of the intermediate spaces and provides the solution of intermediate problems. The convergence of the computational process to the problem solution in all its components is proved, including strong

© Springer Nature Switzerland AG 2021
N. N. Olenev et al. (Eds.): OPTIMA 2021, LNCS 13078, pp. 151–164, 2021.
https://doi.org/10.1007/978-3-030-91059-4_11

convergence in phase and conjugate trajectories, as well as in finite-dimensional variables of the intermediate and boundary value problems, and weak convergence in controls.

The proven convergence of the computational process guarantees obtaining a solution to the original problem with a given accuracy, which is determined by the accuracy of specifying the input information when setting the problem.

2   Controlling Intermediate and Boundary Value Problems of Linear Programming

On the time interval $[t_0, t_2]$ we consider a linear controlled differential system. This system assigns to each control $u(t) \in U$ a phase trajectory $x(t)$, which is a solution to the differential system. The situation is complicated by the fact that the time interval at the point $t_1 \in [t_0, t_2]$ is divided into two parts: subsegments $[t_0, t_1]$ and $[t_1, t_2]$. Accordingly, the phase trajectory is also divided into two parts, and each part of the trajectory can be considered independently of each other on its subsegment. In this case, for any fixed t the phase trajectory takes on its values in n-dimensional space $\mathbb{R}^n$. Therefore, in finite-dimensional spaces corresponding to points t_1, t_2 of the segment, we can formulate linear (convex) programming problems. We will call the first of the problems "intermediate" with respect to the entire segment $[t_0, t_2]$, and the second one we will call "the boundary value problem". In what follows, for the sake of brevity, both problems will also be called intermediate.

In view of the above, we can formulate the following formal setting: find the optimal solution $(x_1^*, x_2^*, u^*(t), x^*(t))$, where $t \in [t_0, t_2]$, satisfying the system of problems

$$
\begin{cases}
\dfrac{d}{dt}x(t) = D(t)x(t) + B(t)u(t), \quad x(t) \in \mathrm{AC}^n[t_0, t_2], \ u(t) \in U, \\[2mm]
t_0 \le t \le t_2, \ x(t_0) = x^0, \ x(t_1) = x_1^*, \ x(t_2) = x_2^*, \\[2mm]
x_1^* \in \mathrm{Argmin}\{\langle \varphi_1, x(t_1)\rangle \mid G_1 x(t_1) \le g_1, \ x(t_1) \in \mathbb{R}^n\}, \\[2mm]
x_2^* \in \mathrm{Argmin}\{\langle \varphi_2, x(t_2)\rangle \mid G_2 x(t_2) \le g_2, \ x(t_2) \in \mathbb{R}^n\}.
\end{cases}
\tag{1}
$$

The problem is considered in Hilbert space $\mathbb{L}_2^n[t_0, t_2]$ with scalar product $\langle x, y\rangle$. The dynamics is defined on segment $[t_0, t_2]$, $U \subset \mathbb{R}^r$ is a closed convex set; the inclusion $u(t) \in U$ means that for almost all $t \in [t_0, t_2]$ the points $u(t)$ belong to U. Matrices $D(t), B(t)$ in (1) are given continuous functions of size $n \times n$, $n \times r$; G_1, G_2 are fixed matrices of size $m \times n$. Vectors φ_1, φ_2, g_1, g_2 are also fixed; $x_1 = x(t_1), x_2 = x(t_2)$. Controls $u(t)$ are functions of space $\mathbb{L}_2^r[t_0, t_2]$. The initial value of the trajectory $x(t_0) = x^0$ is also considered given.

Phase trajectory $x(t)$ as a solution of the differential equation for some fixed control $u(t) \in U$ according to the classical theorems of analysis [15,17] is an absolutely continuous function. An absolutely continuous function can be viewed as a generalization of the concept of an antiderivative function, when it is required to restore the original function from its derivative using indefinite integration. Therefore, as a solution to the differential system (1) we mean any pair $(x(t), u(t)) \in \mathbb{L}_2^n \times U$, satisfying identically the condition

$$x(t) = x(t_0) + \int_{t_0}^{t} (D(\tau)x(\tau) + B(\tau)u(\tau))d\tau, \quad t_0 \le t \le t_2. \tag{2}$$

Identity (2) defines a generalized solution to dynamics (1). In [17, Book 1, p. 443] it is shown that any control $u(t) \in U$ in a linear differential system corresponds to a unique trajectory $x(t)$, and this pair satisfies identity (2). In applications, control $u(t)$ is often a piecewise continuous function. In this case, the presence of discontinuity points on control $u(t)$ does not in any way affect the values of trajectory $x(t)$. Moreover, the trajectory will remain unchanged even if the values of function $u(t)$ are changed on the set of measure zero [15]. Condition (2) also excludes from consideration functions of the Cantor ladder type, i. e., functions that transform sets of measure zero into sets of positive measure. The class of absolutely continuous functions is a linear variety, dense everywhere in $\mathbb{L}_2^n[t_0, t_2]$. This class will be denoted as $AC^n[t_0, t_2] \subset \mathbb{L}_2^n[t_0, t_2]$. For any pair of functions $(x(t), u(t)) \in AC^n[t_0, t_2] \times U$, both the Newton–Leibniz formula and, respectively, the formula for integration by parts hold.

If control $u(t)$ runs through entire set of controls U, then the phase trajectory corresponding to each such $u(t)$ forms at points t_1, t_2 the reachable sets $x_1 \in X_1$, $x_2 \in X_2$. It is also assumed that the intersections of reachable sets and admissible sets (defined by inequalities $G_i x(t_i) \le g_i, i = 1, 2$) for the intermediate and boundary value problems from (1) are not empty. In turn, problem (1) can be viewed as a generalization of the terminal control problem with linear controlled dynamics, which develops on the entire segment $[t_0, t_2]$ (but without intermediate loaded problems) with a fixed left end and a movable right end [3, 4].

It follows from the above that problem (1) splits into two independent subproblems, each on its own subsegment. Each of these subproblems is analogous to a linear programming problem formulated in a functional Hilbert space. Each subproblem has the fixed left end and the movable right end. At the right end of the time segment, we have the finite-dimensional linear programming problem. This problem has its dual counterpart in the dual space. Accordingly, the linear differential system from (1) also has its image in the dual (conjugate) space, known as the dual differential equation [14].

Under regularity conditions like the Slater condition, the Lagrange function for problem (1) has a saddle point [2]. All of the above applies equally to any segment of partition $[t_0, t_1]$, $[t_1, t_2]$, and to segment $[t_0, t_2]$ as a whole. Thus, system (1) on each subsegment can work independently if there are terminal conditions. In principle, passing from one subsegment to the adjacent one, one can solve the entire problem on segment $[t_0, t_2]$. In general, system (1) can be interpreted as a problem whose dynamics are loaded with intermediate and boundary-value linear programming problems [3, 4].

3   Problem Statement in Vector-Matrix Form

Taking into account the separable (shared) structure of problem (1), the latter one can be rewritten as

$$\frac{d}{dt}x(t) = D(t)x(t) + B(t)u(t), \quad x(t) \in AC^n[t_0, t_2], \ u(t) \in U,$$

$$t_0 \le t \le t_2, \ x(t_0) = x^0, \ x^*(t_1) = x_1^*, \ x^*(t_2) = x_2^*,$$

$$\begin{pmatrix} x_1^* \\ x_2^* \end{pmatrix} \in \text{Argmin} \left\{ (\varphi_1, \varphi_2) \begin{pmatrix} x(t_1) \\ x(t_2) \end{pmatrix} \middle| \begin{pmatrix} G_1 & 0 \\ 0 & G_2 \end{pmatrix} \begin{pmatrix} x(t_1) \\ x(t_2) \end{pmatrix} \le \begin{pmatrix} g_1 \\ g_2 \end{pmatrix} \right\}. \quad (3)$$

Recall once again that pair $(x(t), u(t))$ forms the solution to the differential equation of this system, defined on segment $[t_0, t_2]$. Accordingly, $x_i(t), u_i(t)$, $i = 1, 2$, are the parts of this solution defined on subsegments $[t_0, t_1]$, $[t_1, t_2]$. In other words, the phase trajectory of system (3) will also be used in the form

$$x(t) = \begin{cases} x_1(t), & t \in [t_0, t_1], \\ x_2(t), & t \in [t_1, t_2]. \end{cases}$$

We emphasize that each phase trajectory $x(t)$ generates a mapping that assigns to any partition of segment $[t_0, t_2]$ a vector with components $x = (x(t_1), x(t_2)) = (x_1, x_2)$, the number of which is equal to the number of dividing points of the segment (in this case, there are two of them). Moreover, each component of this vector, in turn, is a vector, the size of which is equal to n. Thus, in this case we have a space of dimension $\mathbb{R}^{2n}$. The diagonal matrix $G = \begin{pmatrix} G_1 & 0 \\ 0 & G_2 \end{pmatrix}$ is defined in this space, and we can see two components of this matrix in the form of submatrices G_i, $i = 1, 2$, of size $m \times n$. These submatrices are used to form inequality-type constraints with the right-hand side, which is specified by vector $g = (g_1, g_2)$. The linear objective function of finite-dimensional problem from (3) is determined by its normal vector $\varphi = (\varphi_1, \varphi_2)$ and variable $(x(t_1), x(t_2))$, and has the form $\langle \varphi, x \rangle$.

Using the introduced notation for matrices and vectors, we can represent problem (3) in a compact vector-matrix form

$$\begin{cases} \dfrac{d}{dt}x(t) = D(t)x(t) + B(t)u(t), \ t_0 \le t \le t_2, \ x(t_0) = x^0, x(t_1) = x_1^*, \ x(t_2) = x_2^*, \\ x^* \in \text{Argmin}\{\langle \varphi, x \rangle \mid Gx \le g, \ x \in \mathbb{R}^{2n}\}, \ u(t) \in U, \end{cases} \quad (4)$$

where $D(t), B(t)$ are continuous matrices, respectively, of size $n \times n$, $n \times r$; $x(t_0) = x^0$ is the initial condition, $x^* = (x_1^*, x_2^*) = (x^*(t_1), x^*(t_2))$. The values of control $u(t)$ for each $t \in [t_0, t_2]$ belong to set U, which is a convex compact set from $\mathbb{R}^r$.

Note that macrosystem (4) is obtained as a result of linear convolution of intermediate problems (1), and, both in form and in essence, coincides with the scalar terminal control problem from [3,4]. The authors consider the proposed problem as a dynamic version of the linear programming problem. Therefore, the approaches to the construction of methods for solving the problems under consideration and the proof of their convergence in general repeat logic of reasoning [3,4].

The geometric picture here is as follows. When control $u(t)$ runs through its set of controls U, then the right ends of phase subtrajectories (each on its own intermediate subspace corresponding to the sections at points t_1 and t_2, respectively) describe their own reachable sets. Linear programming problems are specified in each of these subspaces. The minimum point of these problems must be found at the intersection of their reachable set and the feasible set of the problem, the last of which is some polyhedron. Note that in the intermediate and boundary value problems from (1) and (4), the reachability sets X_1, X_2 as separate constraints in an explicit form are absent, since they are automatically taken into account as the ends of the phase trajectory on time subsegments.

4   Classical Lagrangian

We consider the regular case when all polytopes (inequality-type constraints) for intermediate finite-dimensional problems satisfy the Slater conditions. These conditions (the existence of interior points for sets $G_1 x < g_1$, $G_2 x < g_2$) guarantee the existence of saddle points for small Lagrangians of the corresponding intermediate problems [2].

Recall that in problem (1) it is required to choose control $u(t)$ so that the corresponding phase trajectory satisfies the following conditions: at point t_0 the value of phase trajectory is fixed by the initial condition $x(t_0) = x^0$; at point t_1 the value of phase trajectory coincides with solution of the intermediate problem $x^*(t_1) = x_1^*$, and at point t_2 the value of trajectory coincides with the solution of the boundary value problem $x^*(t_2) = x_2^*$.

Problem (4) is nothing more than problem (1) written in macro format. Accordingly, the Lagrange function for this problem in this format has the form

$$\mathcal{L}(x, x(t), u(t); p, \psi(t))$$

$$= \int_{t_0}^{t_2} \langle \psi(t), D(t)x(t) + B(t)u(t) - \frac{d}{dt}x(t) \rangle dt + \langle \varphi, x \rangle + \langle p, Gx - g \rangle \quad (5)$$

for all $(x(t), u(t)) \in \mathrm{AC}^n[t_0, t_2] \times \mathrm{U}$, $\psi(t) \in \Psi^n[t_0, t_2]$, where $\Psi^n[t_0, t_2]$ is the linear variety of absolutely continuous functions from the space dual to the space of primal variables $\mathbb{L}_2^n[t_0, t_2]$. The Lagrange function is defined for all $p = (p_1, p_2)$, $x = (x_1, x_2)$, $p_i \in \mathbb{R}_+^m$, $x_i \in \mathbb{R}^n$, $i = 1, 2$, where $x_1 = x(t_1)$, $x_2 = x(t_2)$. Here $(x, x(t), u(t))$ are primal variables, and $(p, \psi(t))$ are dual variables.

Note that the small Lagrangian of function (5) splits into two small Lagrangians of problem (1), that is,

$$\langle \varphi, x \rangle + \langle p, Gx - g \rangle = \langle \varphi_1, x_1 \rangle + \langle p_1, G_1 x_1 - g_1 \rangle + \langle \varphi_2, x_2 \rangle + \langle p_2, G_2 x_2 - g_2 \rangle.$$

The latter is due to the fact that the finite-dimensional optimization problem from (4) is the optimization problem for a linear function on a parallelepiped, i.e. a separable problem. In other words, the finite-dimensional problem from (4) is split into the system of problems from (1). By virtue of the analogue of the Kuhn–Tucker theorem [2], formulated in Hilbert space for problem (1), we can assert that if the problem has a solution $(x^*, x^*(t), u^*(t))$, then there is a set of corresponding dual variables $(p^*, \psi^*(t))$ such that together these variables form a saddle point of the Lagrange function. Thus, the problem can be reduced to finding the saddle point of the Lagrangian. The converse is also true: primal components $(x^*, x^*(t), u^*(t))$ of the saddle point of Lagrange function (5) are a solution to original problem (4) and, accordingly, a solution to problem (1) [5].

The system of saddle-point inequalities for problem (4) in macro format will have the form

$$\int_{t_0}^{t_2} \langle \psi(t), D(t)x^*(t) + B(t)u^*(t) - \frac{d}{dt}x^*(t)\rangle dt + \langle \varphi, x^* \rangle + \langle p, Gx^* - g \rangle$$

$$\leq \int_{t_0}^{t_2} \langle \psi^*(t), D(t)x^*(t) + B(t)u^*(t) - \frac{d}{dt}x^*(t)\rangle dt + \langle \varphi, x^* \rangle + \langle p^*, Gx^* - g_1 \rangle$$

$$\leq \int_{t_0}^{t_2} \langle \psi^*(t), D(t)x(t) + B(t)u(t) - \frac{d}{dt}x(t)\rangle dt + \langle \varphi, x \rangle + \langle p^*, Gx - g \rangle \quad (6)$$

for all $(x, x(t), u(t)) \in \mathbb{R}^{2n} \times AC^n[t_0, t_2] \times U$, $(p, \psi(t)) \in \mathbb{R}^{2m}_+ \times \Psi^n[t_0, t_2]$.

So, if problem (4) has primal and dual solutions, then this pair is the saddle point of the Lagrange function. Let us show that the converse is true: saddle point (6) of the Lagrange function is primal and dual solutions to (4).

The left inequality of (6) is the problem of maximizing a linear function with respect to variables $(p, \psi(t))$ on the whole space $\mathbb{R}^{2m}_+ \times \Psi^n[t_0, t_2]$:

$$\int_{t_0}^{t_2} \langle \psi(t) - \psi^*(t), D(t)x^*(t) + B(t)u^*(t) - \frac{d}{dt}x^*(t)\rangle dt$$

$$+ \langle p - p^*, Gx^* - g \rangle \leq 0, \quad (7)$$

where $p \in \mathbb{R}^{2m}_+$, $\psi(t) \in \Psi^n[t_0, t_2]$. Inequality (7) implies that

$$D(t)x^*(t) + B(t)u^*(t) - \frac{d}{dt}x^*(t) = 0, \quad x^*(t_0) = x^0, \quad (8)$$

$$\langle p - p^*, Gx^* - g \rangle \leq 0,$$

for all $p \in \mathbb{R}^{2m}_+$. Setting first $p = 0$ and then $p = 2p^*$, we get

$$D(t)x^*(t) + B(t)u^*(t) - \frac{d}{dt}x^*(t) = 0, \quad x^*(t_0) = x^0, \quad (9)$$

$$\langle p^*, Gx^* - g \rangle = 0, \quad Gx^* - g \leq 0.$$

Moving from vector notation to coordinate notation according to (3), the lower finite-dimensional system in (9) can be rewritten as

$$\langle p_1^*, G_1 x_1^* - g_1 \rangle = 0, \quad G_1 x_1^* - g_1 \leq 0,$$

$$\langle p_2^*, G_2 x_2^* - g_2 \rangle = 0, \quad G_2 x_2^* - g_2 \leq 0, \tag{10}$$

which matches (1) and (3).

The right inequality of (6) is the problem of minimizing the Lagrange function with respect to variables $x, x(t), u(t)$ for fixed values $p = p^*$, $\psi(t) = \psi^*(t)$. Let us show that $(p^*, \psi^*(t); x^*, x^*(t), u^*(t))$ is a solution to (6). Taking into account (10), from the right inequality of (6) we have

$$\langle \varphi, x^* \rangle \leq \langle \varphi, x \rangle + \langle p^*, Gx - g \rangle + \int_{t_0}^{t_2} \langle \psi^*(t), D(t)x(t) + B(t)u(t) - \frac{d}{dt}x(t) \rangle dt \tag{11}$$

for all $x \in \mathbb{R}^{2n}$, $(x(t), u(t)) \in AC^n[t_0, t_1] \times U$.

Consider inequality (11) with additional scalar constraints

$$\langle p^*, Gx - g \rangle \leq 0, \quad \int_{t_0}^{t_2} \langle \psi^*(t), D(t)x(t) + B(t)u(t) - \frac{d}{dt}x(t) \rangle dt = 0.$$

Then we get the optimization problem

$$\langle \varphi, x^* \rangle \leq \langle \varphi, x \rangle$$

under constraints

$$\langle p^*, Gx - g \rangle \leq 0, \quad \int_{t_0}^{t_2} \langle \psi^*(t), D(t)x(t) + B(t)u(t) - \frac{d}{dt}x(t) \rangle dt = 0 \tag{12}$$

for all $x \in \mathbb{R}^{2n}$, $(x(t), u(t)) \in AC^n[t_0, t_2] \times U$. Taking into account the inequality and the equation from (9), we obtain that the solution $(x^*(t), u^*(t))$ belongs to a narrower set than (11). Therefore, the indicated point remains a minimum on the subset of solutions of the resulting system

$$\frac{d}{dt}x(t) = D(t)x(t) + B(t)u(t), \quad t_0 \leq t \leq t_2, \ x(t_0) = x^0, \ x(t_1) = x_1^*, \ x(t_2) = x_2^*,$$

$$\langle \varphi, x^* \rangle \leq \langle \varphi, x \rangle, \quad Gx \leq g$$

for all $x \in \mathbb{R}^{2n}$, $(x(t), u(t)) \in AC^n[t_0, t_2] \times U$. Thus, if Lagrange function (5) has a saddle point, then the vector of primal components of the saddle point is a solution to the boundary value (intermediate) linear programming problem (4). The last problem, due to the diagonality of matrix G from (3), splits into two independent linear programming problems (1), each in its own coordinate.

5 Dual Lagrangian

The Lagrange function in finite-dimensional and dynamic problems allows one to pass from the original problem formulated in the space of primal variables to the dual problem formulated in the dual space. Let us show how this can be done using system (4) as an example. We write out formulas for the transition to adjoint linear operators

$$\langle p, Gx \rangle = \langle G^{\mathrm{T}} p, x \rangle,$$

$$\langle \psi(t), D(t)x(t) \rangle = \langle D^{\mathrm{T}}(t)\psi(t), x(t) \rangle, \quad \langle \psi(t), B(t)u(t) \rangle = \langle B^{\mathrm{T}}(t)\psi(t), u(t) \rangle \quad (13)$$

and formulas for integration by parts on intervals $[t_0, t_1]$ and $[t_1, t_2]$

$$\langle \psi(t_1), x(t_1) \rangle - \langle \psi(t_0), x(t_0) \rangle = \int_{t_0}^{t_1} \langle \tfrac{d}{dt}\psi(t), x(t) \rangle dt + \int_{t_0}^{t_1} \langle \psi(t), \tfrac{d}{dt}x(t) \rangle dt,$$

$$\langle \psi(t_2), x(t_2) \rangle - \langle \psi(t_1), x(t_1) \rangle = \int_{t_1}^{t_2} \langle \tfrac{d}{dt}\psi(t), x(t) \rangle dt + \int_{t_1}^{t_2} \langle \psi(t), \tfrac{d}{dt}x(t) \rangle dt.$$

Let us turn to the union of segments $[t_0, t_1]$ and $[t_1, t_2]$ into one large segment $[t_0, t_2]$. To do this, add the last two equalities and get

$$\langle \psi(t_2), x(t_2) \rangle - \langle \psi(t_0), x(t_0) \rangle = \int_{t_0}^{t_2} \langle \frac{d}{dt}\psi(t), x(t) \rangle dt + \int_{t_0}^{t_2} \langle \psi(t), \frac{d}{dt}x(t) \rangle dt. \quad (14)$$

Here, terms $\langle \psi(t_1), x(t_1) \rangle$ with different signs cancel each other out, and the sum of the integrals, due to its additivity, is represented as a single integral, but on the union of these segments. Term $\langle \psi(t_0), x(t_0) \rangle$ in the resulting expression can be omitted: $\psi(t_0) = 0$, since $\psi(t_0)$ has the meaning of a gradient (reference plane normal) for the adjoint equation, which acts as a constraint. For simplicity of calculations, we will assume that the gradient at the moment $t = t_0$ is equal to zero.

Note also that the structure of formula (14) does not depend on the number of dividing points of segment $[t_0, t_2]$ into subsegments. There may be more than two of them. Formula (14) will be correct.

Next, we use the formulas for transition from primal variables to dual variables (13) and (14). For this purpose, all components (terms) with respect to primal variables in (5), and then in (6), are replaced by (13) and (14) with their equal components with respect to dual variables. All terms containing matrices are replaced by transposed ones, and the differential operator under the scalar product is transformed using formula (14). Then we carry out the corresponding transformations and obtain first the dual Lagrange function and then the dual system of saddle point inequalities

$$\mathcal{L}^{\mathrm{T}}(p, \psi(t); x, x(t), u(t))$$

$$= \int_{t_0}^{t_2} \langle D^{\mathrm{T}}(t)\psi(t) + \frac{d}{dt}\psi(t), x(t) \rangle dt + \int_{t_0}^{t_2} \langle B^{\mathrm{T}}(t)\psi(t), u(t) \rangle dt$$

$$+ \langle \varphi + G^{\mathrm{T}} p - \psi, x \rangle + \langle -g, p \rangle + \langle \psi(t_0), x^0 \rangle \qquad (15)$$

for all $(x, x(\cdot), u(\cdot)) \in \mathbb{R}^{2n} \times AC^n[t_0, t_2] \times U$, $(p, \psi(\cdot)) \in \mathbb{R}_+^{2m} \times \Psi^n[t_0, t_2]$. Here

$$\langle \varphi + G^{\mathrm{T}} p - \psi, x \rangle + \langle -g, p \rangle + \langle \psi(t_0), x^0 \rangle = \langle \varphi_1 + G_1^{\mathrm{T}} p_1 - \psi_1, x_1 \rangle + \langle -g_1, p_1 \rangle$$

$$+ \langle \varphi_2 + G_2^{\mathrm{T}} p_2 - \psi_2, x_2 \rangle + \langle -g_2, p_2 \rangle + \langle \psi(t_0), x^0 \rangle.$$

For simplicity of calculations, we can always assume that term $\langle \psi(t_0), x^0 \rangle$ is equal to zero.

The saddle point system of inequalities, dual to (6), has the form:

$$\int_{t_0}^{t_2} \langle D^{\mathrm{T}}(t) \psi(t) + \frac{d}{dt} \psi(t), x^*(t) \rangle dt + \int_{t_0}^{t_2} \langle B^{\mathrm{T}}(t) \psi(t), u^*(t) \rangle dt$$

$$+ \langle \varphi + G^{\mathrm{T}} p - \psi, x^* \rangle + \langle -g, p \rangle$$

$$\leq \int_{t_0}^{t_2} \langle D^{\mathrm{T}}(t) \psi^*(t) + \frac{d}{dt} \psi^*(t), x^*(t) \rangle dt + \int_{t_0}^{t_2} \langle B^{\mathrm{T}}(t) \psi^*(t), u^*(t) \rangle dt$$

$$+ \langle \varphi + G^{\mathrm{T}} p^* - \psi^*, x^* \rangle + \langle -g, p^* \rangle$$

$$\leq \int_{t_0}^{t_2} \langle D^{\mathrm{T}}(t) \psi^*(t) + \frac{d}{dt} \psi^*(t), x(t) \rangle dt + \int_{t_0}^{t_2} \langle B^{\mathrm{T}}(t) \psi^*(t), u(t) \rangle dt$$

$$+ \langle \varphi + G^{\mathrm{T}} p^* - \psi^*, x \rangle + \langle -g, p^* \rangle. \qquad (16)$$

From right inequality (16) we obtain

$$\int_{t_0}^{t_2} \langle D^{\mathrm{T}}(t) \psi^*(t) + \frac{d}{dt} \psi^*(t), x^*(t) - x(t) \rangle dt + \int_{t_0}^{t_2} \langle B^{\mathrm{T}}(t) \psi^*(t), u^*(t) - u(t) \rangle dt$$

$$+ \langle \varphi + G^{\mathrm{T}} p^* - \psi^*, x^* - x \rangle \leq 0 \qquad (17)$$

for all $(x, x(t), u(t)) \in \mathbb{R}^{2n} \times AC^n[t_0, t_2] \times U$. Due to the independent change of each of variables $(x, x(t), u(t))$ within its admissible subspaces (sets), the last inequality is decomposed into three independent inequalities

$$\int_{t_0}^{t_2} \langle D^{\mathrm{T}}(t) \psi^*(t) + \frac{d}{dt} \psi^*(t), x^*(t) - x(t) \rangle dt \leq 0, \quad x(t) \in AC^n[t_0, t_2],$$

$$\int_{t_0}^{t_2} \langle B^{\mathrm{T}}(t) \psi^*(t), u^*(t) - u(t) \rangle dt \leq 0, \quad u(t) \in U.$$

$$\langle \varphi + G^{\mathrm{T}} p^* - \psi^*, x^* - x \rangle \leq 0, \quad x \in \mathbb{R}^{2n}.$$

The linear functional reaches a finite extremum on the entire subspace only if its gradient vanishes, and this leads to system of problems

$$D^{\mathrm{T}}(t) \psi^*(t) + \frac{d}{dt} \psi^*(t) = 0, \quad \varphi + G^{\mathrm{T}} p^* - \psi^* = 0, \qquad (18)$$

$$\int_{t_0}^{t_2} \langle B^{\mathrm{T}}(t)\psi^*(t), u^*(t) - u(t)\rangle dt \leq 0, \quad \forall u(t) \in U. \tag{19}$$

From left inequality (16), taking into account (18) and (19), we have

$$\int_{t_0}^{t_2} \langle D^{\mathrm{T}}(t)\psi(t) + \frac{d}{dt}\psi(t), x^*(t)\rangle dt + \int_{t_0}^{t_2} \langle B^{\mathrm{T}}(t)\psi(t), u^*(t)\rangle dt + \langle \varphi + G^{\mathrm{T}}p - \psi, x^*\rangle$$

$$+\langle -g, p\rangle \leq \langle -g, p^*\rangle + \int_{t_0}^{t_2} \langle B^{\mathrm{T}}(t)\psi^*(t), u^*(t)\rangle dt.$$

Considering this inequality subject to the scalar constraints

$$\langle \varphi + G^{\mathrm{T}}p - \psi, x^*\rangle = 0,$$

$$\int_{t_0}^{t_2} \langle D^{\mathrm{T}}(t)\psi(t) + \frac{d}{dt}\psi(t), x^*(t)\rangle dt = 0,$$

we arrive at the problem of maximizing the scalar function

$$\langle -g, p\rangle + \int_{t_0}^{t_2} \langle B^{\mathrm{T}}(t)\psi(t), u^*(t)\rangle dt \leq \langle -g, p^*\rangle + \int_{t_0}^{t_2} \langle B^{\mathrm{T}}(t)\psi^*(t), u^*(t)\rangle dt,$$

where $(p, \psi(t)) \in \mathbb{R}_+^{2m} \times \Psi^n[t_0, t_2]$.

Combining this problem with (18) and (19), we get a problem dual to (1):

$$\begin{cases} (p^*, \psi^*(t)) \in \mathrm{Argmax}\left\{\langle -g, p\rangle + \int_{t_0}^{t_2} \langle B^{\mathrm{T}}(t)\psi(t), u^*(t)\rangle dt \ \middle| \right. \\ \qquad\qquad\qquad D^{\mathrm{T}}(t)\psi(t) + \frac{d}{dt}\psi(t) = 0, \ \ \psi = \varphi + G^{\mathrm{T}}p \Big\}, \\ \left. \int_{t_0}^{t_2} \langle B^{\mathrm{T}}(t)\psi^*(t), u^*(t) - u(t)\rangle dt \leq 0, \ \ u(t) \in U, \right. \end{cases} \tag{20}$$

where $\psi = \varphi + G^{\mathrm{T}}p$ is the transversality condition.

6   Saddle Point Differential System (Sufficient Conditions for Optimality)

Combining in one system the main elements of primal and dual problems, we finally arrive at the saddle point differential system. With respect to original problem (1), this system plays the role of a necessary and sufficient optimality condition

$$\frac{d}{dt}x^*(t) = D(t)x^*(t) + B(t)u^*(t), \quad x^*(t_0) = x^0, \tag{21}$$

$$\langle Gx^* - g, p - p^*\rangle \leq 0, \quad p \geq 0, \tag{22}$$

$$\frac{d}{dt}\psi^*(t) + D^{\mathrm{T}}(t)\psi^*(t) = 0, \quad \psi^* = \varphi + G^{\mathrm{T}}p^*, \tag{23}$$

$$\int_{t_0}^{t_2} \langle B^{\mathrm{T}}(t)\psi^*(t), u^*(t) - u(t)\rangle dt \le 0, \quad u(t) \in \mathrm{U}. \tag{24}$$

It is interesting that if differential system (21) is loaded with intermediate problems [1], then its dual system (23) belongs to the class of switched differential systems [16].

The variational inequalities of system (21)–(24) can be rewritten in equivalent form of operator equations with operators of projection onto the corresponding convex closed sets:

$$\frac{d}{dt}x^*(t) = D(t)x^*(t) + B(t)u^*(t), \quad x^*(t_0) = x^0,$$

$$p^* = \pi_+(p^* + \alpha(Gx^* - g)),$$

$$\frac{d}{dt}\psi^*(t) + D^{\mathrm{T}}(t)\psi^*(t) = 0, \quad \psi^* = \varphi + G^{\mathrm{T}}p^*,$$

$$u^*(t) = \pi_U(u^*(t) - \alpha B^{\mathrm{T}}(t)\psi^*(t)),$$

where $\pi_+(\cdot), \pi_U(\cdot)$ are projection operators, respectively, to positive orthant $\mathbb{R}_+^{2m}$ and to set of controls U $(\alpha > 0)$.

Saddle point optimality conditions open up great opportunities and prospects for the development of the theory of methods for solving terminal control problems with boundary value problems at the ends of time interval [3–13]. Terminal control problems in this case are transferred to the class of saddle point (game) problems.

7   Saddle Point Method of Extragradient Type

Based on the saddle point differential system, we construct an iterative process (parameter $\alpha > 0$ characterizes the size of iteration step):

$$\frac{d}{dt}x^k(t) = D(t)x^k(t) + B(t)u^k(t), \quad x(t_0) = x^0,$$

$$p^{k+1} = \pi_+(p^k + \alpha(Gx^k - g)),$$

$$\frac{d}{dt}\psi^k(t) + D^{\mathrm{T}}(t)\psi^k(t) = 0, \quad \psi^k = \varphi + G^{\mathrm{T}}p^k,$$

$$u^{k+1}(t) = \pi_U(u^k(t) - \alpha B^{\mathrm{T}}(t)\psi^k(t)), \quad k = 0, 1, 2...,$$

where $x^k = x^k(t)$.

This process is a simple iteration method. This method converges to solutions of optimization problems, where gradient generates vector fields with fixed points of "stable focus" type. For saddle point problems, the gradient method generates vector fields of rotation (around fixed points of "center" type), and in this case the gradient process, generally speaking, will not converge to this point. In order to ensure convergence to a point of "center of rotation" type, we must split the iterative step into two halfsteps. One halfstep implements the movement of the computational process along one closed curve. This is followed by a jump of the trajectory to another closed trajectory. Formally, this jump is realized by splitting one iteration into two half-iterations. Thus, one saddle point iteration is split into two half-iterations. The set of such iterations forms a spiral trajectory that twists around a fixed point of the vector field (saddle point). Therefore, in the considered saddle point situation, we use the saddle point extragradient method to solve the problem.

The extragradient method is a controlled simple iteration method, each iteration of which is split into two half-steps:

1) *predictive half-step*

$$\frac{d}{dt}x^k(t) = D(t)x^k(t) + B(t)u^k(t), \quad x^k(t_0) = x^0, \quad t_0 \leq t \leq t_2, \qquad (25)$$

$$\bar{p}^k = \pi_+(p^k + \alpha(Gx^k - g)), \qquad (26)$$

$$\frac{d}{dt}\psi^k(t) + D^{\mathrm{T}}(t)\psi^k(t) = 0, \quad \psi^k = \varphi + G^{\mathrm{T}}p^k, \qquad (27)$$

$$\bar{u}^k(t) = \pi_U(u^k(t) - \alpha B^{\mathrm{T}}(t)\psi^k(t)); \qquad (28)$$

2) *main half-step*

$$\frac{d}{dt}\bar{x}^k(t) = D(t)\bar{x}^k(t) + B(t)\bar{u}^k(t), \quad \bar{x}^k(t_0) = x^0, \, t_0 \leq t \leq t_2, \qquad (29)$$

$$p^{k+1} = \pi_+(p^k + \alpha(G\bar{x}^k - g)), \qquad (30)$$

$$\frac{d}{dt}\bar{\psi}^k(t) + D^{\mathrm{T}}(t)\bar{\psi}^k(t) = 0, \quad \bar{\psi}^k = \varphi + G^{\mathrm{T}}\bar{p}^k, \qquad (31)$$

$$u^{k+1}(t) = \pi_U(u^k(t) - \alpha B^{\mathrm{T}}(t)\bar{\psi}^k(t)), \quad k = 0, 1, 2... \qquad (32)$$

Note that variables x, p and others in this paper have the dimensions of macro variables with respect to the variables in (3). The considered multi-agent problem (in this case, two agents) admits a finite number of agents. The problem under consideration was formally scalarized in such a way that it does not differ from the scalar formulations [3,4]. Therefore, the theorem on the convergence of the method considered above is not given here due to the limited volume of the publication. The logic of this proof can be easily restored by analogy with [3,4].

The following theorem is proved for the computational process (25)–(32).

8   Theorem on Convergence

Theorem. If set of solutions $(x_1^*, x_2^*, x^*(t), u^*(t); p_1^*, p_2^*, \psi^*(t))$ to problem (1), or (4), is not empty, then sequence $\{(x_1^k, x_2^k, x^k(t), u^k(t); p_1^k, p_2^k, \psi^k(t))\}$ generated by method (25)–(32) with step length α (chosen from some special condition) contains a subsequence $\{(x_1^{k_i}, x_2^{k_i}, x^{k_i}(t), u^{k_i}(t); p_1^{k_i}, p_2^{k_i}, \psi^{k_i}(t))\}$, which converges to the solution of the problem, including: weak convergence in controls, strong convergence in phase trajectories, conjugate trajectories, as well as in variables of finite-dimensional (intermediate and terminal) spaces. In particular, sequence

$$\left\{|p_1^k - p_1^*|^2 + |p_2^k - p_2^*|^2 + \|u^k(t) - u^*(t)\|^2\right\}$$

decreases monotonically on the Cartesian product $\mathbb{R}_+^m \times \mathbb{R}_+^m \times U$.

9   Conclusions

The article investigates the terminal control problem with additional finite-dimensional intermediate and boundary value problems. The problem belongs to the class of multi-agent linear systems and has a convex structure. The latter makes it possible, within the framework of duality theory, to use the saddle point property of the Lagrangian, in particular, to develop the theory of saddle point methods for solving multi-agent terminal control problems.

The proposed approach makes it possible to recalculate the phase trajectories by using the iterative saddle point gradient method, and to solve linear programming problems in intermediate finite-dimensional spaces. The trajectories are gradually "pulled up" to the optimal point. Limit points of finite-dimensional saddle points processes are the points of optimality for phase trajectories. We emphasize once again that only evidence-based optimization transforms mathematical models into a tool for making guaranteed decisions. Evidence-based optimization is, in fact, a beautiful continuation of the ideas of Lyapunov stability.

References

1. Abdullayev, V.M., Aida-zade, K.R.: Approach to the numerical solution of optimal control problems for loaded differential equations with nonlocal conditions. Comput. Math. Math. Phys. **59**(5), 696–707 (2019). https://doi.org/10.1134/S0965542519050026

2. Alekseev, V.M., Tikhomirov, V.M., Fomin, S.V.: Optimal Control. Springer, Boston (1987). https://doi.org/10.1007/978-1-4615-7551-1

3. Antipin, A.S., Khoroshilova, E.V.: Linear programming and dynamics. Trudy Inst. Mat. Mech. Ural Branch Russian Acad. Sci. **19**(2), 7–25 (2015) (in Russian)

4. Antipin, A.S., Khoroshilova, E.V.: Linear programming and dynamics. Ural Math. J. **1**(1), 3–19 (2015)

5. Antipin, A.S., Khoroshilova, E.V.: Optimal control with connected initial and terminal conditions. Proc. Steklov Inst. Math. **289**(1), 9–25 (2015). https://doi.org/10.1134/S0081543815050028

6. Antipin, A.S., Khoroshilova, E.V.: Saddle-point approach to solving problem of optimal control with fixed ends. J. Global Optim. **65**(1), 3–17 (2016)

7. Antipin, A., Khoroshilova, E.: On methods of terminal control with boundary-value problems: lagrange approach. In: Goldengorin, B. (ed.) Optimization and Its Applications in Control and Data Sciences. SOIA, vol. 115, pp. 17–49. Springer, Cham (2016). https://doi.org/10.1007/978-3-319-42056-1_2

8. Antipin, A.S., Khoroshilova, E.V.: Feedback synthesis for a terminal control problem. Comput. Math. Math. Phys. **58**(12), 1903–1918 (2018). https://doi.org/10.1134/S0965542518120035

9. Antipin, A.S., Khoroshilova, E.V.: Lagrangian as a tool for solving linear optimal control problems with state constraints. Optimal control and differential games. In: Proceedings of the International Conference Dedicated to the 110th Anniversary of the Birth of Lev Semenovich Pontryagin, pp. 23–26 (2018)

10. Antipin, A., Khoroshilova, E.: Controlled dynamic model with boundary-value problem of minimizing a sensitivity function. Optim. Lett. **13**(3), 451–473 (2017). https://doi.org/10.1007/s11590-017-1216-8

11. Antipin, A.S., Khoroshilova, E.V.: Dynamics, phase constraints, and linear programming. Comput. Math. Math. Phys. **60**(2), 184–202 (2020)

12. Antipin, A., Jacimovic, V., Jacimovic, M.: Dynamics and variational inequalities. Comp. Maths. Math. Phys. **57**(5), 784–801 (2017)

13. Antipin, A., Vasilieva, O.: Dynamic method of Multipliers in terminal control. Comp. Maths. Math. Phys. **55**(5), 766–787 (2015)

14. Dmitruk, A.V.: Convex Analysis. Moscow, MAKS-PRESS, Elementary Introductory Course (2012).(in Russian)

15. Kolmogorov, A.N., Fomin, S.V.: Elements of the Theory of Functions and Functional Analysis. FIZMATLIT, Moscow (2009).(in Russian)

16. Makarenko, A.V.: Intelligent control. In: Novikov, D.A. (ed.) Control Theory (Additional Chapters), 552p. LENAND, Moscow (2019) (in Russian)

17. Vasilyev, F.P.: Optimization methods. In: 2 Books. Moscow Center for Continuous Mathematical Education (2011) (in Russian)

Existence of Bounded Soliton Solutions for a Finite Difference Analogue of the Wave Equation with a Nonlinear Potential of General Form

Levon A. Beklaryan[1] and Armen L. Beklaryan[2(✉)]

[1] Central Economics and Mathematics Institute RAS, Nachimovky Prospect 47, 117418 Moscow, Russia
beklar@cemi.rssi.ru
[2] National Research University Higher School of Economics, 26-28 Ulitsa Shabolovka, 119049 Moscow, Russia
abeklaryan@hse.ru
http://www.hse.ru/en/staff/beklaryan

Abstract. In the presented work, the existence of a family of bounded soliton solutions for a finite difference wave equation with a nonlinear potential of general form is established. The proof is carried out within the framework of a formalism establishing a one-to-one correspondence between soliton solutions of an infinite-dimensional dynamical system and solutions of a family of functional differential equations of point-wise type. The key fact for the considered class of equations is also the presence of a number of symmetries.

Keywords: Wave equation · Soliton solutions · Nonlinear potential

1 Introduction

In the theory of plastic deformation, the following infinite-dimensional dynamical system is studied

$$m\ddot{y}_i = y_{i-1} - 2y_i + y_{i+1} + \phi(y_i), \quad i \in \mathbb{Z}, \quad y_i \in \mathbb{R}, \quad t \in \mathbb{R}, \qquad (1)$$

where the potential $\phi(\cdot)$ is given by a smooth periodic function. The Eq. (1) is a system with the Frenkel-Kontorova potential [8]. Such a system is a finite difference analogue of a nonlinear wave equation, simulates the behavior of a countable number of balls of mass m, placed at integer points of the real line, where each pair of adjacent balls is connected by an elastic spring, and describes the propagation of longitudinal waves in an infinite homogeneous absolutely elastic rod.

The reported study was partially funded by RFBR according to the research project 19-01-00147.

For equations of mathematical physics, an important class of solutions are traveling wave solutions (soliton solutions) [11,13]. In a number of models, such solutions are well approximated by traveling wave solutions for finite difference analogs of the original equations, which, in place of a continuous environment, describe the interaction of clumps of a environment placed at lattice sites [8,13]. Emerging systems belong to the class of infinite-dimensional dynamical systems. The most widely considered classes of such problems are infinite systems with Frenkel-Kontorova potentials (periodic and slowly growing potentials) and Fermi-Pasta-Ulam (potentials of exponential growth), a broad survey of which is given in the paper [12].

Definition 1. $\{y_i(\cdot)\}_{-\infty}^{+\infty}$ *is called a solution to the system* (1) *if for any* $i \in \mathbb{Z}$ *the function* $y_i(\cdot)$ *is continuously differentiable and its derivative is an absolutely continuous function, and almost everywhere satisfies the system* (1). ■

The study of such systems with different potentials is one of the intensively developing directions in the theory of dynamical systems. For them, the central task is to study soliton solutions (solutions of the traveling wave type) as one of the observed classes of waves.

Definition 2. *We can state that the solution* $\{y_i(\cdot)\}_{-\infty}^{+\infty}$ *of system* (1), *defined for all* $t \in \mathbb{R}$, *has a traveling wave type if there exists* $\tau > 0$ *that does not depend on* t *and* i, *such that for all* $i \in \mathbb{Z}$ *and* $t \in \mathbb{R}$, *the following equality holds:*

$$y_i(t + \tau) = y_{i+1}(t).$$

The constant τ *will be known as characteristic of a traveling wave.* ■

Thus, for the considered finite difference analogue of the wave equation, the study of soliton solutions is reduced to the study of the space of solutions of the following boundary value problem with linear nonlocal boundary conditions.

$$m\ddot{y}_i(t) = y_{i-1}(t) - 2y_i(t) + y_{i+1}(t) + \phi(y_i(t)), \, i \in \mathbb{Z}, \, y_i \in \mathbb{R}, \, t \in \mathbb{R}, \quad (2)$$
$$y_i(t + \tau) = y_{i+1}(t), \quad \tau \geq 0. \quad (3)$$

The presented work uses the capabilities of the formalism developed in [1–4]. Within the framework of this formalism, the localization of soliton solutions is used by specifying their asymptotics both in space and in time. This approach is based on the existence of a one-to-one correspondence of soliton solutions for infinite-dimensional dynamical systems with solutions of the family of induced functional differential equations of pointwise type [1–4,9,10,14,15].

The phase space of the system (2) is the space of infinite sequences

$$\mathcal{K}_{\mathbb{Z}}^2 = \overline{\prod}_{q \in \mathbb{Z}} \mathbb{R}_q^2, \, \mathbb{R}_q^n = \mathbb{R}^2, \, \varkappa \in \mathcal{K}_{\mathbb{Z}}^2, \, \varkappa = \{\mathbf{x}_i\}_{i \in \mathbb{Z}}, \, \mathbf{x}_i = (x_{i1}, x_{i2})'.$$

with the standard Tikhonov topology. In the space $\mathcal{K}_{\mathbb{Z}}^2$ we define the family of Hilbert subspaces $\mathcal{K}_{\mathbb{Z}2\mu}^2, \mu \in (0, 1)$

$$\mathcal{K}_{\mathbb{Z}2\mu}^2 = \left\{ \varkappa : \varkappa \in \mathcal{K}_{\mathbb{Z}}^2; \sum_{i \in \mathbb{Z}} \|\mathbf{x}_i\|_{R^2}^2 \mu^{2|i|} < +\infty \right\}$$

with the norm

$$\|\varkappa\|_{\mathbb{Z}2\mu} = \Big[\sum_{i\in\mathbb{Z}} \|\mathbf{x}_i\|_{\mathbb{R}^2}^2 \mu^{2|i|}\Big]^{\frac{1}{2}}.$$

We define a linear operator $\mathbb{A}$, a shift operator $\mathbb{T}$, and a nonlinear operator $\mathbb{F}$, acting continuously from the space K^2 into itself according to the following rule: for any $i \in \mathbb{Z}$, $\varkappa \in K^2$

$$(\mathbb{A}\varkappa)_i = (x_{i2}, m^{-1}[x_{(i+1)1} - 2x_{i1} + x_{(i-1)1}])',$$

$$(\mathbb{T}\varkappa)_i = (\varkappa)_{i+1}, \quad (\mathbb{F}(\varkappa))_i = (0, m^{-1}\phi(x_{i1}))'.$$

Note that the shift operator $\mathbb{T}$ *commutes* with the operators $\mathbb{A}$ and $\mathbb{F}$, which is typical of models describing processes in homogeneous environments.

The system (2)–(3), which defines soliton solutions, can be rewritten in the following operator form

$$\dot{\varkappa} = \mathbb{A}\varkappa + \mathbb{F}(\varkappa), \quad t \in \mathbb{R}, \tag{4}$$

$$\varkappa(t+\tau) = \mathbb{T}\varkappa(t), \tag{5}$$

which is a boundary value problem with linear nonlocal boundary conditions. The boundary conditions (5) mean that *the time shift of the solution is equal to the space shift*.

In the case of the problem under consideration, the soliton solutions, the solutions of the system (2)–(3), are in one-to-one correspondence with the solutions of the family of induced functional differential equations of pointwise type

$$\dot{z}_1(t) = z_2(t), \quad (z_1, z_2)' \in \mathbb{R}^2, \quad t \in \mathbb{R}, \tag{6}$$

$$\dot{z}_2(t) = m^{-1}[z_1(t-\tau) - 2z_1(t) + z_1(t+\tau) + \Phi(z_1(t))]. \tag{7}$$

The correspondence between these solutions is as follows

$$z_1(t) = y_0(t), \quad z_2(t) = \dot{y}_0(t), \quad t \in \mathbb{R}. \tag{8}$$

To study the existence and uniqueness of soliton solutions, it is proposed to localize solutions of induced functional differential equations of pointwise type in spaces of functions majorized by functions of a given exponential growth with exponent as a parameter

$$\mathcal{L}_\mu^n C^{(k)}(\mathbb{R}) = \Big\{z(\cdot) : z(\cdot) \in C^{(k)}(\mathbb{R}, \mathbb{R}^n), \max_{0\le r\le k}\sup_{t\in\mathbb{R}} \|z^{(r)}(t)\mu^{|t|}\|_{\mathbb{R}^n} < +\infty\Big\},$$

$$\|z(\cdot)\|_\mu^{(k)} = \max_{0\le r\le k}\sup_{t\in\mathbb{R}} \|z^{(r)}(t)\mu^{|t|}\|_{\mathbb{R}^n}, \quad k = 0, 1, \ldots, \quad \mu \in (0, +\infty).$$

This approach turns out to be especially successful for systems with Frenkel-Kontorova potentials. When describing processes in inhomogeneous environments, the commutativity condition for the right-hand side of the system in operator form and the shift operator is violated. In this case, the space of soliton solutions turns out to be trivial. At the same time, within the framework of

the developed formalism, it is possible to obtain the "correct" extension of the
concept of a traveling wave (soliton solution) in the form of solutions of the type
of quasi-traveling waves [5,6].

Under minimal restrictions on the potential $\phi(\cdot)$ in the form of the presence
of the Lipschitz condition (quasilinear potentials), this problem was studied in
the monograph [4]. The corresponding Lipschitz constant for the potential $\phi(\cdot)$
we denote by L_ϕ.

Let us consider a transcendental equation in two variables $\tau \in (0, +\infty)$ and
$\mu \in (0,1)$

$$C_\phi \tau \left(2\mu^{-1} + 1\right) = \ln \mu^{-1}, \tag{9}$$

where

$$C_\phi = \max\left\{1; 2m^{-1}\sqrt{L_\phi^2 + 2}\right\}.$$

The set of solutions to the Eq. (9) is described by the functions $\mu_1(\tau), \mu_2(\tau)$
given in Fig. 1. The value $\hat{\tau}$ has some absolute estimate $\hat{\tau} \le (2C_\phi)^{-1}$ and, in
particular, $\hat{\tau} \le \frac{1}{2}$.

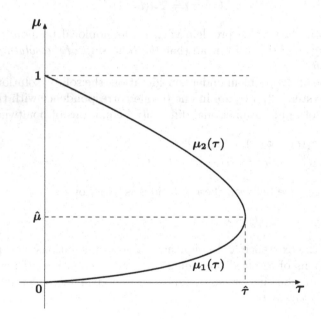

Fig. 1. Graphs of the functions $\mu_1(\tau), \mu_2(\tau)$.

Let us formulate a theorem on the existence and uniqueness of a solution for
the induced functional differential equation (6)–(7).

Theorem 1 ([4]). *Let the potential Φ satisfy the Lipschitz condition with the
constant L_Φ. Then for any initial data $a, b \in \mathbb{R}, \bar{t} \in \mathbb{R}$ and characteristic $\tau > 0$
satisfying the condition*

$$0 < \tau < \hat{\tau},$$

in the space $\mathcal{L}_\mu^2 C^{(0)}(\mathbb{R})$, $\mu^\tau \in (\mu_1(\tau), \mu_2(\tau))$ for the system of functional differential equations (6)–(7) exists and, moreover, a unique solution $(z_1(t), z_2(t))'$, $t \in \mathbb{R}$ such that it satisfies the initial conditions $z_1(\bar{t}) = a$, $z_2(\bar{t}) = b$. Such a solution as an element of the space $\mathcal{L}_\mu^2 C^{(0)}(\mathbb{R})$ continuously depends on the initial data $a, b \in \mathbb{R}$, as well as on the mass m, the characteristic τ, and the potential $\Phi(\cdot)$.
■

Theorem 1 not only guarantees the existence of a solution, but also sets a limit on its possible growth in time t. Obviously, for each $0 < \tau < \hat{\tau}$ the spaces $\mathcal{L}^2_{(\sqrt[\tau]{\mu_2(\tau)}-\varepsilon)} C^{(0)}(\mathbb{R})$ for small $\varepsilon > 0$ are much narrower than the spaces $\mathcal{L}^2_{(\sqrt[\tau]{\mu_1(\tau)}+\varepsilon)} C^{(0)}(\mathbb{R})$. The theorem guarantees the existence of a solution in narrower spaces and uniqueness in wider spaces.

Theorem 1 can be reformulated in terms of traveling wave solutions (soliton solutions) for the original wave equation (in terms of the system (2)–(3)).

Theorem 2 ([4]). *Let the potential Φ satisfy the Lipschitz condition with the constant L_Φ. Then for any initial data $\bar{i} \in \mathbb{Z}$, $a, b \in \mathbb{R}$, $\bar{t} \in \mathbb{R}$ and characteristic $\tau > 0$ satisfying the condition*

$$0 < \tau < \hat{\tau},$$

for the initial system of differential equations (2) there is a unique solution $\{y_i(\cdot)\}_{-\infty}^{+\infty}$ of traveling wave type (soliton solution) with characteristic τ such that it satisfies the initial conditions $y_{\bar{i}}(\bar{t}) = a$, $\dot{y}_{\bar{i}}(\bar{t}) = b$. For any parameter $\mu, \mu^\tau \in (\mu_1(\tau), \mu_2(\tau))$, the values of the vector function

$$\omega(t) = \{(y_i(t), \dot{y}_i(t))'\}_{-\infty}^{+\infty}$$

for any $t \in \mathbb{R}$ belong to the space $\mathcal{K}^2_{\mathbb{Z}2\mu}$, and the function

$$\rho(t) = \|\omega(t)\|_{2\mu}$$

belongs to the space $\mathcal{L}_\mu^1 C^{(0)}(\mathbb{R})$. Such a solution continuously depends on the initial data $a, b \in \mathbb{R}$, as well as on the mass m, the characteristic τ, and the potential $\Phi(\cdot)$. ■

Theorem 2 not only guarantees the existence of a solution, but also sets a restriction on its possible growth both in time t and in coordinates $i \in \mathbb{Z}$ (in space). Obviously, for each $0 < \tau < \hat{\tau}$ the spaces $\mathcal{K}^2_{\mathbb{Z}2(\sqrt[\tau]{\mu_2(\tau)}-\varepsilon)}$ for small $\varepsilon > 0$ are much narrower than the spaces $\mathcal{K}^2_{\mathbb{Z}2(\sqrt[\tau]{\mu_1(\tau)}+\varepsilon)}$. The theorem guarantees the existence of a solution in narrower spaces and uniqueness in wider spaces.

Within the framework of this approach, families of bounded solutions are described in the paper [7] for a finite-difference analog of the wave equation with a quadratic potential. Most of the preliminary results presented there were obtained without taking into account the specific form of the nonlinear potential. Therefore, the presented approach is universal in nature and can be applied to the study of a wide class of systems with nonlinear potential, which will be demonstrated below.

2 Some Preliminary Results for a System with a General Nonlinear Potential

We are going to study soliton solutions of the system (solutions of the system (2)–(3)) with a general nonlinear potential. Within the framework of such a task, a number of universal properties inherent in such systems are established.

Recall that the corresponding induced functional differential equation of pointwise type has the form

$$\dot{z}_1(t) = z_2(t), \quad (z_1, z_2)' \in \mathbb{R}^2, \quad t \in \mathbb{R}, \tag{10}$$
$$\dot{z}_2(t) = m^{-1}\big[z_1(t - \tau) - 2z_1(t) + z_1(t + \tau) + \Phi(z_1)\big]. \tag{11}$$

The Correspondence is Valid: a bounded soliton solution (the solution of the system (2)–(3)) corresponds to the solution of the induced functional differential equation of pointwise type (10)–(11) with bounded first coordinate $z_1(t), t \in \mathbb{R}$ and vice versa.

Remark 1. Points of a closed set

$$S = \{z : z = (z_1, z_2)', \quad (z_1, z_2)' \in \mathbb{R}^2; \quad z_2 = 0, \Phi(z_1) = 0\}$$

and only they are fixed points for the induced functional differential equation of pointwise type (10)–(11). ∎

To study other solutions of the induced functional differential equation with the condition of boundedness in the first coordinate, a family of auxiliary functional differential equations of pointwise type is constructed.

For any $\Delta > 0$, we define the potential Φ_Δ

$$\Phi(\xi) = \begin{cases} \Phi(-\Delta), & \text{if } \xi < -\Delta, \\ \Phi(\xi), & \text{if } \xi \in [-\Delta, \Delta], \\ \Phi(\Delta), & \text{if } \xi > \Delta. \end{cases}$$

The Lipschitz constant for such a function Φ_Δ is equal to the Lipschitz constant for the function $\Phi|_{[-\Delta,\Delta]}$, as restrictions of the potential Φ on the interval $[-\Delta, \Delta]$, and is denoted by L_{Φ_Δ}. Obviously, L_{Φ_Δ} is monotonically increasing in the parameter $\Delta > 0$.

Consider an auxiliary functional differential equation of pointwise type

$$\dot{z}_1(t) = z_2(t), \quad (z_1, z_2)' \in \mathbb{R}^2, \quad t \in \mathbb{R}, \tag{12}$$
$$\dot{z}_2(t) = m^{-1}\big[z_1(t - \tau) - 2z_1(t) + z_1(t + \tau) + \Phi_\Delta(z_1(t))\big]. \tag{13}$$

By analogy with the Eq. (9), we consider the transcendental equation in two variables $\tau \in (0, +\infty)$ and $\mu \in (0, 1)$

$$C_{\Phi_\Delta}\tau\left(2\mu^{-1} + 1\right) = \ln \mu^{-1}, \tag{14}$$

where

$$C_{\Phi_\Delta} = \max\left\{1; 2m^{-1}\sqrt{L^2_{\Phi_\Delta} + 2}\right\}$$

and C_{Φ_Δ} is monotonically increasing in the parameter $\Delta > 0$.

The solution to the Eq. (14) is described by the functions $\mu_{\Delta1}(\tau)$ and $\mu_{\Delta2}(\tau)$. The qualitative behavior of the functions $\mu_{\Delta1}(\tau), \mu_{\Delta2}(\tau)$ is the same as the behavior of the functions $\mu_1(\tau), \mu_2(\tau)$ in Fig. 1, and the value $\hat{\tau}$ is replaced by the corresponding value $\hat{\tau}_\Delta$, which is monotonically decreasing in the parameter $\Delta > 0$.

For the auxiliary functional differential equation of pointwise type (12)–(13), the theorem 1 is also true, in which the potential Φ should be replaced by Φ_Δ, the functions $\mu_1(\tau), \mu_2(\tau)$ replace with functions $\mu_{\Delta1}(\tau), \mu_{\Delta2}(\tau)$, and the value $\hat{\tau}$ by $\hat{\tau}_\Delta$. Any solution $(z_1(t), z_2(t))', t \in \mathbb{R}$ of the auxiliary functional differential equation (12)–(13) with the property of being bounded in the first coordinate $|z_1(t)| \leq \Delta, t \in \mathbb{R}$ is a solution of the induced functional differential equation of pointwise type (10)–(11) with nonlinear potential and the same boundedness condition along the first coordinate. Thus, by virtue of the matching rule formulated above, it suffices to establish the existence of solutions of the auxiliary functional differential equation of pointwise type (12)–(13) satisfying the property of being bounded in the first coordinate $|z_1(t)| \leq \Delta, t \in \mathbb{R}$.

The presence of symmetries, as well as the behavior of the vector field for the considered initial equation and auxiliary equations, allows one to describe a family of bounded solutions.

Let us define the sets

$$B = \{a : \Phi(a) = 0\}, \quad D = \mathbb{R}\backslash B,$$
$$B^\Delta = \{a : \Phi_\Delta(a) = 0\}, \quad D^\Delta = \mathbb{R}\backslash B^\Delta.$$

Obviously, the set S coincides with the natural embedding of the set B into $\mathbb{R}^2$. Due to the continuity of the potentials $\Phi(\cdot), \Phi_\Delta(\cdot)$, the sets B, B^Δ are closed, and the sets D, D^Δ are open. Then the open sets D, D^Δ consist of the union of at most countably many open intervals, that is,

$$D = \bigcup_{i \in I} d_i, \quad d_i = (\alpha_i, \beta_i), \quad i \in I,$$

$$D^\Delta = \bigcup_{i \in I^\Delta} d_i^\Delta, \quad d_i^\Delta = (\alpha_i^\Delta, \beta_i^\Delta), \quad i \in I^\Delta,$$

where I, I^Δ are finite or countable sets of indices. Obviously, the values of the potentials $\Phi(\cdot), \Phi_\Delta(\cdot)$, respectively, on each of the intervals $d_i, i \in I, d_i^\Delta, i \in I^\Delta$, are constant sign. Depending on the sign of the potential value, all indices I, I^Δ can be divided into two subsets $I_-, I_+, I_-^\Delta, I_+^\Delta$ and, accordingly, $I = I_- \cup I_+, I^\Delta = I_-^\Delta \cup I_+^\Delta$. Elements of the sets of indices $I_-, I_-^\Delta, I_+, I_+^\Delta$ we denote by $i-, i+$, respectively. Note that any finite interval $d_{i-}, i- \in I_-$ $(d_{i+}, i+ \in I_+)$, starting from some large value $\Delta > 0$, coincides with one from the intervals $d_{i-}^\Delta, i \in I_-^\Delta, (d_{i+}^\Delta, i \in I_+^\Delta)$.

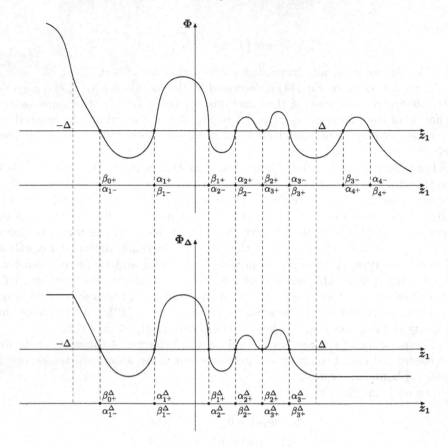

Fig. 2. Graphs of the potentials Φ, Φ_Δ.

Let us comment on Fig. 3. Small circles denote stationary solutions of hyperbolic type. The crosses denote stationary solutions that are sinks and sources for the same solutions and also have a hyperbolic type. Dotted lines denote separatrices, and green color denotes solutions unbounded in the first coordinate z_1.

3   Main Result

Let us formulate a result on the existence of a solution bounded in the first coordinate of the induced functional differential equation.

Theorem 3. *Let $\Delta > 0$ be given, and $B \cap [-\Delta, \Delta] \neq \emptyset$. Then for any $\tau \in (0, \hat\tau_\Delta)$ and $\mu, \mu^\tau \in (\mu_{\Delta 1}(\tau), \mu_{\Delta 2}(\tau))$ the solution $(z_1(\cdot), z_2(\cdot))' \in \mathcal{L}^2_\mu C^{(0)}(\mathbb{R})$ of the induced functional differential equation of pointwise type (10)–(11) with nonlinear potential and initial conditions $(z_1(0), z_2(0))' = (a, 0)'$ exists and is* *unique for all a such that*

$$a \in [(\bigcup_{i-\in I_-^\Delta, -\Delta \le \alpha_{i-}^\Delta} (\alpha_{i-}^\Delta, \beta_{i-}^\Delta)) \bigcup (\bigcup_{i+\in I_+^\Delta, \beta_{i+}^\Delta \le \Delta} (\alpha_{i+}^\Delta, \beta_{i+}^\Delta)) \bigcup B] \bigcap [-\Delta, \Delta]$$

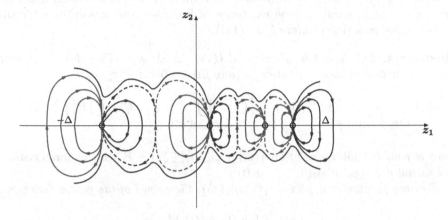

Fig. 3. Qualitative picture of bounded solutions of an auxiliary functional differential equation of pointwise type with a quasilinear right-hand side.

and such a solution satisfies the boundedness condition $|z_1(t)| \le \Delta, t \in \mathbb{R}$ *in the first coordinate* z_1. ∎

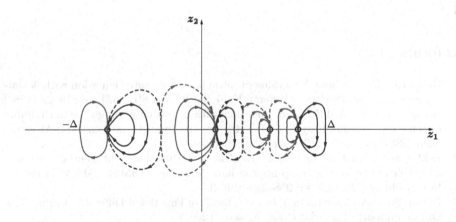

Fig. 4. Qualitative picture of bounded solutions for an induced functional differential equation of pointwise type with a nonlinear potential.

Let us comment on Fig. 4. In Fig. 4, in comparison with Fig. 3, there are only those solutions bounded in the coordinate z_1 that belong to the cylinder

$\{(z_1, z_2) : z_1 \in [-\Delta, \Delta]\}$, since only such solutions of the auxiliary functional differential equation of pointwise type are simultaneously solutions of the induced functional differential equation of pointwise type.

Now we can give an equivalent reformulation of Theorem 3 on the existence of a solution to an induced functional differential equation of pointwise type (system (12)–(13)) bounded in the first coordinate. This reformulation takes the form of the main result on the existence of bounded soliton solutions for the original wave equation (system (10)–(11)).

Theorem 4. *Let $\Delta > 0$ be given, and $B \cap [-\Delta, \Delta] \neq \emptyset$. Then for the system (10)–(11) with nonlinear potential for any fixed $\tau, 0 < \tau < \hat{\tau}_\Delta$ and*

$$a \in [(\bigcup_{i- \in I_-^\Delta, -\Delta \leq \alpha_{i-}^\Delta} (\alpha_{i-}^\Delta, \beta_{i-}^\Delta)) \bigcup (\bigcup_{i+ \in I_+^\Delta, \beta_{i+}^\Delta \leq \Delta} (\alpha_{i+}^\Delta, \beta_{i+}^\Delta)) \bigcup B] \bigcap [-\Delta, \Delta]$$

there is bounded soliton solution $\{(y_i(t), \dot{y}_i(t))'\}_{-\infty}^{+\infty}, t \in \mathbb{R}$ with characteristic τ and initial data $(y_0(0), \dot{y}_0(0))' = (a, 0)'$.

For any parameter $\mu, \mu^\tau \in (\mu_1(\tau), \mu_2(\tau))$, the values of the vector function

$$\omega(t) = \{(y_i(t), \dot{y}_i(t))'\}_{-\infty}^{+\infty}$$

for any $t \in \mathbb{R}$ belong to the space $\mathcal{K}_{\mathbb{Z}2\mu}^2$, and the function

$$\rho(t) = \|\omega(t)\|_{2\mu}$$

belongs to the space $\mathcal{L}_\mu^1 C^{(0)}(\mathbb{R})$. This is the only solution. Moreover, the boundedness condition for such a soliton solution has the form $|y_0(t)| \leq \Delta, t \in \mathbb{R}$.
∎

References

1. Beklaryan, L.A.: A boundary value problem for a differential equation with deviating argument. In: Doklady Akademii Nauk SSSR, vol. 291, no. 1, pp. 19–22 (1986)
2. Beklaryan, L.A.: A differential equation with deviating argument as an infinite-dimensional dynamical system. In: Reports in Applied Mathematics, Akademii Nauk SSSR, Vychislitel, Tsentr, Moscow (1989)
3. Beklaryan, L.A.: Introduction to the theory of functional differential equations and their applications. Group approach. J. Math. Sci. **135**(2), 2813–2954 (2006). https://doi.org/10.1007/s10958-006-0145-3
4. Beklaryan, L.A.: Introduction To the Theory of Functional Differential Equations. Group approach. Factorial Press, Moscow (2007)
5. Beklaryan, L.A.: Quasitravelling waves. Sbornik: Math. **201**(12), 1731–1775 (2010). https://doi.org/10.1070/SM2010v201n12ABEH004129
6. Beklaryan, L.A.: Quasi-travelling waves as natural extension of class of traveling waves. In: Tambov University Reports, Series Natural and Technical Sciences, vol. 19, no. 2, pp. 331–340 (2014)

7. Beklaryan, L.A.: The question of the existence of bounded soliton solutions in the problem of longitudinal vibrations of an elastic infinite rod in a field with a strongly nonlinear potential, to appear in Computational Mathematics and Mathematical Physics (2021)
8. Frenkel, Y.I., Contorova, T.A.: On the theory of plastic deformation and twinning. J. Exp. Theor. Phys. **8**(1), 89–95 (1938)
9. Keener, J.P.: Propagation and its failure in coupled systems of discrete excitable cells. SIAM J. Appl. Math. **47**(3), 556–572 (1987). https://doi.org/10.1137/0147038
10. Mallet-Paret, J.: The Fredholm alternative for functional differential equations of mixed type. J. Dyn. Diff. Equat. **11**(1), 1–47 (1999). https://doi.org/10.1023/A:1021889401235
11. Miwa, T., Jimbo, M., Date, E.: Solitons: Differential Equations, Symmetries and Infinite Dimensional Algebras. Cambridge University Press, Cambridge (2000)
12. Pustyl'nikov, L.D.: Infinite-dimensional non-linear ordinary differential equations and the KAM theory. Russ. Math. Surv. **52**(3), 551–604 (1997). https://doi.org/10.1070/RM1997v052n03ABEH001810
13. Toda, M.: Theory of nonlinear lattices. Springer, Heidelberg (1989)
14. Van Vleck, E.S., Mallet-Paret, J., Cahn, J.W.: Traveling wave solutions for systems of ODEs on a two-dimensional spatial lattice. SIAM J. Appl. Math. **59**(2), 455–493 (1998). https://doi.org/10.1137/S0036139996312703
15. Zinner, B.: Existence of traveling wavefront solutions for the discrete Nagumo equation. J. Differ. Equ. **96**(1), 1–27 (1992). https://doi.org/10.1016/0022-0396(92)90142-A

Attraction Domains in the Control Problem of a Wheeled Robot Following a Curvilinear Path over an Uneven Surface

Alexey Generalov[1,3](✉) (iD), Lev Rapoport[1,3](✉) (iD), and Mikhail Shavin[2](✉) (iD)

[1] Institute of Control Sciences, Moscow 117997, Russia
[2] Skolkovo Institute of Science and Technology, Moscow 121205, Russia
m.shavin@skoltech.ru
[3] Topcon Positioning Systems, Moscow 115114, Russia
{generalov.alexey,lbrapoport}@gmail.com

Abstract. The problem of motion control of a wheeled robot is considered. The robot is supposed to be moving without lateral slippage along an arbitrary, sufficiently smooth three-dimensional surface. The target path of the robot is defined by a curve with constrained curvature on a given surface. The rear wheels are assumed to be driving while the front wheels are responsible for the rotation of the robot's platform. A control law is synthesized based on the feedback linearization approach [5]. The purpose of the paper is to construct an estimate of the invariant attraction domain in the space "cross-track deviation - angular deviation" taking into account the constraints on the maximum steering angle. This problem has received much attention in connection with precision farming applications [14]. The control goal is to drive the specified target point, taken as a middle of the rear axle, to the target path and to stabilize its motion. The system is presented in the so called Lurie form [1,15] and embedded in the class of systems with nonlinearities constrained by the sector condition. The method of attraction domain estimation in the state space of the system is proposed. The negativity condition for the derivative of the Lyapunov function with respect to the system's dynamics under sector conditions is formulated in terms of solvability of the linear matrix inequality (LMI) [2]. The LMI, the left side of which depends on the matrix of the quadratic form, gives the constraints of the considered optimization problem. The cost function of the optimization problem is the trace of the matrix. Such a cost function is widely used in control theory to optimize the volume of invariant sets [2]. Numerical results are presented.

Keywords: Wheeled robot · Curvilinear target path · Absolute stability · Attraction domain

© Springer Nature Switzerland AG 2021
N. N. Olenev et al. (Eds.): OPTIMA 2021, LNCS 13078, pp. 176–190, 2021.
https://doi.org/10.1007/978-3-030-91059-4_13

1   Introduction

In this paper considered is the motion of the wheeled robot following the curvilinear path laid on uneven surface. More precisely, the problem of stabilization of such a motion is considered. It is supposed that the all four wheels move without cross-track slippage i.e. motion is subjected to non-holonomic constraints. The rear wheels are assumed to be driving while the front wheels are responsible for the rotation of the robot's platform. The control goal is to drive the target point (a specific point selected on the platform) to the target path and to stabilize its motion. In this paper, we don't restrict ourselves to the simple case of motion along a straight path (see [8,10,14]). Instead, most general case of feasible target path is considered. Here the feasibility means that the path can be followed by the robot with constrained steering wheels angle. The control law is obtained by the feedback linearization approach and is subject to the two-sided constraints as described in the earlier papers [10,11]. To estimate attraction domain of the closed loop system the quadratic Lyapunov function will be used. Parameters to be chosen are entries of the positive definite square matrix. The motion is described by the system of nonlinear ordinary differential equations.

One of the fundamental problems of nonlinear control systems theory is a description of the attraction domain of the equilibrium state. Usually the equilibrium refers to nominal system's operation, which should be stabilized. Complete solution of this problem in general form is extremely hard. Therefore, an internal (by inclusion) estimate of the attraction domain is important for applications. Usually engineers are looking for an estimate of a domain that combines two properties: it must provide asymptotic attraction and it must be invariant. By invariance we mean such a property of the domain, that the trajectory of the system, once getting inside, will no longer leave it. Standard approach for the construction of such estimates consists in using Lyapunov functions from certain parametric classes. Let $z \in R^n$ be the system state. Given the Lyapunov function $V(z)$, the estimate of the attraction domain is constructed as a level set $\{z : V(z) \leq c\}$, provided the time derivative with respect to the system dynamics is negative: $\dot{V} < 0$. Candidates for use as a Lyapunov function are selected from some parametric class. Thus, the more general is the parametric class of Lyapunov functions, the more is freedom of choice, and the less conservative is the resulting estimate of the attraction domain. Desire to maximize the volume of the attraction domain leads to the problem of optimal parameters choice.

As for design of the target path, it is supposed that the optimal field coverage problem was already solved for the particular working site and desired path was obtained as a result. The site is in general hilly and has curvilinear boundary.

The goal of this study is to propose a method of attraction domain estimation in the state space of the system. Description of the attraction domain inscribed into the band of certain width around the target path and guaranteeing prescribed exponential convergence rate is considered. Results of the paper are illustrated by example.

Note that the absolute stability approach used by authors for stability analysis of systems with constrained control is not the only possible approach, see for example [12, 13] for other approaches.

2   Problem Statement, Motion Equations and Change of Variables

Starting with the description of the coordinate frames we then define the parametrization of the target path and the kinematic model of the wheeled robot. Then ordinary differential equations describing motion of the robot are given. For convenience, the middle point of the rear axle is taken as the target point. Everywhere in the paper vectors are assumed to be columns and the symbol ^{T} stands for the matrix transpose.

2.1   Coordinate Frames

Position of the wheeled robot r is described by position of the origin of a mobile coordinate frame B associated with the robot in some fixed coordinate frame I.

The coordinate frame B has the origin fixed at the target point. The x_B axis is directed along the centerline of the robot's platform, the y_B axis lies in the plane of the platform and directed left orthogonal to x_B, and z_B completes the right triple. Hereinafter subscripts B and I mean that a quantity is expressed in the B or I-frame respectively. Double subscripts x, y, z are used to express appropriate entry of the vector.

The subscript $_{IB}$ is used to denote transformations between coordinate frames. The robot's platform orientation is defined by a quaternion q in such a way, that an arbitrary vector, expressed in the robot's frame B is transformed into the frame I as $r_I = q_{IB} \circ r_B \circ \tilde{q}_{IB}$, where $\circ$ stands for a Hamilton product, the quaternion $\tilde{q}_{IB}$ is inverse to q_{IB}, the fourth zero entry is added to three dimensional vectors to complete quaternion operations.

The unit vectors e_{x_B}, e_{y_B}, and e_{z_B} are aligned with appropriate axes of the B-frame. Taking into account these notations we have for example $e_{x_{B_x}} = 1$ and $e_{x_{B_y}} = e_{x_{B_z}} = 0$.

2.2   Surface and Target Path

We assume that the surface on which the robot moves is given by a continuous function
$$f_s(x, y, z) = 0,$$
for which the first and second derivatives are everywhere defined and continuous. The inverse of the curvature of the surface is much larger than the robot's size, and the third axis of the coordinate system associated with the robot e_{z_B} remains collinear with the normal to the surface.

The target path is given by a continuous function $p(s)$, for which the first and second derivatives are everywhere defined and continuous and which belongs to the surface f_s. Hereinafter the symbol $'$ is used to denote differentiation with respect to the variable s. Function $p(s)$ is naturally parameterized, that is, for any two values of the path parameter s_a and s_b, arc length $l(p(s_a), p(s_b)) = |s_b - s_a|$. Thus, the derivative of the parameter

$$p'(s) = 1,$$

and the second derivative is everywhere perpendicular to the tangent to the curve

$$p'(s)p''(s)^T = 0,$$

and defines the curvature vector of the curve at a given point

$$c_v = p''(s).$$

We assume that the length of the projection of the curvature vector c_v on the tangent plane to the surface f_s is bounded by the value

$$|c_s| \leq c_{max}$$

at each point of the path.

2.3   Motion Equations

The robot's motion with time is defined by its orientation and a scalar velocity of the target point $v > 0$ (hereinafter the symbol $\dot{}$ is used to denote differentiation with respect to the time):

$$\dot{r}_B = vc_{x_B} = (v \quad 0 \quad 0)^T,$$
$$\dot{r}_I = v(q_{IB} \circ e_{x_B} \circ \tilde{q}_{IB}).$$

An angular velocity vector expressed in the B-frame $\Omega_B = (\omega_x \quad \omega_y \quad \omega_z)^T$ connects the quaternion with its time derivative by the Euler–Poisson equation

$$\dot{q}_{IB} = \frac{1}{2} q_{IB} \circ \Omega_B. \tag{1}$$

The current orientation of the robot is defined by the steering wheel angle which in turn defines the instant curvature u of the target point and the normal vector to the surface n_s, the robot is moving on. The orientation quaternion can be expressed in the following form:

$$q_{IB} = q_\gamma \circ q_\alpha,$$
$$q_\gamma = (\cos \tfrac{\gamma}{2} \quad n_s \sin \tfrac{\gamma}{2})^T,$$
$$\dot{\gamma} = vu = \omega_z,$$
$$q_\alpha = (\cos \tfrac{\alpha}{2} \quad n_b \sin \tfrac{\alpha}{2})^T.$$

Here the angle γ defines the rotation of the robot around the vector n_s, α is the angle between n_s and e_z, the vector n_b is defined as a common normal to the vector n_{s_B} and $e_z = (0 \quad 0 \quad 1)^T$. In the case of n_s and e_z are collinear, $q_\alpha = (1 \quad 0 \quad 0 \quad 0)^T$. Since the surface is defined as $f_s(x, y, z) = 0$, then

$$n_s = \frac{\nabla f_s}{\|\nabla f_s\|}.$$

Let Δ be the vector of the lateral deviation of the robot's target point from the target path

$$\Delta_I = r_I - p(s^*),$$

where $p(s^*)$ defines the nearest point of the target path for which the following equality holds:

$$\Delta_I^T p'(s^*) = 0. \tag{2}$$

Then

$$\dot{\Delta}_I = v e_I - p'(s^*)\dot{s}^*. \tag{3}$$

Let δ be the distance to the target path

$$\delta = \sqrt{\Delta_I^T \Delta_I}.$$

Then

$$\dot{\delta} = \frac{\Delta_I^T \dot{\Delta}_I}{\delta} = \frac{\Delta_I^T (v e_I - p'(s^*)\dot{s}^*)}{\delta} = \frac{v \Delta_I^T e_I}{\delta}. \tag{4}$$

Let us choose ξ - the path length as a natural independent variable. Then

$$\dot{\xi} = v. \tag{5}$$

Let us define $z_1 = \delta$, be the distance to the target path. Then, taking into account (4) and (5):

$$(z_1)_\xi = \frac{\dot{\delta}}{\dot{\xi}} = \frac{\Delta_I^T}{\delta} e_I = \cos \varphi,$$

where φ is an angle between a direction to the nearest point of the target path

$$d_I = \frac{\Delta_I}{\delta},$$

and the direction of the velocity of the target point e_I. Let $z_2 = \cos \varphi$. Then

$$(z_2)_\xi = \frac{d_I^T \dot{e}_I + \dot{d}_I^T e_I}{\dot{\xi}},$$

where

$$\dot{e}_I = \Omega_I \times e_I = [\Omega_I]_\times e_I, \tag{6}$$

and

$$\dot{d}_I = \frac{\dot{\Delta}_I}{\delta} - \frac{\Delta_I}{\delta^2}\dot{\delta} = \frac{\dot{\Delta}_I}{\delta} - \frac{v \Delta_I \cos \varphi}{\delta^2}.$$

In the Eq. (6) the matrix $[\Omega_I]_\times$ is defined as

$$[\Omega_I]_\times = \begin{bmatrix} 0 & -\Omega_{I,3} & \Omega_{I,2} \\ \Omega_{I,3} & 0 & -\Omega_{I,1} \\ -\Omega_{I,2} & \Omega_{I,1} & 0 \end{bmatrix}.$$

Thus, the operation $[\cdot]_\times$ converts the vector to the skew-symmetric matrix. Further,

$$(z_2)_\xi = \frac{d_I^T[\Omega_I]_\times + \frac{\dot{\Delta}_I^T e_I}{\delta}}{v} e_I - \frac{\cos^2\varphi}{\delta} = \frac{d_I^T[\Omega_I]_\times e_I}{v} + \frac{\dot{\Delta}_I^T e_I}{\delta v} - \frac{\cos^2\varphi}{\delta}, \quad (7)$$

where $\dot{\Delta}_I$ is defined in (3). Then,

$$(z_2)_\xi = \frac{d_I^T[\Omega_I]_\times e_I}{v} + \frac{\sin^2\varphi}{\delta} - \frac{(p'\dot{s}^*)^T e_I}{\delta v}. \quad (8)$$

The expression for $\dot{s}^*$ can be obtained by differentiation of (2):

$$\dot{s}^* = \frac{v e_I^T p'}{\|p'\|^2 - \Delta_I^T p''}. \quad (9)$$

Substitution of (3, 9) into (7) gives:

$$(z_2)_\xi = \frac{d_I^T[\Omega_I]_\times e_I}{v} + \frac{1}{\delta} - \frac{p'^T e_I^T p' e_I}{\delta(\|p'\|^2 - \Delta_I^T p'')} - \frac{\cos^2\varphi}{\delta}, \quad (10)$$

where

$$p'^T e_I^T p' c_I = \|p'\|^2 \sin^2\varphi.$$

Farther,

$$(z_2)_\xi = \frac{1}{\delta}\left(1 - \cos^2\varphi - \frac{\|p'\|^2 \sin^2\varphi}{\|p'\|^2 - \delta d_I^T p''}\right) + \frac{d_I^T[\Omega_I]_\times e_I}{v}, \quad (11)$$

where

$$\frac{d_I^T[\Omega_I]_\times e_I}{v} = \frac{(e_I \times d_I)^T \Omega_I}{v} = \frac{(e_B \times d_B)^T \Omega_B}{v}. \quad (12)$$

If the closest point of the target path is in the plane of the robot's motion, the cross product in the last equation can be expressed as:

$$e_B \times d_B = e_z \sin\varphi \, S(d_{By}), \quad (13)$$

where $S(d_{By})$ is the sign of the second entry of the vector d_B. The expression (11) can be rewritten as:

$$(z_2)_\xi = \sin\varphi \left(\frac{d_I^T p'' \sin\varphi}{\delta d_I^T p'' - \|p'\|^2} + u \, S(d_{By})\right). \quad (14)$$

Assuming that vector d_I is in the plane of the closest segment of the target path, $d_I^T p'' = c_s(\xi)$. Summing up results of this section, arrive at the following system describing wheeled robot's motion along the target path with curvature $c_s(\xi)$:

$$(z_1)_\xi = \cos\varphi,$$
$$(z_2)_\xi = \sin\varphi \left(\frac{c_s(\xi)\sin\varphi}{c_s(\xi)\delta - 1} + u\, S(d_{By}) \right). \tag{15}$$

2.4 Change of Variables

The angle φ between the velocity vector of the target point (collinear to the centerline if a lateral slippage is absent) and the direction to the nearest point of the target path experiences a singularity as the robot approaches it. To avoid singularity it is convenient to use the angle ψ between the velocity vector and the tangent to the target path at the nearest point (s^*). Then $z_2 = \cos\varphi \equiv \sin\psi$.

The asymptotic stability of the zero solution of the system (15) closed by the control u will be analized using the quadratic Lyapunov functions [7,10]. To simplify analysis let us introduce the new variables $\tilde{z}_1$ and $\tilde{z}_2$ as follows:

$$\tilde{z}_1 = S(d_{By})z_1,$$
$$\tilde{z}_2 = (\tilde{z}_1)_\xi = S(d_{By})z_2.$$

The second equation in (15) can be rewritten in the form:

$$(\tilde{z}_2)_\xi = \cos\psi \left(u + \frac{\tilde{c}_s(\xi)\cos\psi}{\tilde{c}_s(\xi)\tilde{z}_1 - 1} \right),$$

where $\tilde{c}_s(\xi) = S(d_{By})c_s(\xi)$.

Now we arrive at the following system describing motion along the target path with variable curvature $\tilde{c}_s(\xi)$ dependent on ξ:

$$(\tilde{z}_1)_\xi = \tilde{z}_2,$$
$$(\tilde{z}_2)_\xi = u\sqrt{1 - \tilde{z}_2^2} + \frac{\tilde{c}_s(\xi)(1 - \tilde{z}_2^2)}{\tilde{c}_s(\xi)\tilde{z}_1 - 1}. \tag{16}$$

Similar expression was derived in [6]. The following notation must be taken. Using the expression $\sqrt{1 - \tilde{z}_2^2}$ for $\cos\psi$ we look like loosing possibility of having a negative sign for it. In fact, our goal is to preserve a positive sign of $\cos\psi$ along the whole trajectory of the closed - loop system, provided we started with orientation with $\cos\psi > 0$ in the very beginning. This means, that the robot will never have the "wrong" ($\cos\psi \leq 0$) orientation to the target path and will not go along it the opposite direction corresponding to $\dot{\xi} < 0$, if the orientation was "right" at the beginning of operation. To reach this goal we introduce the invariant attraction domains in the following sections.

3   Control Law Synthesis

Using the feedback linearization approach, we choose the control u in (16) in the form

$$u = \frac{-\sigma - \dfrac{\tilde{c}_s(\xi)(1 - \tilde{z}_2^2)}{\tilde{c}_s(\xi)\tilde{z}_1 - 1}}{\sqrt{1 - \tilde{z}_2^2}}, \tag{17}$$

for some desired rate of exponential decrease $\lambda > 0$ and

$$\sigma = 2\lambda\tilde{z}_2 + \lambda^2\tilde{z}_1.$$

The closed loop system from (16), (17) takes the form:

$$(\tilde{z}_1)_\xi = \tilde{z}_2,$$
$$(\tilde{z}_2)_\xi = -\sigma.$$

This leads to $(\tilde{z}_1)_{\xi\xi} + 2\lambda(\tilde{z}_1)_\xi + \lambda^2\tilde{z}_1 = 0$, which implies the exponential stability with the rate $-\lambda$ of all components of vector $\tilde{z} = (\tilde{z}_1, \tilde{z}_2)^T$. However, in general, control (17) does not satisfy the two-sided constraints:

$$-\overline{u} \leq u \leq \overline{u}.$$

Taking constrained control with saturation in the form

$$u = s_{\overline{u}}\left(\frac{-\sigma - \dfrac{\tilde{c}_s(\xi)(1 - \tilde{z}_2^2)}{\tilde{c}_s(\xi)\tilde{z}_1 - 1}}{\sqrt{1 - \tilde{z}_2^2}} \right), \tag{18}$$

where

$$s_{\overline{u}}(u) = \begin{cases} -\overline{u} & \text{for } u \leq -\overline{u}, \\ u & \text{for } |u| < \overline{u}, \\ \overline{u} & \text{for } u \geq \overline{u}, \end{cases}$$

may not guarantee that vector z decrease exponentially with given rate of exponential stability and undesirable overshoot in variations of the variables is possible. In what follows, the problem of attraction domain estimation inscribed into the band of certain width and guaranteeing prescribed exponential convergence rate is considered.

4   Attraction Domain Estimation

In this section, we are estimating the set of initial conditions $\tilde{z}^0$ guaranteeing that a) along the trajectories of the system (16) $\tilde{z}(\xi)$ decreases exponentially with the rate $-\mu$ $(0 < \mu \leq \lambda)$, and b) entries of the vector $\tilde{z}(\xi)$ satisfy the two-sided constraints $|\tilde{z}_1(\xi)| \leq \alpha_1$, $|\tilde{z}_2(\xi)| \leq \alpha_2$. More specific, we want to guarantee

$$\frac{dV(\tilde{z}(\xi))}{d\xi} + 2\mu V(\tilde{z}(\xi)) \leq 0, \quad \xi \geq 0. \tag{19}$$

for some quadratic form

$$V(\tilde{z}) = \tilde{z}^T P \tilde{z} \tag{20}$$

with P being a positively definite matrix $P \succ 0, P^T = P$. In what follows the symbols $\succ$ ($\prec$) and $\succeq$ ($\preceq$) will be used to denote the positive (negative) definiteness and positive (negative) semi-definiteness of symmetric real valued matrices respectively.

Following the absolute stability approach introduced in [7,9,10], denote

$$\Omega(P) = \{\tilde{z} : V(\tilde{z}) \leq 1\}.$$

Given positive values α_1 and α_2, we are looking for the matrix P satisfying the linear matrix inequalities (LMI's)

$$P \succeq \begin{bmatrix} \frac{1}{\alpha_1^2} & 0 \\ 0 & 0 \end{bmatrix}, P \succeq \begin{bmatrix} 0 & 0 \\ 0 & \frac{1}{\alpha_2^2} \end{bmatrix}, \tag{21}$$

meaning that the desired domain $\Omega(P)$ is inscribed into the rectangle

$$\Pi(\alpha_1, \alpha_2) = \{\tilde{z} : -\alpha_1 \leq \tilde{z}_1 \leq \alpha_1, -\alpha_2 \leq \tilde{z}_2 \leq \alpha_2\}.$$

We now characterize the values of μ, α_1 and α_2 that guarantee

$$\Omega(P) \subseteq \Pi(\alpha_1, \alpha_2) \tag{22}$$

for some matrix $P \succ 0$.

Rewrite the last equation in (16) taking the control u as (18):

$$(\tilde{z}_2)_\xi = s_{\overline{u}} \left(\frac{-\sigma - \dfrac{\tilde{c}_s(\xi)(1 - \tilde{z}_2^2)}{\tilde{c}_s(\xi)\tilde{z}_1 - 1}}{\sqrt{1 - \tilde{z}_2^2}} \right) \sqrt{1 - \tilde{z}_2^2} + \frac{\tilde{c}_s(\xi)(1 - \tilde{z}_2^2)}{\tilde{c}_s(\xi)\tilde{z}_1 - 1} \doteq -\Phi(\tilde{z}, \sigma). \tag{23}$$

Then

$$\Phi(\tilde{z}, \sigma) = s_{\overline{u}\sqrt{1-\tilde{z}_2^2}} \left(\sigma + \frac{\tilde{c}_s(\xi)(1 - \tilde{z}_2^2)}{\tilde{c}_s(\xi)\tilde{z}_1 - 1} \right) - \frac{\tilde{c}_s(\xi)(1 - \tilde{z}_2^2)}{\tilde{c}_s(\xi)\tilde{z}_1 - 1}, \tag{24}$$

and the system (16) takes the form:

$$\begin{aligned} (\tilde{z}_1)_\xi &= \tilde{z}_2, \\ (\tilde{z}_2)_\xi &= -\Phi(\tilde{z}, \sigma). \end{aligned} \tag{25}$$

Denote $d = (\lambda^2, \lambda)^T$ and

$$\bar{c} = \sup |\tilde{c}_s(\xi)|. \tag{26}$$

The following auxiliary assertion holds:

Lemma 1. *Assume that for a matrix P satisfying (21) and numbers α_1, α_2 satisfying the inequality*

$$u_0 \doteq \left(\bar{u} - \frac{\bar{c}}{1 - \bar{c}\alpha_1}\right)\sqrt{1 - \alpha_2^2} > 0, \tag{27}$$

the inclusion

$$\tilde{z} \in \Omega(P) \tag{28}$$

holds. Then the following inequalities hold

$$-\sigma_0 \leq \sigma \leq \sigma_0, \tag{29}$$

$$\begin{aligned}
s_{u_0}(\sigma) \leq \Phi(\tilde{z}, \sigma) \leq \sigma & \quad \text{for } \sigma \geq 0, \\
\sigma \leq \Phi(\tilde{z}, \sigma) \leq s_{u_0}(\sigma) & \quad \text{for } \sigma \leq 0,
\end{aligned} \tag{30}$$

where

$$\sigma_0 = \sqrt{d^T P^{-1} d}. \tag{31}$$

Proof. From conditions (28) and (21) it follows that

$$\tilde{z}_1^2 \leq \alpha_1^2, \quad \tilde{z}_2^2 \leq \alpha_2^2. \tag{32}$$

From (24) we have

$$\Phi(\tilde{z}, \sigma) = \begin{cases} \sigma_1 & \text{for } \sigma > \sigma_1, \\ \sigma & \text{for } -\sigma_2 \leq \sigma \leq \sigma_1, \\ -\sigma_2 & \text{for } \sigma < -\sigma_2, \end{cases} \tag{33}$$

where

$$\sigma_1 = \bar{u}\sqrt{1 - \tilde{z}_2^2} - \frac{\tilde{c}_s(\xi)(1 - \tilde{z}_2^2)}{\tilde{c}_s(\xi)\tilde{z}_1 - 1},$$

$$\sigma_2 = \bar{u}\sqrt{1 - \tilde{z}_2^2} + \frac{\tilde{c}_s(\xi)(1 - \tilde{z}_2^2)}{\tilde{c}_s(\xi)\tilde{z}_1 - 1}.$$

Keeping (32), (26) and (27) in mind, we obtain for $j = 1, 2$

$$\sigma_j \geq u_0. \tag{34}$$

Combination (33) and (34) gives (30). Next, consider the convex optimization problem $\sigma \to$ max subject to constraints (28). Necessary and sufficient conditions for the extremum have the form

$$2\nu P\tilde{z} = d, \tag{35}$$

where $\nu > 0$ is the Lagrange multiplier. Multiplying the last equation by $\tilde{z}^T$ and accounting for the fact that the extremum in this case is attained at the boundary of the domain (28), we obtain

$$\nu = \frac{1}{2}\sigma^*, \tag{36}$$

where σ^* is the solution of the optimization problem. Multiplying (35) by $d^T P^{-1}$, we arrive at

$$\nu\sigma^* = \frac{1}{2} d^T P^{-1} d. \tag{37}$$

Combining (36) with (37), we obtain $\sigma^* = \pm\sqrt{d^T P^{-1} d}$, so that at the maximum, the equality $\sigma^* = \sigma_0 = \sqrt{d^T P^{-1} d}$ holds. Now, formulating the minimization problem and using similar reasonings, we obtain that at the minimum point, the equality $\sigma^* = -\sigma_0$ holds, which yields (29). Proof of Lemma 1 is complete.

Along with the function $\Phi(\tilde{z}, \sigma)$ in the formulation of system (25), introduce the function

$$\phi(\xi, \sigma) = \beta(\xi)\sigma,$$

where $\beta(\xi)$ satisfies the conditions

$$k_0 \le \beta(\xi) \le 1, \quad k_0 = \min\{\frac{u_0}{\sigma_0}, 1\}. \tag{38}$$

The graph of the function $\Phi(\tilde{z}, \sigma)$, satisfying the conditions (30), is inscribed into a "sector" on the plane $\sigma - \Phi$ for values σ satisfying conditions (29). Conditions (38) define the size of the sector. We next expand the class of systems (25) by considering systems of the form

$$\begin{aligned}(\tilde{z}_1)_\xi &= \tilde{z}_2, \\ (\tilde{z}_2)_\xi &= -\beta(\xi)\sigma.\end{aligned} \tag{39}$$

We now require that the function $\beta(\xi)$ satisfy the existence conditions of absolutely continuous solution of system (39). If system (39) possesses property (19) for all functions $\beta(\xi)$ satisfying conditions (38) then property (19) also holds along the trajectories of system (25) satisfying (28). Consider the matrices

$$A_\beta = \begin{bmatrix} 0 & 1 \\ -\beta\lambda^2 & -2\beta\lambda \end{bmatrix}.$$

Theorem 1. *Assume that given numbers $\bar{u} > 0$, $\bar{c} > 0$ and $\alpha_1 > 0$, $\alpha_2 > 0$ satisfying (27) there exist numbers $\mu > 0$ and $0 < \beta \le 1$ such that the following LMI's in the variable P are feasible:*

$$\begin{aligned} PA_1 + A_1^T P + 2\mu P &\preceq 0, \\ PA_\beta + A_\beta^T P + 2\mu P &\preceq 0, \end{aligned} \tag{40}$$

$$\begin{bmatrix} P & d \\ d^T & \frac{u_0^2}{\beta^2} \end{bmatrix} \succ 0, \tag{41}$$

$$P \succeq \begin{bmatrix} \frac{1}{\alpha_1^2} & 0 \\ 0 & 0 \end{bmatrix}, P \succeq \begin{bmatrix} 0 & 0 \\ 0 & \frac{1}{\alpha_2^2} \end{bmatrix}, \tag{42}$$

then the domain $\Omega(P)$ is an attraction domain of system (16) under control (18); moreover, the condition (22) holds.

Proof. The stability of the zero solution of system (39) for all possible functions $\beta(\xi)$ satisfying (38) implies the stability of the zero solution of system (25) and hence, the stability of the zero solution of system (16) under control (18) and initial conditions satisfying (28). Moreover, the existence of the Lyapunov function of the form (20) satisfying condition (19) ensures the exponential decay of z with rate μ. In order that condition (19) be fulfilled for all functions $\beta(\xi)$, satisfying (38), it is necessary and sufficient that the conditions

$$
\begin{aligned}
PA_1 + A_1^T P + 2\mu P \preceq 0, \\
PA_{k_0} + A_{k_0}^T P + 2\mu P \preceq 0,
\end{aligned}
\tag{43}
$$

be satisfied, where k_0 is defined by the second condition in (38). For a given value of k_0, conditions (43) are fulfilled for a certain matrix $P \succ 0$, provided that for some $0 < \beta \leq k_0$, the linear matrix inequalities (40) in $P \succ 0$ are feasible. By (31) and (38), the condition $0 < \beta \leq k_0$ writes

$$
\beta^2 \leq \frac{u_0^2}{\sigma_0^2}
$$

or, equivalently,

$$
d^T P^{-1} d \leq \frac{u_0^2}{\beta^2}.
$$

Together with the condition $P \succ 0$ and (27), the last inequality means that the matrix (41) is positive semi-definite. In combination with Lemma 1, this assertion leads to the following result. Under condition (28) and the conditions of Theorem 1, the solution of system (25) is exponentially decaying with rate $-\mu$. Moreover, condition (28) holds along the whole trajectory of system (25). This completes the proof of Theorem 1.

5   Numerical Example

To illustrate the proposed method, numerical example was considered. We used CVX, a Matlab package for solving convex optimization problems [4].

As for design of the target path, it is supposed that the optimal field coverage problem was already solved for the particular working site (Fig. 1, left) and target paths for two robots to operate simultaneously were obtained as a result (Fig. 1, right).

The desired paths were constructed with the reasonable limitation on maximum turning radius of the robot. In the following example this limitation is 3 m, which corresponds to the curvature constraint $|\tilde{c}_s(\xi)| < 0.33$.

The formulation of the Theorem 1 implies not unique choice of parameters. Therefore, possibility of the optimal choice arises. Given the $\bar{u} > 0$, $\bar{c} > 0$, $\lambda > 0$, $0 < \mu \leq \lambda$ one needs to find such a matrix P satisfying assumptions of Theorem 1 for which the volume of the ellipsoid $\Omega(P)$ takes the maximum value. The last demand can be formulated in the form of SDP:

$$
\min \operatorname{tr}(P)
$$

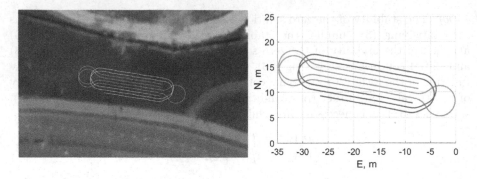

Fig. 1. Left: optimal field coverage for the particular working site. Right: target paths with constrained curvature for two robots operating simultaneously.

under the LMI constraints (40)–(42). Then, we decrease the β satisfying (38) and increase the α_1, α_2 satisfying (27) and check the feasibility of the SDP problem until CVX finds a solution.

To illustrate the behavior of the system (16) closed by the control (18) and to show its trajectories with respect to the estimate of the attraction domain we combine both the phase portrait and the attraction domain. Thus, Fig. 2 shows the attraction domain estimate obtained with the quadratic Lyapunov function based on the formulation of Theorem 1. We set $\bar{u} = 0.5$ and $\bar{c} = 0.33$. The blue ellipsoidal contour line denotes the attraction domain estimate. Green lines show the trajectories of the system (16) for the initial conditions from the boundary of ellipsoid and for the curvature $\tilde{c}_s(\xi)$ taken from previously planned path. The trajectories of the system (16) practically coincide for both target paths shown in the Fig. 1.

The elliptic estimation of the attraction domain is inscribed into the band of a certain width around any state space variables of the system. Here optimal $\alpha_1 = 0.4$, $\alpha_2 = 0.97$ and $\sigma_0 = 0.145$ were achieved for $\mu = 0.1\lambda$ with $\lambda = 0.5$.

The ellipsoid is invariant for the system: once getting inside, the system trajectories are not leave it. Moreover, prescribed exponential convergence rate is guaranteeing for the system.

6   Applications to the Autonomous Robotics

The use of attraction domains in automatic control systems of wheeled robots is introduced in papers [7,8,10] as assistance to operators. On the other hand, the autonomous robots concept is extensively developed today. The autonomy concept excludes the presence of people at all. Therefore, the behavior of robots should be very predictable. There are a lot of potential areas for autonomous robotic operations. For example, small wheeled robots are used for precision farming, golf course lawn mowing, and so on. The common challenge for these areas is to make the robot behavior reliable and fully predictable, including safe and secure operation [3] excluding collisions.

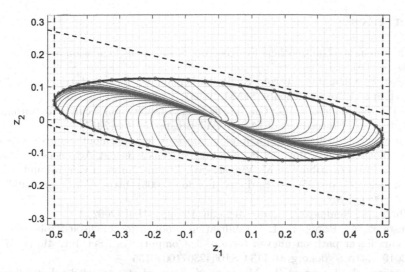

Fig. 2. Attraction domain estimate for given $\bar{u} = 0.5$, $\bar{c} = 0.33$, $\mu = 0.1\lambda$ with $\lambda = 0.5$, $\alpha_1 = 0.5$, $\alpha_2 = 0.97$ and $\sigma_0 = 0.145$.

One can assume, that the robot is delivered first to the neighborhood of the working site and left in arbitrary position near the beginning of the target path. The autonomous control goal is to bring the robot to the desired start of the path and stabilize the motion of a target point along it [10,11,14]. Also, it should be ensured that the control algorithm will not cause a system stability loss. It means, that the robot should get inside the attraction domain of the dynamic system closed by a synthesized control law, thus guaranteeing the asymptotic stability. In addition, even starting sufficiently close to the target path, the robot may perform large oscillations going far enough even if the closed loop system is stable. Thus, the estimate of the attraction domain must be invariant for the closed system and satisfy reasonable geometric constraints.

7   Conclusion

To stabilize the wheeled robot following a curvilinear path over an uneven surface, the control law based on the feedback linearization was used. Using the saturation function for implementation of the constrained control can destroy the exponential stability. To guarantee the exponential stability, a method of attraction domain estimation based on absolute stability approach was proposed. The numerical example illustrates the proposed approach. The results obtained are used in practice for design of the autonomous wheeled robot control system. Of further interest is attraction domain estimation taking into account the internal dynamics of the system (e.g. steering actuator).

References

1. Aizerman, M., Gantmacher, F.: Absolute Stability of Regulation Systems. Holden Day (1964)
2. Boyd, S., Ghaoui, L., Feron, E., Balakrishnan, V.: Linear Matrix Inequalities in System and Control Theory. SIAM (1994)
3. Cui, J., Sabaliauskaite, G.: On the alignment of safety and security for autonomous vehicles. In: Proceedings of IARIA CYBER, pp. 59–64 (2017)
4. Grant, M., Boyd, S.: Graph implementations for nonsmooth convex programs. In: Blondel, V.D., Boyd, S.P., Kimura, H. (eds.) Recent Advances in Learning and Control (a tribute to M. Vidyasagar), Lecture Notes in Control and Information Sciences, pp. 95–110. Springer, London (2008). https://doi.org/10.1007/978-1-84800-155-8_7
5. Khalil, H.: Nonlinear Systems. 3rd edn. Prentice-Hall (2002)
6. Pesterev, A., Rapoport, L.: Stabilization problem for a wheeled robot following a curvilinear path on uneven terrain. J. Comput. Syst. Sci. Int. **49**(4), 672–680 (2010). https://doi.org/10.1134/S1064230710040155
7. Pesterev, A., Rapoport, L., Morozov, Y.: Control of a wheeled robot following a curvilinear path. In: Proceedings of the 6th EUROMECH Nonlinear Dynamics Conference (ENOC 2008), pp. 1–7 (2008)
8. Rapoport, L., Generalov, A.: Lurie systems stability approach for attraction domain estimation in the wheeled robot control problem. In: Olenev, N., Evtushenko, Y., Khachay, M., Malkova, V. (eds.) OPTIMA 2020. LNCS, vol. 12422, pp. 224–238. Springer, Cham (2020). https://doi.org/10.1007/978-3-030-62867-3_17
9. Rapoport, L., Morozov, Y.: Estimation of attraction domains in wheeled robot control using absolute stability approach. In: 17th World Congress the International Federation of Automatic Control, vol. 41, pp. 5903–5908 (2008). https://doi.org/10.1134/S0005117906090062
10. Rapoport, L.: Estimation of attraction domains in wheeled robot control. Autom. Remote Control **67**(9), 1416–1435 (2006). https://doi.org/10.1134/S0005117906090062
11. Rapoport, L.: The periodic solution of two-dimensional linear nonstationary systems and estimation of the attraction domain boundary in the problem of control of a wheeled robot. Autom. Remote Control **72**(11), 2339–2347 (2011). https://doi.org/10.1134/S0005117911110087
12. Tarbouriech, S., Garcia, G., Gomes da Silva, J.J., Queinnec, I.: Stability and Stabilization of Linear Systems with Saturating Actuators. Springer, London (2011). https://doi.org/10.1007/978-0-85729-941-3
13. Tarbouriech, S., Turner, M.: Anti-windup design: an overview of some recent advances and open problems. IET Control Theory Appl. **3**(1), 1–19 (2009). https://doi.org/10.1049/IET-CTA:20070435
14. Thuilot, B., Cariou, C., Martinet, P., Berducat, M.: Automatic guidance of a farm tractor relying on a single CP-DGPS. Auton. Robots **13**(1), 53–71 (2002). https://doi.org/10.1023/A:1015678121948
15. Yakubovich, V.A., Leonov, G.A., Gelig, A.K.: Stability of Stationary Sets in Control Systems with Discontinuous Nonlinearities. World Scientific, Series on Stability, Vibration and Control of Systems (2004)

Optimizing Coefficients of a Controller in the Point Stabilization Problem for a Robot-Wheel

Alexander Pesterev$^{(\boxtimes)}$ ⓘ and Yury Morozov ⓘ

Institute of Control Sciences, Moscow 117997, Russia

Abstract. The problem of stabilizing a robot-wheel at a target point on a straight line subject to control and phase constraints is considered. The phase and control constraints are met by applying an advanced feedback law in the form of nested saturation functions. The selection of the feedback coefficients is discussed that optimizes the performance of the controller. An optimal controller is defined to be that that ensures the greatest convergence rate near the target point, while preserving a node-like phase portrait of the nonlinear system. The paper continues the work reported at the Optima 2020 conference [1], where an estimate of the greatest rate was obtained. The goal of this paper is to improve the results obtained in that work by considering a curvilinear asymptote and to get the exact value of the greatest rate.

Keywords: Robot-wheel · Optimal feedback coefficients · Point stabilization problem · Phase and control constraints · Nested saturators

1 Introduction

This study is a sequel of the work [1] devoted to optimizing a controller stabilizing a wheel at a point. The problem of a wheel rolling on a plane or an uneven terrain is of importance in many practical applications. A rising tide of interest to this classical problem is due to appearance of robotic systems of a new type—ball-shaped or spherical robots and robot–wheels—and search for new actuators for such systems [2–6]. The problem of motion control for mobile robots of this type that move owing to displacements of masses (pendulums) inside the shell (wheel) is discussed in many publications (see, for example, [2,4,6,7]). In this paper, we consider the simplest model of a robot-wheel assuming that it is driven by a control torque applied to the wheel axis. We do not go into detail of implementation of the actuator assuming only that the control torque is constrained, with the limit value being determined by physical parameters of the robot [2,7]. On the one hand, such a model, in spite of its simplicity, is of interest by itself in the study of advanced control strategies, including optimal ones. On the other hand, this model can be used as a reference one, in studying more complicated

© Springer Nature Switzerland AG 2021
N. N. Olenev et al. (Eds.): OPTIMA 2021, LNCS 13078, pp. 191–202, 2021.
https://doi.org/10.1007/978-3-030-91059-4_14

models, with the solutions obtained for the reference model being taken to be a set of target trajectories for the original system [8].

We set the problem of synthesizing a control law in the form of feedback that brings the wheel from an arbitrary initial position on a straight line to a given one, with the velocity of motion being limited. To meet the phase and control constraints, an advanced feedback law in the form of nested saturation functions depending on four coefficients was suggested in [1]. Feedback laws of this type were studied in [9,10]. The basic advantage of such laws is that they ensure global stability of the closed-loop system and guarantee the fulfilment of the phase and control constraints under appropriate choice of feedback coefficients.

Two of the four feedback coefficients are uniquely determined by the limit value of the control torque and the maximum allowed wheel velocity, while the selection of the other two coefficients can be used to optimize the performance of the controller. The optimality criterion employed in this study, as well as in [1], is similar to that in [11], where the selection of feedback coefficients of a saturated linearizing feedback for a wheeled robot with constrained control resource was discussed. The optimality is meant in the sense that the phase portrait of the nonlinear closed-loop system is similar to that of a linear system with a stable node, with the asymptotic rate of approaching the target point being as high as possible. The problem statement in this study differs from that in [1] by the definition of the concept of the node-like phase portrait. While in [1] it was defined only for the domain of the phase plane satisfying the phase constraints and the asymptote dividing the domain into two invariant sets was assumed straight, in this work, the definition is extended to the entire phase plane and the asymptote is allowed to be curvilinear. The optimal value of the asymptotic convergence rate in terms of the new definition to be derived in this work is considerably greater than that in terms of the definition introduced in [1].

The paper is organized as follows. In Sect. 2, the wheel stabilization problem statement is given, the governing equations are reduced to a dimensionless form, and some earlier obtained results from [1] are presented. The optimization problem statement is formulated in Sect. 3, and the solution of the optimization problem is presented in Sect. 4. Section 5 summarizes the results of the study and discusses prospects for future research.

2   Stabilization Problem Statement

We consider a wheel rolling without slipping on a plane along a straight line (Fig. 1). The dynamics of the wheel are described by the equation [1]

$$M\ddot{x} = R, \ Mr^2\ddot{\theta} = rR - f\dot{\theta} - U,$$

where M and r are mass and radius of the wheel, x is the coordinate of the wheel center, θ is the rotation angle, R is the reaction force, f is the viscous friction coefficient, and U is the control torque.

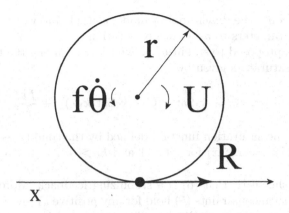

Fig. 1. Schematic of the robot-wheel.

Applying the condition of rolling without slipping $\dot{x} + r\dot{\theta} = 0$, we reduce the system equations to one second-order equation

$$\mu\ddot{x} = -\frac{f\dot{x}}{r^2} + \frac{U}{r}, \tag{1}$$

where $\mu = 2M$. In the point stabilization problem, it is required to synthesize a control law U in the form of a feedback that brings the wheel to a given target point on the line. Without loss of generality, we set the target point to be at the origin. The control torque U is assumed to be limited, and we also assume that the velocity of the wheel center cannot exceed a prescribed value:

$$|U| < U_{max}, \; |\dot{x}| \leq V_{max}. \tag{2}$$

The problem is further simplified by going to dimensionless form. Indeed, by introducing the dimensionless time, coordinate, and control

$$\tilde{t} = tV_{max}/r, \; \tilde{x} = x/r, \; \tilde{U} = U/U_{max}, \tag{3}$$

as well as dimensionless parameters

$$\tilde{\mu} = \frac{\mu V_{max}^2}{U_{max}}, \; \tilde{f} = \frac{fV_{max}}{rU_{max}},$$

using the dot notation for the derivatives with respect to the new time, and assuming that $U_{max} - fV_{max}/r > 0$ (see [1] for detail), Eq. (1) turns to the dimensionless form:

$$\tilde{\mu}\ddot{\tilde{x}} = -\tilde{f}\dot{\tilde{x}} + \tilde{U}, \tag{4}$$

where $0 \leq \tilde{f} < 1$, with constraints (2) taking the form

$$|\tilde{U}| \leq 1, \; |\dot{\tilde{x}}| \leq 1. \tag{5}$$

In what follows, only the dimensionless model is used, and we omit tilde over all variables and parameters to avoid messy notation.

In [1], it was proposed to stabilize the wheel by applying the feedback in the form of nested saturators given by

$$U(x, \dot{x}) = -k_4 \text{Sat}(k_3(\dot{x} + k_2\text{Sat}(k_1x))) + \frac{f\dot{x}}{r}, \qquad (6)$$

where $\text{Sat}(x)$ is the saturation function defined by the conditions $\text{Sat}(x) = x$ for $|x| \le 1$ and $\text{Sat}(x) = \text{sign}(x)$ for $|x| > 1$ and $k_i > 0$, $i = 1, 2, 3, 4$, are positive coefficients.

It has been shown [1] that (6) is a stabilizing feedback. Moreover, if $k_2 = 1$ and $k_4 = 1 - f$, then constraints (5) hold for any positive k_1 and k_3. Substituting (6) into (4) with the above-specified coefficients k_2 and k_4, we get the following equation governing the closed-loop system:

$$\ddot{x} = -\eta\text{Sat}(k_3(\dot{x} + \text{Sat}(k_1x))). \qquad (7)$$

where $\eta = (1 - f)/\mu$ is the control resource per unit mass.

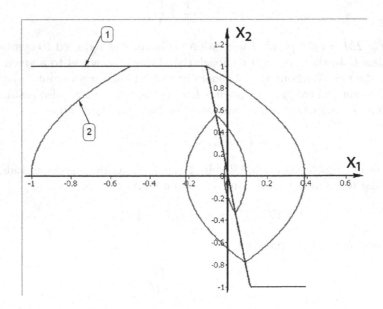

Fig. 2. An example of inappropriate selection of feedback coefficients in (7).

Although feedback (6) with the coefficients $k_2 = 1$ and $k_4 = 1 - f$ stabilizes the system and respects the constraints, inappropriate selection of the other two coefficients can result in poor performance of the control system and great overshooting. Figure 2 illustrates this. It shows a phase trajectory (curve 2) of the wheel with $\mu = 1$ and $f = 0$. Because of inappropriate selection of the fecdback coefficients (here, $k_1 = 9$ and $k_3 = 100$), the wheel missed the target point several times, with the overshootings being quite large. The phase portrait of the system in this case reminds that of a focus, with the overshootings being quite large, which does not sound good. The broken blue line (marked by 1) shows the curve $x_2 + \text{Sat}(k_1 x_1) = 0$. Hence, it follows that the freedom in selection of k_1 and k_3 can be employed to optimize the performance of the controller, which is discussed in the remainder of the paper.

3 Optimization Problem Statement

Intuitively, speaking of desirable behavior, we want to have fast asymptotic convergence to the origin in the time domain and the phase portrait of the nonlinear system to look like that of a linear system with a node, when any trajectory approaches the origin monotonically, or has at most one overshooting. Recall that, in the linear case, the phase plane is divided into two invariant half-planes by a straight line, which is the asymptote for all (but two if the node is not a degenerate one) phase trajectories of the system. The concept of a node-like phase portrait for a nonlinear system can formally be defined in terms of a *curvilinear asymptote* dividing the phase plane into two invariant sets, which is a generalization of the straight asymptote for a linear system.

Definition 1. *We will say that the phase portrait of a nonlinear system is of the node-like type if there is a curvilinear asymptote lying completely in the second and fourth quadrants.*

The property of being node-like defined above is a global one. It means that not only the origin is a node of the linearized system but also that the behavior of the phase trajectories in the entire phase plane is similar to the behavior of the phase trajectories of a linear system. Likc the straight asymptote in the linear casc, the curvilinear asymptote divides the phase plane into two invariant sets such that any phase trajectory passes through only two quadrants of the phase plane.

Now, the problem to be solved in this study can be formulated as follows.

Problem. *Determine feedback coefficients k_1 and k_3 for which the asymptotic rate of approaching the target point is maximal under the condition that the phase portrait of system (7) is of the node-like type.*

4 Solution of the Optimization Problem

First, we establish the general form a curvilinear asymptote (further, simply *asymptote*) for system (7) and, then, will determine under what conditions the asymptote passes only through the second and forth quadrants.

Let us introduce the notation $x_1 = x$ and $x_2 = \dot{x}$ and rewrite (7) in the state-space form as

$$\begin{aligned} \dot{x}_1 &= x_2 \\ \dot{x}_2 &= -\eta \text{Sat}(k_3(x_2 + \text{Sat}(k_1 x_1))). \end{aligned} \tag{8}$$

It is easy to see that the closed-loop system (8) is piecewise linear. Figure 3 shows the partitioning of the phase plane. Here, the dashed lines depict the boundaries between different linearity regions where one linear system switches to another. The solid broken line

$$x_2 + \text{Sat}(k_1 x_1) = 0 \tag{9}$$

is the set of points where the right-hand side of the second equation in (7) vanishes. The control reaches saturation outside the broken strip bounded by the two dashed lines parallel to (9).

In the intersection of the sets $|x_1| \leq 1/k_1$ and $|x_2 + k_1 x_1| \leq 1/k_3$, which includes the origin, Eq. (7) takes the form

$$\ddot{x} + \eta k_3 \dot{x} + \eta k_1 k_3 x = 0. \tag{10}$$

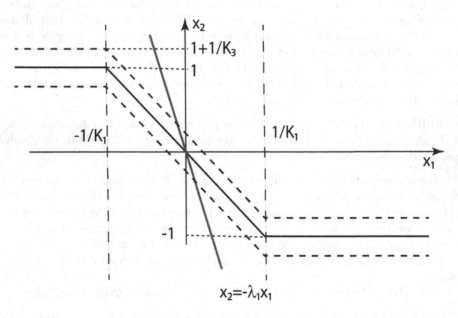

Fig. 3. Partition of the phase plane for system (8).

To simplify the following calculations, we confine our consideration in this paper to the case of a degenerate node (repeated root of the characteristic equation) of the linearized system, which is governed by the equation

$$\ddot{x} + 2\lambda \dot{x} + \lambda^2 x = 0, \ \lambda > 0, \tag{11}$$

where λ is the rate of the asymptotic convergence. Comparing (10) and (11), we find that the coefficients k_1 and k_3 are to be selected from the one-parameter family

$$k_1 = \frac{\lambda}{2}, \ k_3 = \frac{2\lambda}{\eta} \tag{12}$$

parameterized by the exponent λ, and will seek for the maximal λ for which the phase portrait of system (7) is of the node-like type.

Clearly, being a curve dividing the phase plane into two invariant sets, any asymptote must be an integral curve of the system [12]. It was proved in [1, Lemma 1] that any trajectory of equation (7) beginning in the strip $|x_2| \leq 1$ never leaves it, i.e., the strip is an invariant set of the system. It is easy to prove that any trajectory beginning outside the strip cannot intersect the horizontal segment of line (7) either. This follows from the facts that the horizontal segments of line (9) are negative half-trajectories with the initial points $(-1/k_1, 1)$ and $(1/k_1, -1)$, respectively, and that no trajectories can intersect [12]. Indeed, let $x_1(0) = -1/k_1, x_2(0) = 1$. Since the right-hand side of the second equation in (7) is zero, $x_2(t) \equiv 1$. Then, by virtue of the first equation, $x(t) = x_1(0) + t < 0$ and, for $t \leq 0$, the negative half-trajectory beginning at the point $(-1/k_1, 1)$ is the left horizontal segment of line (9). Similarly, it is proved that the right horizontal segment is the negative half-trajectory beginning at the point $(1/k_1, -1)$. Note also that the positive half-trajectories of the system beginning at the same points asymptotically approach the origin by virtue of the fact that the origin is the equilibrium point of the system. This brings us at the following lemma.

Lemma 1. *The asymptote of system (7) is an integral curve consisting of the singular phase trajectory (equilibrium point) $x(t) \equiv 0$ and two pairs of the half-trajectories beginning at the points $(-1/k_1, 1)$ and $(1/k_1, -1)$.*

Thus, solving the Problem reduces to finding the maximal exponent λ for which the positive half-trajectories beginning at the points $(-1/k_1, 1)$ and $(1/k_1, -1)$ completely lie in the second and fourth quadrants. Taking into account the symmetry of the phase portrait with respect to the origin, it will suffice to consider only one of these half-trajectories, say, that beginning in the fourth quadrant.

In [1], the estimate $\tilde{\lambda} = \eta$ for the maximal λ was obtained by seeking for a straight asymptote that divides the strip $|x_2| \leq 1$ into two invariant sets. Further in this section, we will show that, allowing the asymptote to be curvilinear, an exact value of maximal λ can be obtained, which is considerably greater than the estimate from [1]. Moreover, the curvilinear asymptote is shown to divide the entire phase plane, rather than the strip, into two invariant sets.

In view of the system symmetry, we may confine our consideration to the trajectories beginning in the left half-plane. It is evident that the positive half-trajectory beginning in the corner of the broken line (9) completely lies in the fourth quadrant if and only if it does not intersect the straight asymptote $x_2 = -\lambda x_1$ of the linear system (10). The latter may hold in the following two cases. First, this obviously happens when the trajectory does not intersect the dotted line $x_2 + k_1 x_1 = 1/k_3$ (i.e., when the control does not reach saturation). The other case takes place when the system does reach saturation but the trajectory still does not intersect the straight asymptote. Whether the second case is possible will further be verified.

Consider the first case. Solution of the linear equation (11) is given by

$$x_1(t) = -\frac{1}{\lambda}(\lambda t + 2)\exp(-\lambda t), \quad x_2(t) = (\lambda t + 1)\exp(-\lambda t), \tag{13}$$

from which it follows that

$$\frac{x_1}{x_2} = -\frac{1}{\lambda}\frac{\lambda t + 2}{\lambda t + 1}. \tag{14}$$

Equation (11) in the half-plane $x_2 > 0$ can be written as

$$\frac{dx_2}{dx_1} = -2\lambda - \lambda^2\frac{x_1}{x_2} = -2\lambda - \lambda\frac{\lambda t + 2}{\lambda t + 1} = -\frac{\lambda^2 t}{\lambda t + 1}.$$

Taking into account that $k_1 = \lambda/2$, the slope of the saturation (switching) line $x_2 + k_1 x_1 = 1/k_3$ is $-\lambda/2$.

Let us find t_* for which the trajectory (13) has the same slope,

$$-\frac{\lambda^2 t}{\lambda t + 1} = -\frac{\lambda}{2}.$$

This yields $\lambda t_* = 1$ and $t_* = 1/\lambda$; the corresponding trajectory point is given by $x_1(t_*) = -3e^{-1}/\lambda$, $x_2(t_*) = 2e^{-1}$. Substituting these into the equation of the saturation line, we get

$$2e^{-1} - \frac{3}{2}e^{-1} = \frac{\eta}{2\lambda},$$

from which it follows that $\lambda_* = \eta e$. The estimate obtained is by e times greater than the estimate obtained with the help of a straight asymptote in [1]. The coordinates of the touching point are $(-3e^{-2}/\eta, 2e^{-1})$.

Thus, in the considered case, the desired asymptote is defined parametrically by Eq. (13) for $\lambda = \eta e$ as t varies from 0 to ∞. On the asymptote, the system is linear and the control reaches saturation at the single point.

Now, let us check whether the exponent λ can be increased if we permit saturation on the positive half-trajectory. After intersecting the saturation line, system (8) turns to

$$\dot{x}_1 = x_2, \quad \dot{x}_2 = -\eta.$$

Rewriting these equations as

$$\frac{dx_1}{dx_2} = -\frac{x_2}{\eta},$$

and integrating the resulting equation, we find that the system trajectory is the parabola

$$x_1(t) = -\frac{1}{2\eta}x_2^2(t) + C(x_{10}, x_{20}), \tag{15}$$

where x_{10} and x_{20} are the coordinates of the point where the trajectory intersects the saturation line and

$$C(x_{10}, x_{20}) = x_{10} + \frac{x_{20}^2}{2\eta} = \frac{1}{8\eta}\left(\lambda^2 x_{10}^2 + 6x_{10}\eta + \frac{\eta^2}{\lambda^2}\right).$$

Equating the slope of the parabola to that of the straight asymptote, we find that the tangent line to the parabola is parallel to the asymptote when $x_2 = \eta/\lambda$. The condition that the parabola touches the asymptote is that it passes through the point with the coordinates $x_1^{**} = -\eta/\lambda^2$, $x_2^{**} = \eta/\lambda$. Note also that it is at this point where the asymptote and the saturation line intersect. Substituting these into the parabola equation, we obtain

$$-\frac{\eta}{\lambda^2} = -\frac{\eta}{2\lambda^2} + C$$

from which it follows that $C = -\eta/2\lambda^2$. Equating the two expressions for C, we get the following second-order algebraic equation in x_{10}:

$$\frac{\lambda^2 x_{10}^2}{\eta^2} + \frac{6x_{10}}{\eta} + \frac{5}{\lambda^2} = 0.$$

Two solutions of this equation are $-5\eta/\lambda^2$ and $-\eta/\lambda^2$. The former is the abscissa of the first intersection point where the trajectory leaves the strip and the control reaches saturation, and the latter is the abscissa of the second point where the trajectory enters again the strip.

The above implies that, in order that the trajectory return to the strip at the right point (x_1^{**}, x_2^{**}), the first intersection with the saturation line must be at the point with the coordinates $x_1^* = -5\eta/\lambda^2$, $x_2^* = 3\eta/\lambda$. The value of λ and the corresponding time t^* are found by equating solutions (13) at $t = t^*$ to the coordinates obtained

$$-\frac{1}{\lambda}(\lambda t^* + 2)e^{-\lambda t^*} = -\frac{5\eta}{\lambda^2}$$

and

$$(\lambda t^* + 1)e^{-\lambda t^*} = \frac{3\eta}{\lambda}.$$

Dividing the first equation by the second one and solving the equation obtained, we get

$$\lambda t^* = \frac{1}{2}, \ \lambda = 2\eta\sqrt{e}.$$

As can be seen, the exponential rate of approaching the origin obtained is by $2/\sqrt{e} \approx 1.2$ times greater than that in the previous case and is by $2\sqrt{e} \approx 3.3$ times greater than the estimate obtained in [1].

Moreover, this is the exact value of the maximal rate λ: $\lambda_{max} = 2\eta\sqrt{e}$. Indeed, for any $\lambda > \lambda_{max}$, the positive half-trajectory emerging from the corner of the broken line (9) necessarily intersects the straight asymptote and, being the trajectory of the linear system (10), will intersect the x_2-axis and enter the first quadrant.

The above results are summarized in the following theorem.

Theorem 1. *The greatest exponential rate λ of the deviation x decrease for which the phase portrait of the nonlinear system (8) is of the node-like type is*

$$\lambda_{max} = 2\eta\sqrt{e}.$$

The corresponding coefficients k_1 and k_3 are given by

$$k_1 = \eta\sqrt{e}, \quad k_3 = 4\sqrt{e}. \tag{16}$$

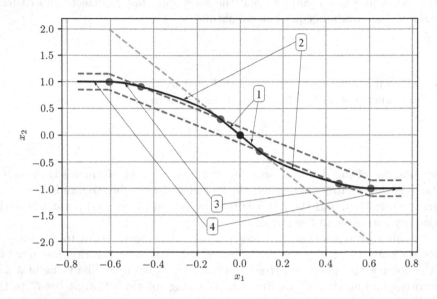

Fig. 4. Optimal asymptote for system (8) with $\mu = 1$ and $f = 0$.

The optimal curvilinear asymptote for the system with $\mu = 1$ and $f = 0$ (and, hence, $\eta = 1$), corresponding to the optimal value of λ is depicted in Fig. 4 by the bold black curve. For this system, $\lambda_{max} = 2\sqrt{e}$, $k_1 = \sqrt{e}$, and $k_3 = 4\sqrt{e}$. The two broken dashed lines in the figure are boundaries of the region where the control does not reach saturation. The straight dashed line is the straight asymptote $x_2 = -\lambda_{max}x_1$ of the linear system (10).

Let us describe the part of the asymptote lying in the fourth quadrant. As noted earlier, it consists of the negative and positive half-trajectories beginning at the point $(-1/k_1, 1)$. The former is the straight line (marked by 4) given parametrically by $x_1(t) = -1/k_1 + t$, $x_2(t) \equiv 1$, $-\infty < t \le 0$. The latter, in turn, consists of the three segments: the first segment (curve 3) is the trajectory of the linear equation (11) given by (13), where $0 \le t < t^* = 1/2\lambda$; the second segment (curve 2) is a piece of parabola (15), $t^* \le t < t^{**} = 5/2\lambda$; and the third segment (line 1) is a piece of the straight asymptote of (11), $t^{**} \le t < \infty$. The other part of the asymptote in the second quadrant is symmetric to this one with respect to the origin.

Figure 5 shows the phase portrait of system (8) with $\mu = 1$ and $f = 0.25$ ($\eta = 0.75$) for the optimal value of $\lambda = \lambda_{max}$. The black bold line is the asymptote of the system. The green broken lines are the boundaries of the region where the control is not saturated. As can seen, any trajectory beginning below (above)

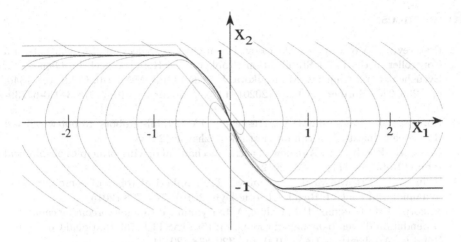

Fig. 5. Phase portrait of system (8) with $\mu = 1$ and $f = 0.25$.

the asymptote completely lies below (above) the asymptote, and any trajectory intersects the x_2-axis at most once. If we further increase λ, the asymptote will intersect the x_2-axis and will pass through all quadrants, which means that the property of the phase portrait being node-like will be violated.

5   Conclusions

In the paper, the problem of optimizing a controller stabilizing a robot-wheel at a target point on a straight line subject to phase and control constraints has been discussed. The controller implementing an advanced feedback law in the form of nested saturation functions was suggested in [1]. The feedback depends on four coefficients two of which ensure the fulfillment of the phase and control constraints, while the other two can be adjusted to optimize the performance of the controller. An optimal controller has been defined to be that that ensures the greatest convergence rate near the target point, while preserving a node-like phase portrait of the nonlinear system. Optimal values of the feedback coefficients have been found, and the corresponding asymptote dividing the phase plane into two invariant sets has been constructed. The use of the new definition of the node-like phase portrait relying on the concept of a curvilinear asymptote made it possible to get a greater value of the asymptotic convergence rate near the target point compared to that in [1].

In the future, we plan to apply the approach developed in this paper to optimizing coefficients of a controller for a more complicated system of a robot–wheel with a pendulum. We also plan to synthesize a hybrid control law where the selection of the feedback coefficients will depend on whether the system is in the neighborhood of the target point or far from it.

References

1. Pesterev, A., Morozov, Y., Matrosov, I.: On Optimal Selection of Coefficients of a Controller in the Point Stabilization Problem for a Robot-Wheel. In: Olenev, N., Evtushenko, Y., Khachay, M., Malkova, V. (eds.) OPTIMA 2020. CCIS, vol. 1340, pp. 236–249. Springer, Cham (2020). https://doi.org/10.1007/978-3-030-65739-0_18

2. Borisov, A.V., Pavlovskii, D.V., Treshchev, D.V.: Mobile Robots: Robot-wheel and Robot-ball. Institute of computer studies, Izhevsk (2013)

3. Chase, R., Pandya, A.: A review of active mechanical driving principles of spherical robots. Robotics **1**(1), 3–23 (2012)

4. Kilin, A.A., Pivovarova, E.N., Ivanova, T.B.: Spherical robot of combined type: dynamics and control. Regul. Chaotic Dyn. **20**(6), 716–728 (2015)

5. Ylikorpi, T.J., Forsman, P.J., Halme, A.J.: Dynamic obstacle overcoming capability of pendulum driven ball-shaped robots. In: 17th IASTED International Conference Robotics Applications (RA 2014), pp. 329–338 (2014)

6. Bai, Y., Svinin, M., Yamamoto, M.: Dynamics-based motion planning for a pendulum-actuated spherical rolling robot. Regul. Chaotic Dyn. **23**(4), 372–388 (2018). https://doi.org/10.1134/S1560354718040020

7. Chernous'ko, F.L., Ananievski, I.M., Reshmin, S.A.: Control of Nonlinear Dynamical Systems: Methods and Applications. Springer, Heidelberg (2008)

8. Matrosov, I.V., Morozov, Y.V., Pesterev, A.V.: Control of the robot-wheel with a pendulum. In: Proceedings of the 2020 International Conference Stability and Oscillations of Nonlinear Control Systems (Pyatntitskiy's Conference), pp. 1–4. IEEE Xplore, Piscataway, NJ (2020)

9. Saberi, A., Lin, Z., Teel, A.: Control of linear systems with saturating actuators. IEEE Trans. Aut. Contr. **41**(3), 368–378 (1996)

10. Olfati-Saber, R.: Global stabilization of a flat underactuated system: the inertia wheel pendulum. IEEE Conf. Decis. Conf. **4**, 3764–3765 (2001)

11. Pesterev, A.V.: Synthesis of a stabilizing feedback for a wheeled robot with constrained control resource. Autom. Remote Control **77**(4), 578–593 (2016). https://doi.org/10.1134/S0005117916040044

12. Andronov, A.A., Vitt, A.A., Haikin, S.E.: Teoriya kolebanii. Fizmatgiz, Moscow (1959)

Algorithm for the Numerical Solution of Optimal Control Problems in Robotic Systems

Pavel Sorokovikov[1]([⊠]) [ID], Alexander Gornov[1] [ID], and Alexander Strelnikov[2] [ID]

[1] Matrosov Institute for System Dynamics and Control Theory of SB RAS,
Lermontov Street, 134, 664033 Irkutsk, Russia
gornov@icc.ru
[2] Irkutsk National Research Technical University,
Lermontov Street, 83, 664074 Irkutsk, Russia
strelnikov077@rambler.ru
http://idstu.irk.ru/en
http://www.eng.istu.edu/

Abstract. The paper discusses the algorithm for the numerical solution of applied optimal control problems in robotics. The proposed algorithm is the Powell method modification, which uses the combined one-dimensional nonlocal search algorithm developed by the authors based on the Strongin and parabolas methods as an auxiliary one. The developed algorithm is implemented in C language and integrated within a single software package. The obtained using the proposed algorithm solutions of two problems are presented: the problem of the optimal control of the mobile robot and the task of the industrial robot arm control. All the obtained solutions found a meaningful interpretation.

Keywords: Optimal control · Global optimization · Numerical methods · Robotics

1 Introduction

The problems of optimal control (optimization of dynamic systems, trajectory optimization) everywhere arise in the consideration and investigation of mechanical systems. One of the recent and relevant classes of trajectory optimization problems is applied tasks from the field of robot control. The optimal control problems in robotic systems have distinct specific features. The most important of them is the indispensable presence of constraints on the trajectory of the controlled system imposed both at finite time instants (terminal restrictions) and throughout the entire time interval (phase, mixed, interval constraints). Many problems of this class are optimal performance tasks, which consist of minimizing the time of transferring a system from one point to another, are non-convex and multi-extreme.

Supported by Russian Foundation for Basic Research, No 19-37-90065.

The finding of a global extremum in non-convex optimal control tasks remains one of the most troublesome optimization problems and, to date, does not have a satisfying solution. The development of theoretical approaches to the investigation of linear problems was carried out by many specialists (R. Bellman, V.A. Baturin, F.L. Chernousko, V.A. Dykhta, Yu.G. Evtushenko, R.P. Fedorenko, C.A. Floudas, C.J. Goh, V.I. Gurman, V.F. Krotov, I.L. Lopez Cruz, E. Polak, A. Schwartz, V.A. Srochko, A.S. Strekalovsky, K.L. Teo, K.H. Wong, et al.). A significant number of publications are devoted to the development of numerical methods for optimizing dynamical systems. Some of them (for example, [4, 8, 9, 14, 25]) focus on the creation of algorithms established on the optimal control theory – the Pontryagin maximum principle, sufficient optimality conditions, et al. But no approaches have been found that guarantee a globally optimal solution for nonlinear systems, based on which it is possible to construct efficiently working algorithms. According to the authors, to date, no theoretical results have been obtained that could become the ideological basis for building guaranteed algorithms for solving nonlinear optimal control problems. The authors of some works (for instance, [1, 2, 17, 20, 22, 23]) proposed to reduce optimal control problems to finite-dimensional optimization tasks and use developed software for mathematical programming. All currently known global extremum search algorithms in nonlinear optimal control problems should be considered heuristic. To date, for example, the following globalized heuristic algorithms for searching the extremum in nonlinear optimal control problems are known: genetic algorithms [5], random multistart method [10], convexity method [24], algorithm of stochastic approximations of the reachable set [11], "curvilinear search" method [12], "stochastic coverings" algorithm [13], et al.

We can argue that in any software implementation of the algorithms, hypotheses are implicitly laid that do not guarantee the finding of a global extremum (we can only talk about increasing the probability of its obtaining). At present, it seems that it is impossible to guarantee a global solution to nonlinear optimal control problems even if there is a reliable theoretical basis. Therefore, it is advisable to use other methodological approaches to the study of non-convex problems. Consequently, the paper proposes the use of "presentation logic": we consider a problem temporarily successful if the algorithm can find an extremum with a sufficiently "good" value of the objective functional. Further, the result obtained is published and is considered a plausible solution to the problem until the "moment of refutation" (the presentation of a better solution by someone). A similar approach has long been used in finite-dimensional optimization (in mathematical programming) when creating collections of test problems (the principle of "best of known").

As you know, the basis of any successful search strategy is the balance between global scanning of the feasible set and local refinement of the approximations obtained. Therefore, numerical methods for solving applied optimal control problems should include separate "global" and "local" stages, oriented accordingly to find an approximation to the solution in the whole space understudy and local refinement of the result obtained at the first stage [29]. The combination of various stages in the algorithms and the order of their alternation determine a specific computational scheme.

A well-known method for finding the global extremum for the optimal control problem is the dynamic programming method (see, for example, [6,16]). Unfortunately, no other methods have been published that would guarantee a global solution for nonlinear systems. The main drawback of the dynamic programming method is the "curse of dimensionality" – its complexity increases dramatically with the increase in the dimension of the problem. Therefore, this method is suitable only for optimization problems of small dimensions. At least 100 sampling nodes (finite-dimensional variables) are usually required to construct adequate approximations of optimal control problems, which doesn't allow using the dynamic programming method in practice. A way out of this situation is the idea of constructing specialized optimization algorithms that take into account the specifics of the optimal control problem as an extremal problem.

The paper proposes a heuristic algorithm for solving non-convex optimal control problems, which is much lighter computationally than the dynamic programming method. It is based on the Powell method and a combined one-dimensional optimization algorithm established on the Strongin and "parabolas" methods. At iterations of the algorithm under consideration, we constructed conjugate descent directions by solving numerous auxiliary problems of finding the global minimum of a univariate function.

2   Statement of the Optimal Control Problem

We described the controlled process by a system of ordinary differential equations with initial conditions

$$\dot{x} = f\left(x\left(t\right), u\left(t\right), t\right), \ \ x\left(t_0\right) = x^0, \tag{1}$$

where t is the time from the interval $[t_0, t_1]$, $x\left(t\right) = \left(x_1\left(t\right), x_2\left(t\right), ..., x_n\left(t\right)\right)$ is the vector of phase coordinates, $u\left(t\right) = \left(u_1\left(t\right), u_2\left(t\right), ..., u_r\left(t\right)\right)$ is the vector of control actions. A vector function $f\left(x\left(t\right), u\left(t\right), t\right)$ is assumed to be continuously differentiable for all arguments except t. Admissible functions are piecewise-continuous control functions $u(t)$ for any time values t belonging to the set U, where

$$U = \left\{u\left(t\right) \in \mathbb{R}^r : \ u_l \leq u\left(t\right) \leq u_g\right\}, \tag{2}$$

$u_l, u_g \in \mathbb{R}^r$ are vectors of lower and upper control constraints. The optimal control problem in the standard-setting is to find an admissible control $u^*\left(t\right)$ that delivers a minimum to the terminal functional

$$I_0\left(u\right) = \varphi_0\left(x\left(t_1\right)\right) \to \min. \tag{3}$$

There are also terminal restrictions

$$I_j\left(u\right) = \varphi_j\left(x\left(t_1\right)\right) = \left(\leq\right)0, \ j = \overline{1, m} \tag{4}$$

and inequality type phase constraints

$$I_j(u) = g_j(x(t), u(t), t) \leq 0, \ j = \overline{m+1, mt}. \tag{5}$$

All functions $\varphi_j(x(t_1))$, $j = \overline{0, m}$ and $g_j(x(t), u(t), t)$, $j = \overline{m+1, mt}$ are assumed to be continuously differentiable for all arguments.

3   Consideration of Phase Constraints

The complexity of problems with phase constraints is associated with the inertia property of a dynamic system, and the inertia of robotic systems is usually quite high. Currently, the following methods of consideration of phase constraints are known [10]: algorithm of external penalty functionals, method of modified Lagrange functionals, parameterization of constraints algorithm, reduced gradient method, linearization algorithm.

In this work, when solving problems, the method of external penalty functionals was used since, by its simplicity, it does not entail any unpredictable computational effects. We reduced all phase constraints to the terminal one $x_{n+1}(t_0) = 0$ by introducing cubic penalty functionals, which made it possible to preserve the continuity property of the second derivatives, as follows:

$$\dot{x}_{n+1} = K \sum_{i=1}^{n+1} \theta(g_i(x)) g_i^3(x), \tag{6}$$

where $\theta(A) = \begin{cases} 1, \ A > 0, \\ 0, \ A \leq 0 \end{cases}$ is the Heaviside function, $K \to \infty$ is the penalty

parameter. After this transformation, the problem assumed the standard form of the optimal control problem with terminal constraints [10]. Thus, the total functional has the following structure: $\bar{I}(u) = I_0(u) + x_{n+1}(t_1) \to \min$.

4   Algorithm for the Numerical Solution of Problems

For the numerical solution of optimal control problems, the discretization of the system of differential equations and approximate methods for solving the Cauchy problem are used. We divided the time change segment into $n_u - 1$ parts and constructed a uniform grid, in the nodes of which the controls and trajectories are stored (n_u is the number of sampling points). The type of control approximation is piecewise linear.

Powell method [15] is a local search algorithm in which, due to the solution of numerous univariate problems on iterations, we constructed the conjugate descent directions.

4.1   Powell Algorithm

1. Choose $u^0(t)$, $t \in T$.
2. Set parameters of the algorithm: $\Delta\tau$ is the sampling rate by τ and t; K_{upd} is the update frequency.
3. Assume $P^0(\tau, t) = 0$, if $\tau \neq t$; $P^0(\tau, t) = 1$, if $\tau = t$; $\tau \in T$, $t \in T$.
   On the k-th iteration $(k \geq 0)$:
4. Perform conversion of variables:
   (a) Assume $d^k(t) = \frac{2(u^k(t) - 0.5(u_l + u_g))}{u_g - u_l}$, $t \in T$.
   (b) To avoid the rounding errors effect, project $d^k(t)$ onto a valid area:
       if $d^k(t) > 1$, assume $d^k(t) = 1$, $t \in T$;
       if $d^k(t) < -1$, assume $d^k(t) = -1$, $t \in T$.
   (c) Calculate the converted control $w^k(t) = \arcsin\left(d^k(t)\right)$, $t \in T$.
5. If k is divisible by K_{upd}, make an update: assume $P^k(\tau, t) = 0$, if $\tau \neq t$; $P^k(\tau, t) = 1$, if $\tau = t$; $\tau \in T$, $t \in T$.
6. Assume $y^k(0, t) = w^k(t)$, $t \in T$.
7. In the cycle of τ with the step $\Delta\tau$, $\tau \in T$:
   (a) Assume $h^k(t) = P^k(\tau, t)$, $t \in T$.
   (b) Search $y^k(\tau + \Delta\tau, t) = \arg\min\left\{I_0\left(y^k(\tau, t) + \alpha h^k(t)\right), -\infty < \alpha < \infty\right\}$.
   (c) Assume $s^k(t) = y^k(\tau + \Delta\tau, t) - w^k(t)$, $t \in T$.
   (d) Assume $P^{k+1}(\tau, t) = P^k(\tau + \Delta\tau, t)$, if $\tau \neq T$;
       $P^{k+1}(T, t) = s^k(t)$, if $\tau = T$.
8. Search
   $w^{k+1}(t) = \arg\min\left\{I_0\left(y^k(T, t) + \alpha\left[w^k(t) - y(T, t)\right]\right), -\infty < \alpha < \infty\right\}$.
9. Perform inverse conversion of variables
   $u^{k+1}(t) = 0.5(u_l + u_g) + (u_g - u_l)\sin\left(w^{k+1}(t)\right)$, $t \in T$.
   The iteration is complete.

In this paper, we proposed a modification of the Powell method, which consists in using the combined one-dimensional nonlocal search algorithm developed by the authors based on the Strongin and "parabolas" methods as an auxiliary one (steps 7(b), 8 of the Powell algorithm). We used the indicated algorithm to solve numerous univariate problems of constructing conjugate descent directions.

The algorithm proposed by R.G. Strongin [21], is one of the most famous and effective methods of univariate global optimization. In this method, between the neighboring points, a search is performed for the interval with the most probable location of the global extremum. On the found interval, we selected a point that corresponds to the mathematical expectation of the minimum position. At a given point, we calculated the function value, added to the set of known values, and the algorithm proceeds to the next iteration. The algorithm stops when the length of the compression interval becomes less than the specified criterion.

One of the methods of univariate search, established on heuristic ideology, is the "parabolas" method [3, 28]. It is based on the idea of finding the minima of all parabolas formed by "convex triples". The algorithm has been successfully used in many applications, including for solving optimal control problems [27].

By "convex triple" for a function $I_0(\alpha)$, we mean a sequentially located triple of points α_i, α_{i+1}, α_{i+2} belonging to $[a, b]$ for which the following inequality holds:

$$I_0(\alpha_i) \geq I_0(\alpha_{i+1}) \leq I_0(\alpha_{i+2}). \tag{7}$$

Since the "parabolas" method is a heuristic algorithm, it cannot guarantee the finding of a global minimum of a function in all possible cases. It seems appropriate hybridization of the "parabolas" method with one of the "guaranteed" algorithms – the Strongin method. Therefore, we proposed a combined algorithm using these methods as the base.

4.2 The Combined Algorithm Based on the Strongin and the "Parabolas" Methods

1. Set the parameter of the algorithm: $N_p \in \mathbb{N}$ is the number of starting samples. Select points $\alpha_1 = a$, $\alpha_{N_p} = b$. Randomly generate points α_i, $2 \leq i \leq N_p - 1$ on the line segment $[a, b]$.
2. Sort points α_i, $1 \leqslant i \leqslant k$: $a = \alpha_1 < \alpha_2 < ... < \alpha_k = b$.
3. If the iteration has an odd number, then
   (a) Assume $I_0^{\max} = -\infty$. Check each triple of points α_i, α_{i+1}, α_{i+2}, $1 \leqslant i \leqslant N_p - 2$ for "convexity". Under the condition (7), if $I_0(\alpha_{i+1}) > I_0^{\max}$, then $I_0^{\max} = I_0(\alpha_{i+1})$.
   (b) For each "convex triple" $\tilde{\alpha}_i$, $\tilde{\alpha}_{i+1}$, $\tilde{\alpha}_{i+2}$ calculate evaluation $\delta_i = |I_0^{\max} - I_0(\alpha_{i+1})|$.
   (c) Based on the calculated estimates, select the index of the convex triple in a probabilistic way.
   (d) Find a minimum α_i^* of the parabola formed by the selected convex triple using a combination of local one-dimensional search methods. Choose the point $\alpha^{k+1} = \alpha_i^*$, the number of samples k increases by 1.
4. If the iteration has an even number, then perform the steps of the Strongin method:
   (a) Calculate the value m that is an estimate of the Hölder constant:
   $$M = \max \frac{|I_0(\alpha_i) - I_0(\alpha_{i-1})|}{|\alpha_i - \alpha_{i-1}|^{\frac{1}{N}}}, \quad m = \begin{cases} 1, & M = 0 \\ M, & M > 0 \end{cases}, \quad 1 \leqslant i \leqslant k, \text{ where } N \geqslant 1 \text{ is}$$
   the Hölder index.
   (b) For each interval (α_{i-1}, α_i), $1 \leqslant i \leqslant k$, calculate characteristics
   $$R(i) = (\alpha_i - \alpha_{i-1}) + \frac{(I_0(\alpha_i) - I_0(\alpha_{i-1}))^2}{m^2 K_c^2 (\alpha_i - \alpha_{i-1})} - 2\frac{I_0(\alpha_i) + I_0(\alpha_{i-1})}{mK_c}, \text{ where } K_c > 1 \text{ is}$$
   the caution ratio.
   (c) Select the interval (α_{t-1}, α_t) to which the maximum characteristic $R(t) = \max\{R(i), 1 \leqslant i \leqslant k\}$ corresponds.
   (d) If the stop criterion is met $(\alpha_t - \alpha_{t-1} \leqslant \varepsilon_\alpha$, where ε_α is the set accuracy), then the algorithm stops working; otherwise, select a point
   $$\alpha^{k+1} = \frac{1}{2}(\alpha_t + \alpha_{t-1}) - sign(I_0(\alpha_i) - I_0(\alpha_{i-1}))\frac{1}{2K_c}\left[\frac{I_0(\alpha_i) - I_0(\alpha_{i-1})}{m}\right],$$
   calculate the value $I_0(\alpha^{k+1})$, the number of samples k increases by 1. The iteration is complete.

The main disadvantage of the Powell method is that it is a local search method. A modification of the Powell method proposed in the paper eliminates this drawback and ultimately represents a globalized algorithm. The globalization of the algorithm is achieved by solving auxiliary problems of nonlocal one-dimensional search.

The above non-convex optimization algorithm is implemented in C using uniform software standards. The next section presents the numerical solutions of several applied problems of optimal control in robotic systems, performed using the proposed algorithm. Since the original Powell method is a local search algorithm, it converges to a local extremum and cannot cope with the solution of the problems presented in Sect. 5.

5 Solving Applied Problems

5.1 The Optimal Control Problem of the Mobile Robot

The mathematical model of the mobile robot is described by the following system of differential equations [7,18,19]:

$$\begin{cases} \dot{x}_1 = 0.5\,(u_1 + u_2)\cos x_3, \\ \dot{x}_2 = 0.5\,(u_1 + u_2)\sin x_3, \\ \dot{x}_3 = 0.5\,(u_1 - u_2). \end{cases}$$

The controls are subject to conditions: $|u_i(t)| \leqslant 10$, $i = \overline{1,2}$. The phase coordinates must satisfy the following inequalities:

$$g_1\,(x) = 1.5 - \sqrt{(x_1 - 2.5)^2 + (x_2 - 2.5)^2} \leqslant 0,$$

$$g_2\,(x) = 1.5 - \sqrt{(x_1 - 7.5)^2 + (x_2 - 7.5)^2} \leqslant 0,$$

$$g_3\,(x) = 3 - \sqrt{(x_1 - 2)^2 + (x_2 - 8)^2} \leqslant 0,$$

$$g_4\,(x) = 3 - \sqrt{(x_1 - 8)^2 + (x_2 - 2)^2} \leqslant 0.$$

The problem is to transfer the system from point $x(t_0) = (10, 10, 0)$ to point $x(t_1) = (0,0,0)$ in the shortest possible time t_1, under all restrictions. The objective functional is $I_0\,(u) = t_1 \to \min$. The selection of the optimum time was performed, minimizing the discrepancy (8) using the proposed algorithm:

$$\sum_{i=1}^{n} [x_i - x_i\,(t_1)]^2 \to \min. \tag{8}$$

A mobile robot is an automatic wheeled vehicle that has a moving chassis with automatically controlled drives. Figure 1 shows a model of this wheeled robot. The target point X_m is in the middle of the rear axis of the robot platform and has

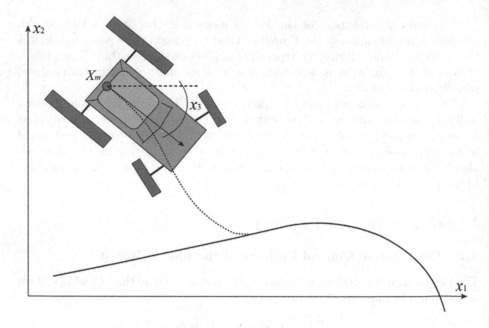

Fig. 1. Kinematic diagram of a wheeled mobile robot.

coordinates x_1, x_2. The angle x_3 formed by the central axis of the platform and the x_1 axis determines the orientation of the robot. The two rear wheels are driving, and the two front wheels are responsible for turning the platform.

In the course of the solution, it was possible to obtain the optimal value of the functional $I_0^* (u) = 2.48$ and the trajectory, which are slightly better compared to the results of computations given in [7] (the value of the functional is 2.51). Figure 2 shows the optimal trajectories of the mobile robot for the solution found: on the plane (left), depending on time (right). On the left graph, in the form of dashed circles, the areas defined by phase constraints are shown – areas beyond which the robot cannot enter. The arrows indicate the direction of movement of the robot.

5.2   The Optimal Control Problem of the Industrial Robot Arm

When industrial robots are introduced into production, a problem arises related to the speed of technological operations. Even with simple actions, the robot can work slower than a skilled worker, which limits the productivity of the work. Robot control is often built on guaranteed but completely non-optimal programs. Thus, there is a significant problem of increasing the speed of an industrial robot when performing simple mechanical operations. On the other hand, the movement of the robot along a given trajectory is traditionally provided by independent drives for different degrees of mobility.

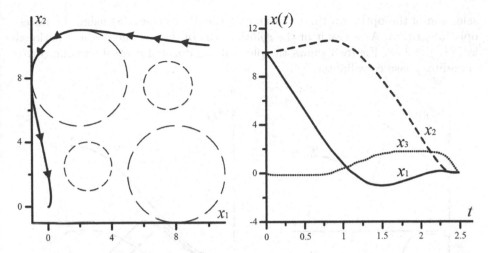

Fig. 2. Optimal trajectories of the mobile robot: on the plane (left), depending on time (right).

Consider the problem of finding the controls and trajectories of the robot with a limited influence of various degrees of mobility on each other. The motion dynamics of a flat two-link robot of an anthropomorphic type is described by the following system of differential equations:

$$
\begin{cases}
\dot{x}_1 = x_2, \\
\dot{x}_2 = \dfrac{(M_1 - F_1) \cdot a_{22} - (M_2 - F_2) \cdot a_{12}}{a_{11} \cdot a_{22} - a_{21} \cdot a_{12}}, \\
\dot{x}_3 = x_4, \\
\dot{x}_4 = \dfrac{(M_2 - F_2) \cdot a_{11} - (M_1 - F_1) \cdot a_{21}}{a_{11} \cdot a_{22} - a_{21} \cdot a_{12}},
\end{cases}
$$

where $M_1 = -C_1(x_1 - u_1)$, $M_2 = -C_2(x_3 - x_1 - u_2)$, $F_1 = -m_1 \cdot l_1 \cdot R_1 \cdot \sin(x_3 - x_1) \cdot x_2^2$, $F_2 = -m_2 \cdot l_2 \cdot R_2 \cdot \sin(x_3 - x_1) \cdot x_4^2$, $a_{11} = m_1 \cdot \rho_1^2 + m_2 \cdot l_1^2$, $a_{12} = a_{21} = m_2 \cdot R_2 \cdot l_1 \cdot \cos(x_3 - x_1)$, $a_{22} = m_2 \cdot \rho_2^2$. Here x_1, x_3 are angles of link rotation, x_2, x_4 are turning velocities, u_1, u_2 are software angle values (controls), l_1, l_2 are link lengths, m_1, m_2 are link masses, ρ_1, ρ_2 are radii of inertia, R_1, R_2 are distances to the center of mass, C_1, C_2 are gear ratios.

The problem variables are subject to conditions associated with the design features of the robot: $|u_i(t)| \leqslant \pi$, $i = \overline{1,2}$, $|M_i(t)| \leqslant 10$, $i = \overline{1,2}$, $\pi/6 \leqslant x_1(t) \leqslant 5/6 \cdot \pi$, $\pi/3 \leqslant x_1(t) - x_3(t) \leqslant 5/6 \cdot \pi$, $t \in [0, t_1]$.

We consider a specific version of the robot ("TUR-10") with the following characteristics: $m_1 = 7.62$, $\rho_1 = 0.968$, $R_1 = 0.239$, $l_1 = 0.50$, $C_1 = 10.0$, $m_2 = 8.73$, $\rho_2 = 0.973$, $R_2 = 0.251$, $l_2 = 0.67$, $C_2 = 10.0$.

The problem is to transfer the system from point $x(t_0) = (\pi/6, 0, -\pi/6, 0)$ to point $x(t_1) = (5/6\pi, 0, \pi/3, 0)$ in the shortest possible time t_1 under all restrictions. The objective functional is $I_0(u) = t_1 \to \min$. We performed the selection of the optimum time, minimizing the discrepancy (8) using the developed algorithm. As a result of the solution, the optimal value of the functional is $I_0^*(u) = 2.88$. Figure 3 shows the plots of the optimal control and the corresponding phase coordinates.

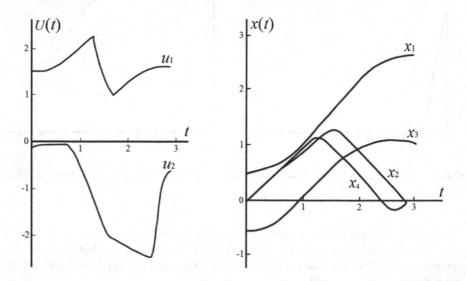

Fig. 3. Optimal control and corresponding trajectories in the problem of the industrial robot arm.

We compared efficiency and reliability of the proposed Powell method modification (PMM) with other heuristic algorithms: genetic algorithm (GA), differential evolution (DE) [5], "stochastic coverings" algorithm (SC) [13], "curvilinear search" method (CS) [12], harmony search (HS), biogeography-based optimization method (BBO) and firefly algorithm (FA) [26]. Table 1 shows the values of the functionals found by each algorithm for 100000 solved Cauchy problems (for both optimal control tasks). The testing of the developed algorithm has shown significantly higher efficiency and reliability in comparison with its competitors.

Table 1. Solutions obtained by algorithms for the optimal control problems of the mobile robot (Value 1) and the industrial robot arm (Value 2).

No.	Algo	Value 1	Value 2	No.	Algo	Value 1	Value 2
1	PMM	2.48	2.88	5	CS	2.61	2.99
2	GA	2.53	2.95	6	HS	2.83	3.24
3	DE	2.54	2.95	7	BBO	2.86	3.25
4	SC	2.54	2.97	8	FA	3.02	3.47

6   Conclusion

An algorithm for the numerical solution of non-convex optimal control problems was proposed. The developed algorithm was implemented in C language and integrated within a single software package. Several applied optimal control problems in robotic systems have been successfully solved using this algorithm. In all the problems considered, a comparison of the obtained solutions with the published computations allows us to conclude that the proposed algorithm has sufficiently high efficiency and reliability. The method proposed in the paper was tested on a variety of nonlinear optimal control problems from different application areas. All the obtained solutions found a meaningful interpretation.

References

1. Betts, J.: Practical methods for optimal control and estimation using nonlinear programming. SIAM, Philadelphia (2010)
2. Blatt, M., Schittkowski, K.: Pdecon: A fortran code for solving control problems based on ordinary, algebraic and partial differential equations. In: Technical report, Department of Mathematics, University of Bayreuth, Bayreuth (1997)
3. Bulatov, V., Hamisov, O.: Cut-off method in e(n+1) for solving global optimization problems on one class of functions. Comput. Math Math. phys. **47**(11), 1830–1842 (2007)
4. Chernousko, F., Banichuk, V.: Variational Problems of Mechanics and Control. Nauka, Moscow (1973)
5. Cruz, I.: Efficient Evolutionary Algorithms for Optimal Control. Wageningen University, Wageningen (2002)
6. Denardo, E.: Dynamic Programming: Models and Applications. Courier Corporation, Chelmsford (2012)
7. Diveev, A., Konstantinov, S.: The study of evolutionary algorithms for solving the optimal control problem. Proceedings of the Moscow Inst. Phys. Technol. **9**(3), 76–85 (2017)
8. Evtushenko, Y.: Methods for solving extremal problems and their application in optimization systems. Nauka, Moscow (1982)
9. Fedorenko, R.: Approximate Solution of Optimal Control Problems. Nauka, Moscow (1978)
10. Gornov, A.: Computational Technologies for Solving Optimal Control Problems. Nauka, Novosibirsk (2009)
11. Gornov, A., Zarodnyuk, T.: Optimal control problem: heuristic algorithm for global minimum. In: Proceedings of the Second International Conference on Optimization and Control, pp. 27–28. National University of Mongolia, Ulanbaatar (2007)
12. Gornov, A., Zarodnyuk, T.: The "curvilinear search" method of global extremum in optimal control problems. Modern techniques. System analysis. Simulation **1**(3), 19–26 (2009)
13. Gornov, A., Zarodnyuk, T.: Method of stochastic coverings for the optimal control problem. Comput. Technol. **17**(2), 31–42 (2012)
14. Gurman, V., Baturin, V., Rasina, I.: Approximate methods of optimal control. Irkutsk University, Irkutsk (1983)
15. Himmelblau, D.: Applied Nonlinear Programming. McGraw-Hill, New York (1972)

16. Kamien, M., Schwartz, N.: Dynamic Optimization: The Calculus of Variations and Optimal Control in Economics and Management. Courier Corporation, Chelmsford (2012)
17. Krotov, V.: Global Methods in Optimal Control Theory. CRC Press, Boca Raton (1995)
18. Pesterev, A.V.: A linearizing feedback for stabilizing a car-like robot following a curvilinear path. J. Comput. Syst. Sci. Int. **52**(5), 819–830 (2013). https://doi.org/10.1134/S1064230713050109
19. Rapoport, L.: Estimation of attraction domains in wheeled robot control. Automation and Remote Control **1**(9), 69–89 (2006)
20. Schwartz, A., Polak, E.: Consistent approximations for optimal control problems based on runge-kutta integration. SIAM J. Control and Optim. **34**(4), 1235–1269 (1996)
21. Strongin, R.: Numerical Methods in Multiextremal Problems. Nauka, Moscow (1978)
22. Stryk, O.V.: Numerical solution of optimal control problems by direct collocation. Optim. Control **1**(111), 129–143 (1993)
23. Teo, K., Goh, C., Wong, K.: A Unified Computational Approach to Optimal Control Problems. John Wiley and Sons, New York (1991)
24. Tolstonogov, A.: Differential Inclusions in a Banach Space. Springer Science and Business Media, Luxembourg (2012)
25. Tyatyushkin, A.: Numerical Methods and Software for Optimizing managed Systems. Nauka, Novosibirsk (1992)
26. Xing, B., Gao, W.: Innovative Computational Intelligence: A Rough Guide to 134 Clever Algorithms. Springer, Cham (2014)
27. Zarodnyuk, T., Gornov, A.: Global extremum search technology in the optimal control problem. Modern techniques. System analysis. Simulation **1**(3), 70–76 (2008)
28. Zhiglyavsky, A.: Mathematical theory of random global search. Leningrad University Publishing Department, Leningrad (1985)
29. Zhigljavsky, A., Zilinskas, A.: Stochastic global optimization. Springer, Luxembourg (2008)

Optimization and Data Analysis

A Derivative-Free Nonlinear Least Squares Solver

Igor Kaporin[✉][iD]

Dorodnicyn Computer Center of FRC CSC RAS, Vavilova 40, Moscow, Russia

Abstract. A nonlinear least squares iterative solver developed earlier by the author is modified to fit the derivative-free optimization paradigm. The proposed algorithm is based on easily parallelizable computational kernels such as small dense matrix factorizations and elementary vector operations and therefore has a potential for a quite efficient implementation on modern high-performance computers. Numerical results are presented for several standard test problems to demonstrate the competitiveness of the proposed method.

Keywords: Nonlinear least squares · Derivative-free optimization · Preconditioned subspace descent

1 Introduction

Application areas of nonlinear least squares are numerous and include, for instance, acceleration of neural network learning processes using Levenberg - Marquardt type algorithms, pattern recognition, signal processing etc. This explains the need in further development of robust and efficient nonlinear least squares solvers.

The present paper is mainly based on the results of [11] specialized to the case of derivative-free optimization. Similar to [5–7], we use the inexact Newton/Krylov subspace framework, however with search subspaces augmented by several previous directions, with different stepsize choice rule and with the use of quasirandom rectangular preconditioner. The latter algorithimic feature is critically important for the nonlinear least square solver proposed in the present paper. Indeed, in general case the residuals cannot be readily used to form search directions as it was done in [5–7] when the number of equations is equal to the number of unknowns.

2 General Description of Nonlinear LS Solver

A standard least squares problem is formulated as

$$x_* = \arg \min_{x \in R^n} \varphi(x), \tag{1}$$

Supported by RFBR grant No.19-01-00666.

where the function $\varphi : R^n \to R$ has the form

$$\varphi(x) = \frac{1}{2}\|f(x)\|^2 \equiv \frac{1}{2}f^\top(x)f(x), \tag{2}$$

and $f(x)$ is a nonlinear mapping

$$f : R^n \to R^m, \qquad m \geq n. \tag{3}$$

Assuming sufficient smoothness of f, an iterative procedure is constructed to find the minimizer x_* numerically. Note that x_* satisfies the equation

$$\operatorname{grad} \varphi(x_*) = 0, \tag{4}$$

where

$$\operatorname{grad} \varphi(x) = J^\top(x)f(x) \in R^n, \tag{5}$$

and

$$J(x) \equiv \frac{\partial f}{\partial x} \in R^{m \times n}, \tag{6}$$

is the Jacobian matrix of f at x.

2.1   Descent Along a Subnormalized Direction

Let $x_0, x_1, \ldots, x_t, \ldots$ be the sequence of approximations to the stationary point x_* constructed in the course of iterations, where t is the outer iteration index. Further on, we will use the notations

$$f_t = f(x_t), \quad J_t = J(x_t), \quad g_t = \operatorname{grad}(x_t) = J_t^\top f_t. \tag{7}$$

The next approximation x_{t+1} to x_* is constructed as

$$x_{t+1} = x_t + \alpha_t p_t, \tag{8}$$

where the stepsize parameter α_t satisfies

$$0 < \alpha_t < 2,$$

and p_t is a direction vector satisfying the *subnormalization* condition

$$(J_t p_t)^\top (f_t + J_t p_t) \leq 0, \tag{9}$$

which can conveniently take into account the inexactness of the Jacobian by a vector products. Inequality (9) is a generalization of the normalization condition

$$(J_t p_t)^\top (f_t + J_t p_t) = 0$$

used earlier in [8–11], where the explicit availability of J_t as an $m \times n$ matrix was assumed. Next we consider sufficient conditions for the descent of $\varphi(x_t)$.

2.2 General Estimate for Residual Norm Reduction

Under rather mild conditions, see, e.g. [9,11], there exists the limiting stepsize $\widehat{\alpha}_t \in (0,2)$ such that for all $0 < \alpha \le \widehat{\alpha}_t$ the estimate

$$\frac{\varphi(x_t + \alpha p_t)}{\varphi(x_t)} \le 1 - \left(\left(\alpha - \frac{\alpha^2}{2}\right)\theta_t^2\right)^2 \tag{10}$$

is valid, where φ is defined in (2), direction p_t is subnormalized by (9), and θ_t is determined as

$$\theta_t = \frac{\|J_t p_t\|}{\|f_t\|}. \tag{11}$$

Note that by the subnormalization condition (9) it holds

$$\vartheta_t = \vartheta(f_t, J_t p_t) \equiv \frac{-(J_t p_t)^\top f_t}{\|f_t\|\|J_t p_t\|} \ge \frac{\|J_t p_t\|}{\|f_t\|} = \theta_t, \tag{12}$$

so that the quantity (11) is a lower bound for the cosine ϑ_t of the Euclidean acute angle between m-vectors f_t and $(-J_t p_t)$. Clearly, estimate (10) shows the importance of finding subnormalized directions p_t with values of θ_t as large as possible.

Remark 1. Similar to [8,11], the proof of (10) is based on the assumption that the limiting stepsize $\widehat{\alpha} = \widehat{\alpha}(f,p)$ along a subnormalized direction p exists such that the limiting stepsize condition

$$\|f(x + \alpha p) - f - \alpha J p\| \le \left(\alpha - \frac{\alpha^2}{2}\right)\frac{\|J p\|^2}{\|f\|} \tag{13}$$

is satisfied for all $0 < \alpha \le \widehat{\alpha}$. (To clarify the notations, further we will omit the iteration index t where possible.) Indeed, (10) can be obtained from (13) and (9) as follows:

$$\|f(x + \alpha p)\| \le \|f + \alpha J p\| + \|f(x + \alpha p) - f - \alpha J p\|$$

$$= \left(\|f\|^2 + 2\alpha f^\top J p + \alpha^2 \|J p\|^2\right)^{1/2} + \|f(x + \alpha p) - f - \alpha J p\|$$

$$\le \left(\|f\|^2 - 2\alpha \|J p\|^2 + \alpha^2 \|J p\|^2\right)^{1/2} + \left(\alpha - \frac{\alpha^2}{2}\right)\frac{\|J p\|^2}{\|f\|}$$

$$= \|f\|\left(\left(1 - (2\alpha - \alpha^2)\frac{\|J p\|^2}{\|f\|^2}\right)^{1/2} + \left(\alpha - \frac{\alpha^2}{2}\right)\frac{\|J p\|^2}{\|f\|^2}\right)$$

$$\le \|f\|\left(1 - \left(\left(\alpha - \frac{\alpha^2}{2}\right)\frac{\|J p\|^2}{\|f\|^2}\right)^2\right)^{1/2},$$

where the latter estimate follows from the inequality

$$\sqrt{1 - \eta} + \frac{\eta}{2} \le \sqrt{1 - \frac{\eta^2}{4}},$$

which holds for any $0 \le \eta \le 1$ and is used with $\eta = \alpha(2 - \alpha)\|J p\|^2/\|f\|^2$, see also (12).

Remark 2. It appears that $\widehat{\alpha}$ characterizes the nonlinearity of f in the neighborhood of x, while θ is related to the precision of approximate solution p of the "Newton equation" $f + Jp = 0$. Note that the latter may not (and often cannot) be solved exactly in the context of our considerations.

2.3   Choosing the Stepsize

Based on estimate (10) one can develop the following Armijo type procedure [1] for evaluating appropriate stepsize α_t providing for a certain decrease of the residual norm. Let p_t be a direction vector satisfying the subnormalization condition (9). The value of stepsize is determined by checking the validity of estimate (26) (see the corresponding Section below; we cannot directly use Jp as in earlier papers) for a decreasing sequence of trial values of $\alpha \in (0,2)$; the standard choice is

$$\alpha^{(l)} = 2^{-l}, \qquad l = 0, 1, \ldots, l_{\max} - 1, \tag{14}$$

with $l_{\max} = 30$, which approximately corresponds to $\alpha^{(l)} > 2 \cdot 10^{-8}$. As soon as (10) be satisfied, one sets $\alpha_t = \alpha^{(l)}$. In numerical testing, the backtracking criterion (10) was often satisfied at once for $l = 0$ with the stepsize $\alpha_t = 1$.

2.4   Approximating Product of Jacobian by a Vector

The derivative-free approximation for products like Jp is obtained using

$$J(x)p \approx \widetilde{J(x)p} = \frac{f(x + \zeta p) - f(x)}{\zeta}, \tag{15}$$

where $\zeta = O(\tau^{1/2})$ and τ is the floating point tolerance, or by the more precise formula

$$J(x)p \approx \widetilde{J(x)p} = \frac{f(x + \zeta p) - f(x - \zeta p)}{2\zeta}, \tag{16}$$

where $\zeta = O(\tau^{1/3})$.

2.5   Choosing Subspace Basis and Descent Direction

Let us choose the direction p_t such as

$$p_0 \widetilde{\in} \operatorname{span}\{K_0^\top f_0, \ K_0^\top J_0 K_0^\top f_0, \ldots, (K_0^\top J_0)^{k-1} K_0^\top f_0\},$$
$$p_1 \widetilde{\in} \operatorname{span}\{K_1^\top f_1, \ K_1^\top J_1 K_1^\top f_1, \ldots, (K_1^\top J_1)^{k-1} K_1^\top f_1, \ p_0\},$$

$$\cdots$$

$$p_t \widetilde{\in} \operatorname{span}\{K_t^\top f_t, \ K_t^\top J_t K_t^\top f_t, \ldots, (K_t^\top J_t)^{k-1} K_t^\top f_t, \ p_{t-1}, \ldots, p_{t-\min(t,l)}\},$$

$$\cdots$$

where $k \geq 1$ and $l \geq 1$ are small integers. Recall that $p_{t-1} = (x_t - x_{t-1})/\alpha_{t-1}, \ldots$ are the previous search directions. Here the rectangular matrix $K_t \in R^{m \times n}$

serves as a kind of preconditioner, see Sect. 2.8 below. Symbol $\widetilde{\in}$ reflects the use of (15) or (16) for the approximation of the Jacobian by a vector products. (The exact meaning of $\widetilde{\in}$ is determined by the orthogonalization procedure given below.) This construction is a generalization of the one proposed in [9] and used in [10, 11].

Further in this section, we omit the iteration index t, for instance, $J = J(x_t)$ and $p_{k-i} = p_{t+k-i}$.

Since we use rather rough approximations for the products of J by a vector, we will use the following Arnoldi-type orthogonalization procedure for constructing the bases of the above defined subspaces:

$$v_1 \chi_{1,0} = f, \qquad v_{i+1} \chi_{i+1,i} = \widetilde{Ju_i} - \sum_{j=1}^{i} v_j \chi_{j,i}, \quad i = 1, \ldots, k+l,$$

where

$$u_i = K^\top v_i, \quad i = 1, \ldots, k,$$

$$u_i = p_{k-i}, \quad i = k+1, \ldots, k+l,$$

and the coefficients $\chi_{1,0} = \|f\|$,

$$\chi_{j,i} = v_j^\top \widetilde{Ju_i}, \quad j = 1, \ldots, i, \qquad \chi_{i+1,i} = \left\| \widetilde{Ju_i} - \sum_{j=1}^{i} v_j \chi_{j,i} \right\|$$

are determined to satisfy the orthonormality condition $v_i^\top v_j = \delta_{i-j}$. As the result, one obtains the factorization

$$JU = VH + Z, \tag{17}$$

where

$$U = [u_1 \mid \ldots \mid u_{k+l}], \qquad V = [v_1 \mid \ldots \mid v_{k+l+1}], \qquad V^\top V = I_{k+l+1},$$

$$U \in R^{n \times (k+l)}, \quad V \in R^{m \times (k+l+1)}, \quad H \in R^{(k+l+1) \times (k+l)}, \quad Z \in R^{m \times (k+l)}.$$

The matrix Z accounts for the errors arising from the approximation of the Jacobian by a vector multiplications so that

$$z_i = \widetilde{Ju_i} - \widetilde{Ju_i}.$$

Therefore, the direction is determined as

$$p = Us, \quad s \in R^{k+l}, \tag{18}$$

where s will be specified below in the next Section.

Remark 3. Setting $\alpha = \zeta$ and combining (13) with (15) one obtains the condition

$$\frac{\|Jp - \widetilde{Jp}\|}{\|Jp\|} \le \left(1 - \frac{\zeta}{2}\right) \frac{\|Jp\|}{\|f\|} = \left(1 - \frac{\zeta}{2}\right) \theta, \tag{19}$$

which means that smaller values of $\theta = \|Jp\|/\|f\|$ impose a more restrictive upper bound on the numerical finite difference error in (15). Since the second order formula (16) typically reduces the left hand part of (19), it can be more appropriate for practical use. Moreover, the use of larger subspace dimensions $k + l$ is preferable in order to increase θ in the right hand side of (19).

2.6   Characterizing Inexactness and Choosing Search Directions

Assume for a moment that $Z = 0$ in (17), which corresponds to exact computations with the Jacobian. In this case, the minimum norm solution of the form (18) for the overdetermined linear equation $f + Jp = 0$ is given by

$$s = -(H^\top H)^{-1} H^\top e_1 \|f\|,$$

where $e_1 = [1\ 0\ \ldots\ 0]^\top \in R^{k+l+1}$ is the first unit vector. Indeed, using

$$f = V e_1 \|f\|,$$

(18), and (17) with $Z = 0$, one has

$$\|f + Jp\| = \|V e_1 \|f\| + JUs\| = \|V(e_1 \|f\| + Hs)\| = \|e_1 \|f\| + Hs\|, \quad (20)$$

and the formula for the least squares solution readily follows. To estimate the effect of inexactness (which corresponds to $Z \neq 0$), the following condition is the most convenient:

$$Z^\top Z \leq \xi^2 H^\top H, \quad (21)$$

where ξ is a (typically small) positive parameter. As is shown in the next section, setting

$$s = -\frac{1}{2}(H^\top H)^{-1} H^\top e_1 \|f\| \quad (22)$$

is a safe choice for s whenever $0 \leq \xi \leq 1/2$.

Remark 4. In actual computations, the use of an approximate pseudo-inversion formula

$$s = -\frac{1}{2}\left(H^\top H + \zeta I\right)^{-1} H^\top e_1 \|f\|, \qquad 0 < \zeta \ll \|H\|^2,$$

is preferable, which relaxes the possible ill-conditioning of H. A comprehensive analysis for the case of nonzero ζ will be given elsewhere.

2.7   Subnormality of Search Directions and the Lower Bound for θ

Denote the orthogonal projector

$$\Pi = H(H^\top H)^{-1} H^\top \in R^{(k+l+1)\times(k+l+1)}, \quad (23)$$

so that

$$Hs = -\frac{1}{2}\Pi e_1 \|f\|. \quad (24)$$

In the general case, similar to (20) one has

$$Jp = VHs + Zs, \qquad f + Jp = V(e_1\|f\| + Hs) + Zs,$$

and therefore

$$(Jp)^\top(f + Jp) = (s^\top H^\top V^\top + s^\top Z^\top)(V(e_1\|f\| + Hs) + Zs)$$
$$= s^\top H^\top(e_1\|f\| + Hs) + s^\top Z^\top V(e_1\|f\| + 2Hs) + s^\top Z^\top Zs$$
$$= -\frac{1}{4}e_1^\top \Pi e_1\|f\|^2 + s^\top Z^\top V(I - \Pi)e_1\|f\| + s^\top Z^\top Zs,$$

where the last equality holds by (24). Further, using (21) one has

$$s^\top Z^\top Zs \le \xi^2 s^\top H^\top Hs = \frac{\xi^2}{4}e_1^\top \Pi e_1\|f\|^2$$

which yields

$$(Jp)^\top(f + Jp) \le \|f\|^2 \left(-\frac{1}{4}e_1^\top \Pi e_1 + \frac{\xi}{2}\sqrt{e_1^\top \Pi e_1(1 - e_1^\top \Pi e_1)} + \frac{\xi^2}{4}e_1^\top \Pi e_1\right).$$

Finally, the right hand side of the latter inequality is negative if

$$\xi \le \frac{\sqrt{e_1^\top \Pi e_1}}{1 + \sqrt{1 - e_1^\top \Pi e_1}}.$$

Therefore, a simple sufficient condition for subnormality (as defined by (9)) of direction (22) is

$$\xi \le \frac{1}{2}\sqrt{e_1^\top \Pi e_1}. \tag{25}$$

It only remains to notice that the quantity $e_1^\top \Pi e_1 \in (0, 1]$ monotonically tends to 1 from below in the progress of the Arnoldi iterations. Thus, the condition (21) turns to be less restrictive (as soon as the minimum singular value of H stays separated from zero).

Using the orthogonality of V and (21), one has then

$$\|Jp\| = \|(VH + Z)s\| = \frac{\|f\|}{2}\|V\Pi e_1 + Z(H^\top H)^{-1}H^\top e_1\|$$

$$\ge \frac{\|f\|}{2}\left(\|V\Pi e_1\| - \|Z(H^\top H)^{-1}H^\top e_1\|\right) \ge \frac{\|f\|}{2}(1 - \xi)\sqrt{e_1^\top \Pi e_1},$$

and, assuming that by (25) it holds $\xi \le 1/2$, we estimate θ from below:

$$\theta = \frac{\|Jp\|}{\|f\|} \ge \frac{1}{4}\sqrt{e_1^\top \Pi e_1}.$$

Therefore, the stepsize can be safely determined from appropriately modified estimate (10):

$$\frac{\|f(x + \alpha p)\|^2}{\|f\|^2} \le 1 - \left(\left(\alpha - \frac{\alpha^2}{2}\right)\frac{e_1^\top \Pi e_1}{16}\right)^2, \tag{26}$$

as soon as conditions (21) and (25) hold.

2.8 Using Quasirandom Preconditioning

As the preconditioner we consider a full column rank matrix $K_t \in R^{m \times n}$ satisfying

$$K_t^{\top} K_t \approx I_n. \tag{27}$$

Clearly, the forming of K_t and multiplying it by a vector $q = K_t v$ must be as cheap as possible. Here we will consider preconditionings having a potential for a quite efficient implementation on modern high-performance computers. In particular, we consider K_t taken as Hankel matrix with quasirandom entries generated by the logistic sequence.

2.9 Description of Computational Algorithm

The above described preconditioned subspace descent algorithm can be summarized as follows. Note that indicating $f(x)$ as an input means the availability of computational module for the evaluation of vector $f(x)$ for any given x.

Algorithm 1.

Key notations:

$$U_t = [u_1| \ldots |u_{i_{\max}+1}] \in R^{m \times (i_{\max}+1)}, \qquad V_t = [v_1| \ldots |v_{i_{\max}}] \in R^{n \times i_{\max}},$$

$$H_i = \begin{bmatrix} \chi_{1,1} & \chi_{1,2} & \chi_{1,3} & \cdots & \chi_{1,i} \\ \chi_{2,1} & \chi_{2,2} & \chi_{2,3} & \cdots & \chi_{2,i} \\ 0 & \chi_{3,2} & \chi_{3,3} & \cdots & \chi_{3,i} \\ \cdots & \cdots & \cdots & \cdots & \cdots \\ 0 & \cdots & 0 & \chi_{i,i-1} & \chi_{i,i} \\ 0 & \cdots & 0 & 0 & \chi_{i+1,i} \end{bmatrix} \in R^{(i+1) \times i}, \quad h_i = \begin{bmatrix} \chi_{1,1} \\ \chi_{1,2} \\ \chi_{1,3} \\ \cdots \\ \chi_{1,i} \end{bmatrix} \in R^i;$$

Input: $f(x) \in R^m$, $x_0 \in R^n$;
Initialization:
$s = k + l \leq n$, $\delta = 10^{-12}$, $\zeta = 5 \cdot 10^{-6}$,
$\varepsilon = 10^{-10}$, $\tau_{\min} = 10^{-8}$, $t_{\max} = 10000$,
$f_0 = f(x_0)$, $\rho_0 = f_0^{\top} f_0$;
Iterations:
for $t = 0, 1, \ldots, t_{\max} - 1$:
    generate quasirandom $K_t \in R^{m \times n}$
    $u_1 := f_t / \sqrt{\rho_t}$
    $w := u_1$
    $i_{\max} := k + \min(l, t)$
    for $i = 1, \ldots, i_{\max}$:
        if $(i \leq k)$ **then**
            $v_i := K_t^{\top} w$
        end if
        $w := (f(x_t + \zeta v_i) - f(x_t - \zeta v_i))/(2\zeta)$
        for $j = 1, \ldots, i$:
            $\chi_{j,i} = u_j^{\top} w$

$$w := w - u_j \chi_{j,i}$$
end for
$$\chi_{i+1,i} = \sqrt{w^\top w}$$
$$w := w/\chi_{i+1,i}$$
$$u_{i+1} = w$$
end for
$$L_t L_t^\top = H_{i_{\max}}^\top H_{i_{\max}} + \delta \operatorname{trace}(H_{i_{\max}}^\top H_{i_{\max}})I$$
$$z_t := (L_t)^{-1} h_{i_{\max}}$$
$$\vartheta_t := z_t^\top z_t$$
$$z_t := (L_t)^{-\top} z_t \rho_t$$
$$p_t = -V_t z_t$$
$$v_{k+1+(t \bmod l)} := p_t$$
$$\alpha^{(0)} = 1$$
for $l = 0, 1, \ldots, l_{\max} - 1$:
$$x_t^{(l)} = x_t + \alpha^{(l)} p_t$$
$$f_t^{(l)} = f(x_t^{(l)})$$
$$\rho_t^{(l)} = (f_t^{(l)})^\top f_t^{(l)}$$
$$\tau = \alpha^{(l)}(2 - \alpha^{(l)})\vartheta_t/16$$
if $(\tau < \tau_{\min})$ **return** x_t
if $(\rho_t^{(l)}/\rho_t > 1 - (\tau/2)^2)$ **then**
$$\alpha^{(l+1)} = \alpha^{(l)}/2$$
$$x_t^{(l+1)} = x_t + \alpha^{(l+1)} p_t$$
else
      go to NEXT
end if
end for
NEXT: $x_{t+1} = x_t^{(l)}$, $f_{t+1} = f_t^{(l)}$, $\rho_{t+1} = \rho_t^{(l)}$;
if $(\rho_{t+1} < c^2 \rho_0)$ **or** $(\rho_{t+1} \geq \rho_t)$ **return** x_{t+1}
end for

Remark 5. In the above algorithm, the notation was changed as $u \leftrightarrow v$ to conform the one used in an earlier code.

Remark 6. The use of quantity ϑ_t can be explained as follows. For simplicity, let us consider $\delta = 0$ and drop the indices t, $i_{\max}$, and (l). Then, by $H^\top H = LL^\top$, $h = H^\top e_1$, and (23), it holds

$$\vartheta = z^\top z = h^\top L^{-\top} L^{-1} h = e_1^\top H(H^\top H)^{-1} H^\top e_1 = e_1^\top \Pi e_1,$$

and therefore

$$\tau/2 = \frac{1}{2}\alpha(2 - \alpha)\vartheta/16 = \left(\alpha - \frac{\alpha^2}{2}\right)\frac{e_1^\top \Pi e_1}{16}.$$

Comparing the latter equality with (26) gives exactly the backtracking condition $\rho_t^{(l)}/\rho_t > 1 - (\tau/2)^2$ used in Algorithm 1 for the refinement of stepsize α.

3   Test Problems and Numerical Results

Below some results of application of Algorithm 1 to several standard hard-to-solve nonlinear test problems are presented. For the test runs, one core of Pentium(R) Dual-Core CPU E6600 3.06 GHz, 3.25 Gbytes RAM desktop PC was used. We will consider sufficiently large subspace dimensions with $k \geq l$ and $k + l \leq n$. For the nonzero residual problems, the iterations typically terminate by the condition $\tau < \tau_{min} = 10^{-10}$, see the corresponding line in Algorithm 1.

Table 1. Performance of Algorithm 1 for medium-size problems

Test name	m	n	$k + l$	#iter	#fun. eval.	opt. value	$\|x\|_C$
Broyden tridiagonal	500	500	50 + 50	15	1726	6.7E−10	0.707
chained Rosenbrock	198	100	50 + 50	664	133538	9.9E−08	1.000
inv. 3D dist. tensor	27000	150	75 + 75	101	25008	0.01113926	0.601
inv. 3D dist. tensor	125000	250	125 + 125	126	47880	0.02763617	0.653
Lennard-Jones 2D	4950	200	180 + 20	74	29258	68.260376	7.915
Lennard-Jones 3D	78	39	20 + 19	41	2861	5.8028612	1.459
Lennard-Jones 3D	1485	165	150 + 15	308	101720	34.762002	12.85
3 × 3 × 3 Brent eq.	729	63	33 + 30	29	2768	4.1E−13	1.144

3.1   Broyden Tridiagonal Function

Following [14], for $n = m$ and $m = 500$ define $f(x)$ as

$$f_i = (3 - 2x_i)x_i - x_{i-1} - 2x_{i+1} + 1, \quad 1 \leq i \leq 500,$$

where $x_0 = x_{n+1} = 0$. The optimum value is $f^\top f = 0$ and the starting point is set as $\tilde{x} = [-1 \ldots -1]^\top$. The results are presented in Table 1. This test can be considered as relatively easy due to the actual closeness of the initial guess $\tilde{x}$ to the solution x_*.

3.2   Chained Rosenbrock Function

This test function was introduced in [18], and we will use its version with $m = 2n - 2$ and essentially variable coefficients:

$$f_{2i-1} = i(x_i - x_{i+1}^2), \qquad f_{2i} = 1 - x_{i+1}, \qquad i = 1, \ldots, n - 1.$$

The optimum value is $f^\top f = 0$ at $x_* = [1 \ldots 1]^\top$ and the starting point is $\tilde{x} = [-1 \ldots -1]^\top$. The results are given in Table 1 for $m = 198$ and $n = 100$. For this test case, the convergence history demonstrated the behavior typical for linear conjugate gradients with fast residual norm decrease at initial steps followed by a near stagnation phase and fast superlinear decrease at the final stage.

3.3 Approximate Canonical Decomposition of Inverse 3D Distance Tensor

This problem was considered, e.g., in [12,16,17]. Since the 3D array under consideration

$$t_{i,j,k} = \left(i^2 + j^2 + k^2\right)^{-1/2}$$

is symmetric, the residual function can be taken as

$$f_{i+(j-1)q+(k-1)q^2} = -\left(i^2 + j^2 + k^2\right)^{-1/2} + \sum_{l=1}^{r} x_{(l-1)q+i} x_{(l-1)q+j} x_{(l-1)q+k},$$

where $1 \leq i, j, k \leq q$, so that $m = q^3$ and $n = qr$. The particular case we consider is $q = 30$ and $r = 5$. The initial guess was set as $\widetilde{x} = [1/2 \ldots 1/2]^\top$. This is rather hard-to-solve nonzero residual problem (especially for $k + l \ll n$).

3.4 Lennard-Jones Potential Minimization

The problem of finding

$$x = [r_1^\top \ldots r_N^\top]^\top = \arg\min_x \sum_{1 \leq i < j \leq N} \left(\|r_i - r_j\|^{-12} - 2\|r_i - r_j\|^{-6}\right),$$

where $r_i \in R^d$ and $d = 2$ or $d = 3$, serves as a popular hard-to-solve benchmark system for optimization algorithms, see, e.g., [2,15,19]. Its reformulation as a nonzero residual nonlinear LS problem with $m = N(N-1)/2$ and $n = 3d$ readily follows if one sets

$$f_{i,j} = \|r_i - r_j\|^{-6} - 1, \qquad 1 \leq i < j \leq N.$$

Clearly, the minimum of the Lennard-Jones potential is expressed as

$$\min \sum_{1 \leq i < j \leq N} \left(\|r_i - r_j\|^{-12} - 2\|r_i - r_j\|^{-6}\right) = \min_x \|f(x)\|^2 - \frac{N(N-1)}{2}.$$

The results for $d = 2$, $N = 100$ and $d = 3$, $N = 13$ or $N = 55$ are shown in Table 1. The obtained minima well agree with that published in the existing literature: for 2D problem $f(x) = 68.26037$ yields -290.521 compared to -293.697 in [2], while for the smaller 3D problem $f(x) = 5.802861$ yields -44.326801 which value exactly coincides with that of [15,19]. For the larger 3D problem with $N = 55$ we have obtained the local minimum -276.603 compared to -279.248 in [15,19].

Note that for such complicated problems with multiple minima, the choice of the initial guess is probably the most important tuning parameter. In our tests with Lennard-Jones and Brent equations, we used 100 quasirandom initial guesses generated by the called logistic sequence (see, e.g., [20] and references cited therein):

$$\xi_0 = 0.2, \qquad \xi_k = 1 - 2\xi_{k-1}^2, \qquad k = 1, 2, \ldots;$$

$$x_0^{(s)}(j) = \xi_{sj}/8, \qquad 1 \leq j \leq n, \qquad s = 1, 2, \ldots, 100;$$

the best results are shown in Table 1.

3.5 $3 \times 3 \times 3$ Brent Equations: Semi-analytical Solution

Let us consider (see, e.g., [11] and references cited therein) a particular case of "Brent Equations" [4] which arise in connection with the development of Fast Matrix Multiplication (FMM) algorithms. The problem setting is specified by the following definition of 3D tensor components

$$t_{i,j,k} = \delta(i_2 - j_1)\delta(j_2 - k_1)\delta(k_2 - i_1), \qquad 1 \le i, j, k \le q^2, \qquad (28)$$

where

$$i = i_1 + (i_2 - 1)q, \quad 1 \le i_1, i_2 \le q,$$
$$j = j_1 + (j_2 - 1)q, \quad 1 \le j_1, j_2 \le q,$$
$$k = k_1 + (k_2 - 1)q, \quad 1 \le k_1, k_2 \le q.$$

Thus, for a general Canonical Decomposition for 3D array (28) we obtain the residual equation

$$f_{i,j,k}(x) = -t_{i,j,k} + \sum_{l=1}^{r} u_{(l-1)q^2+i} v_{(l-1)q^2+j} w_{(l-1)q^2+k}$$

with the vector of unknowns $x^\top = [u^\top \ v^\top \ w^\top]$ and sizes $m = q^6$ and $n = 3q^2r$. This problem can immediately be related to the construction of fast multiplication of two $q \times q$ matrices using r essential multiplications.

Solving Brent equations (more precisely, finding exact solutions with minimum possible r) is known as extremely hard and, in general, unsolved problem, even for small $q \ge 3$. In [3], a specific symmetry was revealed for Brent equations with $q = 3$, which allows to considerably reduce the number of unknowns involved. In view of these results, we propose the following semi-analytical generalization of Brent equations:

$$t_{i,j,k} == \sum_{t=0}^{p-1} \sum_{l=1}^{L} (Q^t X_l Q^{-t})_{i_1,i_2} (Q^t X_{\pi(l)} Q^{-t})_{j_1,j_2} (Q^t X_{\sigma(l)} Q^{-t})_{k_1,k_2} \qquad (29)$$

where $L = l_0 + 3l_1$, π and σ are permutations such that $\sigma = \pi^2$ and π^3 is an identity; without loss of generality one can use

$$l = (1 \ 2 \ 3 \ \dots \ l_0 \ l_0 + 1 \ l_0 + 2 \ l_0 + 3 \ \dots \ l_0 + 3l_1),$$
$$\pi(l) = (1 \ 2 \ 3 \ \dots \ l_0 \ l_0 + 2 \ l_0 + 3 \ l_0 + 1 \ \dots \ l_0 + 3l_1 - 2),$$
$$\sigma(l) = (1 \ 2 \ 3 \ \dots \ l_0 \ l_0 + 3 \ l_0 + 1 \ l_0 + 2 \ \dots \ l_0 + 3l_1 - 1),$$

Q is a fixed matrix parameter such that $Q^p = I_q$ or $Q^p = -I_q$ and $Q^t \ne I_q$ for $1 < t < p$. This equation contains only $n = Lq^2 = q^2r/p$ scalar unknowns but presents a Canonical Decomposition of the rank $r = Lp$. Considering $q = 3$, we can choose

$$l_0 = 3, \quad l_1 = 1, \quad p = 4, \quad Q = \begin{bmatrix} 1 & 0 & 0 \\ 0 & 0 & -1 \\ 0 & 1 & 0 \end{bmatrix},$$

so that $m = 729$, $n = 54$ and $r = 24$. Hence, the number of unknowns is 12 times smaller compared to standard Brent equations. Moreover, Eq. (29) can be very efficiently solved by Algorithm 1 for several quasirandomly generated starting vectors. The results shown in Table 1 correspond to the solution of Eq. (29) with Q^{-1} replaced by $Q^{\top}$ and Q included into the vector of unknowns, so that $n = 63$.

Unfortunately, $r = 24$ is not the optimum rank for this case, since the minimum value for $q = 3$ is known to be at least $r = 23$, see [13]. However, it may happen that (29) can be solved for larger sizes q with better results for r.

4   Concluding Remarks

In the present paper, a nonlinear least squares solver is developed which is based on derivative-free computations and is formally applicable to all types of least squares problems with sufficiently smooth residual function. Key feature of the algorithm is the use of quasirandom rectangular preconditioners for the construction of an approximate Krylov subspaces containing descent directions. Moreover, the proposed version of the algorithm is well suited for an efficient implementation on modern high-performance computers. The results of numerical testing on several hard-to-solve problems have confirmed the efficiency and robustness of the derivatibe-free Preconditioned Subspace Descent method.

Acknowledgement. The author thanks the anonymous referee for insightful comments and suggestions which allow to significantly improve the exposition of the paper.

References

1. Armijo, L.: Minimization of functions having Lipschitz continuous first partial derivatives. Pac. J. Math. **16**(1), 1–3 (1966)
2. Averick, B.M., Carter, R.G., Xue, G.L., More, J.J.: The MINPACK-2 test problem collection (No. ANL/MCS-TM-150-Rev.). Argonne National Lab., IL (United States) (1992)
3. Ballard, G., Ikenmeyer, C., Landsberg, J.M., Ryder, N.: The geometry of rank decompositions of matrix multiplication II: 3 × 3 matrices. J. Pure Appl. Algebra **223**(8), 3205–3224 (2019)
4. Brent, R.P.: Algorithms for matrix multiplication (No. STAN-CS-70-157). Stanford University CA Department of Computer Science, p. 58 (1970)
5. Brown, P.N.: A local convergence theory for combined inexact-Newton/finite-difference projection methods. SIAM J. Numer. Anal. **24**(2), 407–434 (1987)
6. Brown, P.N., Saad, Y.: Hybrid Krylov methods for nonlinear systems of equations. SIAM J. Sci. Stat. Comput. **11**(3), 450–481 (1990)
7. Brown, P.N., Saad, Y.: Convergence theory of nonlinear Newton-Krylov algorithms. SIAM J. Optim. **4**(2), 297–330 (1994)
8. Kaporin, I.E.: Esimating global convergence of inexact Newton methods via limiting stepsize along normalized direction, Report 9329, Department of Mathematics, Catholic University of Nijmegen, Nijmegen, The Netherland, p. 8, July 1993

9. Kaporin, I.E.: The use of preconditioned Krylov subspaces in conjugate gradient type methods for the solution of nonlinear least square problems. (Russian) Vestnik Mosk. Univ. Ser. (Comput. Math. Cybern.) **15**(3), 26–31 (1995)
10. Kaporin, I.E., Axelsson, O.: On a class of nonlinear equation solvers based on the residual norm reduction over a sequence of affine subspaces. SIAM J. Sci. Comput. **16**(1), 228–249 (1994)
11. Kaporin, I.: Preconditioned subspace descent method for nonlinear systems of equations. Open Comput. Sci. **10**(1), 71–81 (2020)
12. Kazeev, V.A., Tyrtyshnikov, E.E.: Structure of the Hessian matrix and an economical implementation of Newton's method in the problem of canonical approximation of tensors. Comput. Math. Math. Phys. **50**(6), 927–945 (2010)
13. Laderman, J.D.: A noncommutative algorithm for multiplying 3×3 matrices using 23 multiplications. Bull. Am. Math. Soc. **82**(1), 126–128 (1976)
14. More, J.J., Garbow, B.S., Hillstrom, K.E.: Testing unconstrained optimization software. Argonne National Laboratory. Appl. Math. Division Tech. Memorandum **324**, 96 (1978)
15. Northby, J.A.: Structure and binding of Lennard-Jones clusters: $13 \leq N \leq 147$. The J. Chem. Phys. **87**(10), 6166–6177 (1987)
16. Oseledets, I.V., Savostyanov, D.V.: Minimization methods for approximating tensors and their comparison. Comput. Math. Math. Phys. **46**(10), 1641–1650 (2006)
17. Sterck, H.D., Miller, K.: An adaptive algebraic multigrid algorithm for low-rank canonical tensor decomposition. SIAM J. Sci. Comput. **35**(1), B1–B24 (2013)
18. Toint, P.L.: Some numerical results using a sparse matrix updating formula in unconstrained optimization. Math. Comput. **32**(143), 839–851 (1978)
19. Wales, D.J., Doye, J.P.K.: Global optimization by basin-hopping and the lowest energy structures of Lennard-Jones clusters containing up to 110 atoms. The J. Phys. Chem. A **101**(28), 5111–5116 (1997)
20. Yu, L., Barbot, J.P., Zheng, G., Sun, H.: Compressive sensing with chaotic sequence. IEEE Sig. Process. Lett. **17**(8), 731–734 (2010)

Max-Min Problems of Searching for Two Disjoint Subsets

Vladimir Khandeev[1]([envelope])[ORCID] and Sergey Neshchadim[2]

[1] Sobolev Institute of Mathematics, 4 Koptyug Avenue, 630090 Novosibirsk, Russia
khandeev@math.nsc.ru
[2] Novosibirsk State University, 2 Pirogova Street, 630090 Novosibirsk, Russia
s.neshchadim@g.nsu.ru

Abstract. The work considers three problems of searching for two disjoint subsets among a finite set of points in Euclidean space. In all three problems, it is required to maximize the minimal size of these subsets so that in each cluster, the total intra-cluster scatter of points relative to the cluster center does not exceed a predetermined threshold. In the first problem, the centers of the clusters are fixed points of Euclidean space and are given as input. In the second one, centers are unknown, but they belong to the initial set. In the last problem, the center of the cluster is the arithmetic mean of all its elements. Earlier works considered problems with constraints on the quadratic intra-cluster scatter.

Quadratic analogs of the first two problems were proven to be NP-hard even in the one-dimensional case. For the third analog, the complexity remains unknown. The main result of the work are proofs of NP-hardness of all considered problems even in the one-dimensional case.

Keywords: Euclidean space · Clustering · Max-min problem · NP-hardness · Bounded scatter

1 Introduction

The subject of research is three problems of finding two disjoint subsets in a finite set of points in Euclidean space. These problems model the applied problems of finding disjoint subsets in a collection of objects so that each subset consists of objects similar in the sense of a certain criterion. Such applied problems are often encountered in many applications such as pattern recognition and machine learning [4], data mining [2], data cleaning [13].

In all three problems, it is required to maximize the size of the minimum cluster in terms of cardinality so that in each cluster, the total intra-cluster scatter of points relative to the center (for each problem the center is determined in its own way) of the cluster does not exceed a predetermined threshold.

Considered problems have a simple interpretation in terms of searching for two groups of similar objects and filtering out foreign objects (see the next section).

The purpose of the research is to substantiate the NP-hardness of all three considered problems.

© Springer Nature Switzerland AG 2021
N. N. Olenev et al. (Eds.): OPTIMA 2021, LNCS 13078, pp. 231–245, 2021.
https://doi.org/10.1007/978-3-030-91059-4_17

2   Formulations and Related Problems

The problems under consideration are formulated as follows.

Problem 1. *Given* an N-element set $\mathcal{Y} = \{y_1, \ldots, y_N\}$ of points in Euclidean space $\mathbb{R}^d$, points z_1, z_2, and a real number $A \in \mathbb{R}_+$.

Find non-empty disjoint subsets $\mathcal{C}_1, \mathcal{C}_2 \subset \mathcal{Y}$ such that the minimal size of a subset is maximal. In other words,

$$\min\left(|\mathcal{C}_1|, |\mathcal{C}_2|\right) \to \max, \tag{1}$$

where

$$F_1\left(\mathcal{C}_i\right) = F\left(\mathcal{C}_i, z_i\right) := \sum_{y \in \mathcal{C}_i} \|y - z_i\|_2 \leq A, \ i = 1, 2, \tag{2}$$

$\|v\|_2 = \sqrt{v_1^2 + \ldots + v_d^2}$ for each $v = (v_1, \ldots, v_d) \in \mathbb{R}^d$.

Problem 2. *Given* an N-element set $\mathcal{Y} = \{y_1, \ldots, y_N\} \subset \mathbb{R}^d$ and a real number $A \in \mathbb{R}_+$.

Find non-empty disjoint subsets $\mathcal{C}_1, \mathcal{C}_2 \subset \mathcal{Y}$ and points $u_1, u_2 \in \mathcal{Y}$ such that (1) holds and $\mathcal{C}_1, \mathcal{C}_2$ satisfy

$$F_2\left(\mathcal{C}_i, u_i\right) = F\left(\mathcal{C}_i, u_i\right) \leq A, \ i = 1, 2. \tag{3}$$

Problem 3. *Given* an N-element set $\mathcal{Y} = \{y_1, \ldots, y_N\} \subset \mathbb{R}^d$ and a real number $A \in \mathbb{R}_+$.

Find non-empty disjoint subsets $\mathcal{C}_1, \mathcal{C}_2 \subset \mathcal{Y}$ such that (1) holds and $\mathcal{C}_1, \mathcal{C}_2$ satisfy

$$F_3\left(\mathcal{C}_i\right) = F\left(\mathcal{C}_i, \bar{y}(\mathcal{C}_i)\right) \leq A, \ i = 1, 2,$$

where $\bar{y}(\mathcal{C}_i) = \frac{1}{|\mathcal{C}_i|} \sum_{y \in \mathcal{C}_i} y$, $i = 1, 2$, are the centroids (geometric centers) of the clusters $\mathcal{C}_i$.

In all three problems, it is required to maximize the size of the minimum cluster such that in each cluster, the total intra-cluster scatter of points relative to the center (for each problem the center is determined in its own way) of the cluster does not exceed the predetermined threshold A.

In Problem 1, the centers are arbitrary but given as input. In Problem 2, the centers are unknown but belong to the input set (that is, the centers are medoids [7]). In Problem 3, the center of the cluster is the centroid (the arithmetic mean of all cluster's elements).

As in the problems considered in [1], as a threshold A for the scatter, one can use α-fraction of the maximum scatter within this instance, i.e., one can put $A = \alpha F_{\max}$, where $F_{\max}$ is the maximum value described above. For example, in Problem 3, one can put $F_{\max} = \sum_{y \in \mathcal{Y}} \|y - \bar{y}(\mathcal{Y})\|_2$. The concentration of points in subsets $\mathcal{C}_i$ can be controlled by adjusting the value of α between 0 and 1.

Problems 1–3 have a fairly simple applied interpretation. There is a table with the data containing the results of measurements of a set of characteristics for a set of objects (in the problem statement, the set $\mathcal{Y}$ corresponds to the table). The set of measured objects contains two disjoint groups of homogeneous objects (subsets $\mathcal{C}_1$, $\mathcal{C}_2$) and a group of outliers ($\mathcal{Y} \setminus (\mathcal{C}_1 \cup \mathcal{C}_2)$). The characteristics measured for elements of a homogeneous group have some scatter due to the measurement technology. The value of this scatter is bounded by the specified threshold value A. The goal is to find two groups of homogeneous objects containing the largest number of admissible elements and separate the group of outliers.

Previously, problems were considered with a restriction on the quadratic intracluster variation. For the quadratic analogs of the first two problems (that is, problems where in (2) and (3) there is the sum of the squared norms), NP-hardness is known [8]. The issue of the complexity of similar problems but with other restrictions on the desired subsets remained open. Also note that the complexity of the quadratic analog of Problem 3 (that is, the problem in which the center of the cluster is its centroid) remains unknown.

The main result of the paper is a proof of NP-hardness of Problems 1–3 even in the one-dimensional case. From this, as is known [5], it follows that there are no known exact polynomial algorithms for these problems unless $P = NP$.

Note that the NP-hardness in the one-dimensional case is not typical for all clustering problems. For example, let's recall the well-known NP-complete M-Variance problem (see, e.g., [9] and references therein).

M-Variance Problem. *Given* a set $\mathcal{Y} = \{y_1, \ldots, y_N\} \subset \mathbb{R}^d$ and a positive integer $M > 1$.

Find a set $\mathcal{C} \subset \mathcal{Y}$ of cardinality M such that

$$\sum_{y \in \mathcal{C}} \|y - \bar{y}(\mathcal{C})\|^2 \to \min.$$

In the one-dimensional case, it is shown [11] that this problem is solvable in polynomial time $\mathcal{O}(N \log N)$.

There is also a more general property. In [10], the following problem was considered.

K-Means and Given J-Centers Problem. *Given* an N-element set $\mathcal{Y} = \{y_1, \ldots, y_N\} \subset \mathbb{R}^d$, positive integers $K \in \mathbb{N}$ and a tuple $\{c_1, \ldots, c_J\}$ of points from $\mathbb{R}^d$.

Find a partition of $\mathcal{Y}$ into $K + J$ clusters $\mathcal{C}_1, \ldots, \mathcal{C}_K, \mathcal{D}_1, \ldots, \mathcal{D}_J$ such that

$$F(\mathcal{C}_1, \ldots, \mathcal{C}_K; \mathcal{D}_1, \ldots, \mathcal{D}_J) = \sum_{k=1}^{K} \sum_{y \in \mathcal{C}_k} \|y - \bar{y}(\mathcal{C}_k)\|^2 + \sum_{j=1}^{J} \sum_{y \in \mathcal{D}_j} \|y - c_j\|^2 \to \min.$$

Its one-dimensional case was proven to be solvable in $\mathcal{O}(KJN^2)$ time, i.e., such multicluster generalization of the M-Variance problem is also polynomially solvable.

One of the properties that allow one to construct efficient algorithms for the one-dimensional case of the above-mentioned problems is that in these problems, the convex hulls of the optimal clusters do not contain elements of the original set that do not belong to the cluster itself. For the one-dimensional case of M-Variance it means that the minimum segment containing all the points of the cluster C, at the intersection with the original set, is equal to the cluster itself.

This property does not hold for Problems 1–3 considered in this paper. We will demonstrate this for Problem 3.

Consider the set $\mathcal{Y} = \{3, 8, 18, 25, 27, 43, 91, 98\}$ and put $A = 111$.

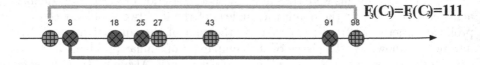

Fig. 1. Admissible solution with $|\mathcal{C}_1| = |\mathcal{C}_2| = 4$.

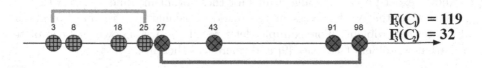

Fig. 2. Invalid solution with disjoint convex hulls of the clusters.

There is an admissible solution in which each cluster contains 4 elements of the original set—Fig. 1. However, a solution that satisfies the non-intersection property of the convex hulls of the clusters is not valid since its scatter is 119—Fig. 2. Therefore, there is no admissible solution satisfying the above property in which $|\mathcal{C}_1| = |\mathcal{C}_2| = 4$, since such a solution, up to the re-assignment of $\mathcal{C}_1, \mathcal{C}_2$, is unique. Thus, for certain values of the threshold A, for any optimal (in the sense of the maximality of the minimum cardinality) solution, the property of non-intersection of the convex hulls of the clusters does not hold.

The NP-hardness in the one-dimensional case makes the problems considered in this paper "more difficult" than the other problems described in this section.

Finally, we note that the problems under consideration are not equivalent to any of the known clustering geometric problems—k-means (k-MSSC) [3], k-median [14], k-center [12], k-center clustering with outliers [6]. One of the main differences between these problems and Problems 1–3 is the previously described property of non-intersection of the convex hulls of the optimal clusters.

3   Computational Complexity Analysis

We will assume that $\mathcal{Y}$, $\mathcal{C}_1$, and $\mathcal{C}_2$ are multisets, which means that it is allowed to include the same element multiple times.

3.1   Problem 1

Problem 1 in the form of a decision problem is given below.

Problem 1A. *Given an N-element set* $\mathcal{Y} = \{y_1, \ldots, y_N\} \subset \mathbb{R}^d$, *points* $z_1, z_2 \in \mathbb{R}^d$, *and numbers* $A \in \mathbb{R}_+$, $M \in \mathbb{N}$. *Question:* are there non-empty disjoint subsets $\mathcal{C}_1, \mathcal{C}_2 \subset \mathcal{Y}$ such that

$$\min\left(|\mathcal{C}_1|, |\mathcal{C}_2|\right) \geq M \tag{4}$$

and

$$F_1\left(\mathcal{C}_i\right) = \sum_{y \in \mathcal{C}_i} \|y - z_i\|_2 \leq A, \; i = 1, 2. \tag{5}$$

Theorem 1. *Problem 1A is NP-complete even in the one-dimensional case.*

Proof. It is obvious that Problem 1A (as well as Problems 2A and 3A, which will be formulated below) in the one-dimensional case belongs to the class *NP*. The NP-completeness is proved by constructing a polynomial reduction of the already known [5] NP-complete Problem **PARTITION** to Problem 1A (a similar scheme of the proof will be applied to subsequent Theorems 2, 3).

Problem **PARTITION** is formulated as follows.

Problem PARTITION. *Given a 2K-element set* $\mathcal{X} \subset \mathbb{N}$. *Question:* is there a partition of $\mathcal{X}$ into subsets $\mathcal{S}_1$, $\mathcal{S}_2$ such that

$$|\mathcal{S}_1| = |\mathcal{S}_2| = K, \; \mathcal{S}_1 \cup \mathcal{S}_2 = \{x_1, \ldots, x_{2K}\}, \; \sum_{x \in \mathcal{S}_1} x = \sum_{x \in \mathcal{S}_2} x. \tag{6}$$

Note that [5] considers the **PARTITION** problem in which there is no condition for equal cardinalities of the subsets $\mathcal{S}_1$, $\mathcal{S}_2$. But it is not difficult to understand that the original **PARTITION** problem is reduced to the one under consideration by adding N zeros, where N is the number of elements in the original problem.

Consider an arbitrary instance of Problem **PARTITION**:

$$\mathcal{X} = \{x_1, \ldots, x_{2K}\}.$$

We construct the following instance of Problem 1A (Fig. 3): $\mathcal{Y} = \{x_1, \ldots, x_{2K}\}$, $M = K$, $A = \frac{1}{2} \sum_{j=1}^{2K} x_j$, $z_1 = z_2 = 0$.

It is obvious that this reduction is polynomial.

Next, we show that in the constructed instance of Problem 1A, multisubsets that satisfy inequalities (4) and (5) exist if and only if in Problem **PARTITION** exists a partition into multisubsets such that equality (6) holds. $\Leftarrow$: Let a solution to the constructed example of Problem **PARTITION** exist. In other words, there are sets of indices $\mathcal{I}_1, \mathcal{I}_2$, such that $\mathcal{S}_i = \{x_k : k \in \mathcal{I}_i\}$, $i = 1, 2$, are the solution to Problem **PARTITION**.

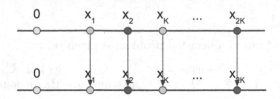

Fig. 3. Reduction for Problem 1A.

Define the following sets:

$$C_i = \{x_k : k \in \mathcal{I}_i\},\ i = 1, 2. \tag{7}$$

Now we show that C_1, C_2 are the solution to Problem 1A. To do this, we should prove that conditions (4) and (5) are satisfied. Condition (4) is satisfied by construction (7) of multisubsets C_i.

Let's compute $F_1(C_i)$, $i = 1, 2$:

$$F_1(C_i) = \sum_{y \in C_i} \|y - z_i\|_2 = \sum_{y \in C_i} |y - z_i| = \sum_{y \in C_i} y = \frac{1}{2} \sum_{j=1}^{2K} x_j = A,\ i = 1, 2.$$

Consequently, condition (5) is also satisfied, and C_1, C_2 are multisubsets required in Problem 1A.

$\Rightarrow$: Suppose that C_1, C_2 are the solution to Problem 1A. It means that there are sets of indices $\mathcal{I}_1, \mathcal{I}_2$ such that $|\mathcal{I}_1| = |\mathcal{I}_2| = K$, $\mathcal{I}_1 \cup \mathcal{I}_2 = \{1 \ldots 2K\}$, and $C_i = \{x_k : k \in \mathcal{I}_i\}$, $i = 1, 2$.

Then the following inequality holds:

$$F_1(C_i) = \sum_{k \in \mathcal{I}_i} x_k \leq A = \frac{1}{2} \sum_{j=1}^{2K} x_j. \tag{8}$$

If we define multisubsets $S_i = \{x_k : k \in \mathcal{I}_i\}$, $i = 1, 2$, then we will get that $\sum_{y \in S_1} y = \sum_{y \in S_2} y$.

This holds because:

$$\sum_{y \in S_1} y + \sum_{y \in S_2} y = \sum_{k \in \mathcal{I}_1} x_k + \sum_{k \in \mathcal{I}_2} x_k = \sum_{k \in \mathcal{I}_1 \cup \mathcal{I}_2} x_k = \sum_{j=1}^{2K} x_j. \tag{9}$$

Combining (8) and (9) gives that multisubsets S_i, $i = 1, 2$, are the solution to Problem **PARTITION**.   □

3.2   Problem 2

Now let's analyze the complexity of Problem 2A. Firstly we formulate Problem 2 in the form of a decision problem.

Problem 2A. *Given an N-clement set* $\mathcal{Y} = \{y_1, \ldots, y_N\} \in \mathbb{R}^d$ *and numbers* $A \in \mathbb{R}_+$, $M \in \mathbb{N}$. *Question: are there non-empty disjoint subsets* $\mathcal{C}_1, \mathcal{C}_2 \subset \mathcal{Y}$ *and points* $u_1, u_2 \in \mathcal{Y}$ *such that*

$$\min\left(|\mathcal{C}_1|, |\mathcal{C}_2|\right) \geq M \tag{10}$$

and

$$F_2\left(\mathcal{C}_i, u_i\right) = \sum_{y \in \mathcal{C}_i} \|y - u_i\|_2 \leq A, \ i = 1, 2. \tag{11}$$

Theorem 2. *Problem 2A is NP-complete even in the one-dimensional case.*

Proof. Given an arbitrary instance $\mathcal{X} = \{x_1, \ldots, x_{2K}\}$ of Problem **PARTITION**, we construct the following instance of Problem 2A (Fig. 4). We put

$$\mathcal{Y} = \{\underbrace{-B, \ldots, -B}_{2S_1}, \underbrace{0, \ldots, 0}_{2S_2}, y_1, \ldots, y_{2K}\}, \ M = S_1 + S_2 + K,$$

where $S_1 = 4K$, $S_2 = 8K + 2$, $D = \max\{\frac{1}{2}S, 2\max_k x_k\} + 1$, $S = \sum_{i=1}^{2K} x_k$, $B = KD + \frac{1}{2}S + 1$, $A = S_1 B + KD + \frac{1}{2}S$, $y_i = x_i + D, i = 1 \ldots 2K$.

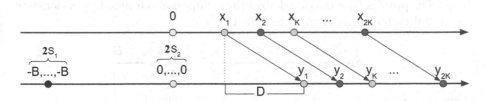

Fig. 4. Reduction for Problem 2A.

We will show that required multisubsets exist in the constructed instance of Problem 2A if and only if Problem **PARTITION** has a solution.

$\Leftarrow$: Suppose that required multisubsets in Problem **PARTITION** exist, i.e., there exist sets of indices $\mathcal{I}_i$, $i = 1, 2$, such that $|\mathcal{I}_1| = |\mathcal{I}_2| = K$, $\mathcal{I}_1 \cup \mathcal{I}_2 = \{1 \ldots 2K\}$ and multisubsets $\mathcal{S}_i = \{x_k : k \in \mathcal{I}_i\}$, $i = 1, 2$, are the solution to Problem **PARTITION**. We define multisubsets for Problem 2A as follows:

$$\mathcal{C}_i = \underbrace{\{-B, \ldots, -B\}}_{S_1} \cup \underbrace{\{0, \ldots, 0\}}_{S_2} \cup \underbrace{\{y_k = x_k + D\}}_{k \in \mathcal{I}_i}, \ i = 1, 2.$$

Firstly notice that $|\mathcal{C}_1| = |\mathcal{C}_2| = S_1 + S_2 + |\mathcal{I}_1| = M$, this means that restriction (10) is satisfied.

Put $u_i = 0$, $i = 1, 2$. Let's compute $F_2(\mathcal{C}_i, u_i)$:

$$F_2(\mathcal{C}_i, u_i) = S_1 B + S_2 \cdot 0 + \sum_{k \in \mathcal{I}_i} y_k = S_1 B + KD + \sum_{k \in \mathcal{I}_i} x_k. \tag{12}$$

Using the fact that S_i, $i = 1, 2$, are the solution to Problem **PARTITION**, we can continue the equality (12):

$$F_2(\mathcal{C}_i, u_i) = S_1 B + KD + \frac{1}{2}S = A.$$

Thus, the constructed sets $\mathcal{C}_i$, $i = 1, 2$, are the solution to Problem 2A.

$\Rightarrow$: Now suppose that the multiset $\mathcal{Y}$ in Problem 2A has required multisubsets $\mathcal{C}_1, \mathcal{C}_2$ which satisfy conditions (10) and (11).

Note that from (10) and the fact that $|\mathcal{Y}| = 2M = 2S_1 + 2S_1 + 2K$, it follows that $|\mathcal{C}_1| = |\mathcal{C}_2| = S_1 + S_2 + K$.

Let us show that the following properties of the sets $\mathcal{C}_1$, $\mathcal{C}_2$ are true.

1. Each set $\mathcal{C}_i$ contains exactly S_1 elements $-B$.
2. Each set $\mathcal{C}_i$ contains exactly S_2 elements 0.
3. Each set $\mathcal{C}_i$ contains exactly K elements from $\{y_1, \ldots, y_{2K}\}$.

Note that it is enough to prove any two of the above statements, and the last one will follow from them. Let us prove the first two facts in the following statements.

Proposition 1. *Each set $\mathcal{C}_i$ contains exactly S_1 elements $-B$.*

Proof. The proof is by reductio ad absurdum. Suppose that quantity of elements $-B$ in multisubset $\mathcal{C}_1$ is greater that S_1, i.e.,

$$\mathcal{C}_1 = \underbrace{\{-B, \ldots, -B\}}_{Q \leq 2S_1} \cup \underbrace{\{0, \ldots, 0\}}_{Z \leq 2S_2} \cup \underbrace{\{y_k = x_k + D\}}_{k \in \mathcal{I}_1, |\mathcal{I}_1| = L \leq 2K},$$

where $Q \geq S_1 + 1$ and $Z + L = S_1 + S_2 + K - Q$ (since $|\mathcal{C}_1| = S_1 + S_2 + K$).

Let's consider all the options that u_1 can take.

If $u_1 \geq 0$:

$$F_2(\mathcal{C}_1, u_1) \geq QB \geq (S_1 + 1)B = S_1 B + B$$

$$= S_1 B + KD + \frac{1}{2}\sum_{j=1}^{2K} x_j + 1 > S_1 B + KD + \frac{1}{2}\sum_{j=1}^{2K} x_j = A.$$

If $u_1 = -B$:

$$F_2(\mathcal{C}_1, u_1) \geq (Z + L)B = (S_1 + S_2 + K - Q)B$$
$$\geq (S_1 + S_2 + K - 2S_1)B = (5K + 2)B > (4K + 1)B > A.$$

We have proved that for all possible values of u_1, $F_2(\mathcal{C}_1, u_1) > A$. In other words, our assumption that was made at the beginning of the proposition proof is false. This implies that each multisubset $\mathcal{C}_i$ contains exactly S_1 element $-B$ (similarly, it can be proved that the case when $\mathcal{C}_2$ contains more than S_1 elements is unrealizable). ∎

Proposition 2. *Each set C_i contains exactly S_2 elements 0.*

Proof. Assume that multisubset C_2 contains more zero elements than C_1. Therefore, taking into account Proposition 1, C_1 has following form:

$$C_1 = \underbrace{\{-B,\ldots,-B\}}_{S_1} \cup \underbrace{\{0,\ldots,0\}}_{Z \leq S_2 - 1} \cup \underbrace{\{y_k = x_k + D\}}_{k \in \mathcal{I}_1, |\mathcal{I}_1| = L \leq 2K},$$

where $Z \leq S_2 - 1$, $L + Z = S_2 + K$ (which follows from Statement 1). It follows from these two relations that $L \geq K + 1$. Now let's consider all options that u_1 can take.
If $u_1 = -B$:

$$F_2(C_1, u_1) = ZB + \sum_{i \in \mathcal{I}_1}(y_i + B)$$

$$= (Z + L)B + \sum_{i \in \mathcal{I}_1} y_i > (Z + L)B + L\min_k y_k \geq (S_2 + K)B + (K+1)D$$

$$= (9K + 2)B + (K+1)D > 4KB + (K+1)D = S_1 B + KD + D$$

$$= S_1 B + KD + \max\{\frac{1}{2}\sum_{j=1}^{2K} x_j, 2\max_k x_k\} + 1 > S_1 B + KD + \frac{1}{2}\sum_{j=1}^{2K} x_j = A.$$

If $u_1 = 0$: $F_2(C_1, 0) \geq S_1 B + (K+1)D = S_1 B + KD + D > A$.
If $u_1 = y_j$:

$$F_2(C_1, u_1) = S_1(B + y_j) + Zy_j + \sum_{i \in \mathcal{I}_1}|y_i - y_j| = S_1 B + (S_1 + Z)y_j + \sum_{i \in \mathcal{I}_1}|y_i - y_j|$$

$$> S_1 B + S_1 D > S_1 B + (K+1)D > A.$$

In this way, for all possible values of u_1, we get that $F_2(C_1, u_1) > A$,—it means that our assumption that multisubset C_2 contains more zero elements than C_1 is false. Therefore, each C_i contains exactly S_2 zero elements. ∎

From Propositions 1 and 2, it follows that multisubsets C_i have the following structure:

$$C_i = \underbrace{\{-B,\ldots,-B\}}_{S_1} \cup \underbrace{\{0,\ldots,0\}}_{S_2} \cup \underbrace{\{y_k = x_k + D\}}_{k \in \mathcal{I}_i, |\mathcal{I}_i| = K}, \quad i = 1, 2.$$

Proposition 3. *Suppose that C is a multisubset of $\mathcal{Y}$, i.e., C has the following structure:*

$$C = \{\underbrace{-B,\ldots,-B}_{N_1}, \underbrace{0,\ldots,0}_{N_2}, \tilde{y}_1 \leq \ldots \leq \tilde{y}_{N_3}\},$$

where $B > 0$, $\tilde{y}_i \geq 0$. If conditions $N_2 + N_3 > N_1$, $N_1 + N_2 > 4N_3$, $\frac{3}{2}\min_k \tilde{y}_k > \max_k \tilde{y}_k$ are satisfied, then $F_2(C, u)$ as a function of $u \in \mathcal{Y}$ achieves its minimum at $u = 0$.

Proof. Firstly, we notice that if $u \notin \mathcal{C}$, then there exists element $u^* \in \mathcal{C}$ such that $F_2(\mathcal{C}, u^*) \le F_2(\mathcal{C}, u)$. It can be obtained from the following relation:

$$F_2(\mathcal{C}, u) = \sum_{z \in \mathcal{C}} |z - u| = \sum_{u < z \in \mathcal{C}} |z - u| + \sum_{u \ge z \in \mathcal{C}} |z - u|$$

$$= \sum_{u < z \in \mathcal{C}} (z - u) + \sum_{u \ge z \in \mathcal{C}} (u - z) = \sum_{z \in \mathcal{C}} z - 2 \sum_{u \ge z \in \mathcal{C}} z - \sum_{z \in \mathcal{C}} u + 2 \sum_{u \ge z \in \mathcal{C}} u$$

$$= \sum_{z \in \mathcal{C}} z - 2 \sum_{u \ge z \in \mathcal{C}} z - (|\mathcal{C}| - 2|\{z \in \mathcal{C} : z \le u\}|) \cdot u.$$

If u is located between two adjacent elements of $\mathcal{C}$ or lies outside the minimum segment containing the entire set $\mathcal{C}$, then all terms except the last one are constant, and the sign of expression in brackets defines to which element of $\mathcal{C}$ point u should be moved for reducing the value of F_2. Therefore, we can assume that $u \in \mathcal{C}$.

Denote by $\tilde{S} = \sum_{k=1}^{N_3} \tilde{y}_k$ the sum of all nonnegative elements, by $\tilde{S}_i = \sum_{k=1}^{i} \tilde{y}_k$— the sum of elements with indices $1, \ldots, i$. Firstly, we compute $F_2(\mathcal{C}, u)$ for all possible values of $u \in \mathcal{C}$.

If $u = -B$: $F_2(\mathcal{C}, -B) = (N_2 + N_3)B + \tilde{S}$.

When $u = 0$ we get that $F_2(\mathcal{C}, 0) = N_1 B + \tilde{S}$.

And the last case $u = \tilde{y}_i$:

$$F_2(\mathcal{C}, \tilde{y}_i) = N_1(B + \tilde{y}_i) + N_2 \tilde{y}_i + \sum_{j=1}^{i-1} (\tilde{y}_i - \tilde{y}_j) + \sum_{j=i+1}^{N_3} (\tilde{y}_j - \tilde{y}_i)$$

$$= N_1 B + (N_1 + N_2)\tilde{y}_i + (i-1)\tilde{y}_i - \tilde{S}_{i-1} + (\tilde{S} - \tilde{S}_i) - (N_3 - i)\tilde{y}_i$$

$$= N_1 B + (N_1 + N_2 + i - 1 - N_3 + i)\tilde{y}_i + \tilde{S} - \tilde{S}_i - \tilde{S}_{i-1}$$

$$= N_1 B + (N_1 + N_2 + 2i - N_3)\tilde{y}_i + \tilde{S} - 2\tilde{S}_i.$$

If we show that $F_2(\mathcal{C}, -B) > F_2(\mathcal{C}, 0)$ and $F_2(\mathcal{C}, \tilde{y}_i) > F_2(\mathcal{C}, 0)$, then the current proposition will be proved.

Indeed, $F_2(\mathcal{C}, -B) - F_2(\mathcal{C}, 0) = (N_2 + N_3 - N_1)B > 0$ since $N_2 + N_3 - N_1 > 0$. The second inequality is proved in a similar way:

$$F_2(\mathcal{C}, \tilde{y}_i) - F_2(\mathcal{C}, 0) = (N_1 + N_2 + 2k - N_3)\tilde{y}_i - 2\tilde{S}_i$$

$$> (N_1 + N_2 - N_3) \min_k \tilde{y}_k - 2N_3 \max_k \tilde{y}_k > (4N_3 - N_3) \min_k \tilde{y}_k - 2N_3 \max_k \tilde{y}_k$$

$$= 2N_3 \left(\frac{3}{2} \min_k \tilde{y}_k - \max_k \tilde{y}_k\right) > 0.$$

Thus, the proposition is proved. ∎

Note that the requirements of Proposition 3 are satisfied for each multisubset $\mathcal{C}_i$, as $S_2 + K = 9K + 2 > 4K = S_1$, $S_1 + S_2 = 12K + 2 > 4K$, and $\frac{3}{2} \min_k y_k \ge \frac{1}{2}D + D > \max_k x_k + D = \max_k y_k$. Consequently, centers u_1, u_2 can be set equal to zero.

Compute $F_2(\mathcal{C}_i, u_i)$:

$$F_2(\mathcal{C}_i, u_i) = S_1 B + \sum_{k \in \mathcal{I}_i} y_k = S_1 B + KD + \sum_{k \in \mathcal{I}_i} x_k \leq S_1 B + KD + \frac{1}{2} S.$$

As a result we get that $\sum_{k \in \mathcal{I}_i} x_k \leq \frac{1}{2} S$, $i = 1, 2$. Recalling that $\sum_{k \in \mathcal{I}_1} x_k + \sum_{k \in \mathcal{I}_2} x_k = S$, we get $\sum_{k \in \mathcal{I}_1} x_k = \sum_{k \in \mathcal{I}_2} x_k = \frac{1}{2} S$. So we showed that $\mathcal{S}_i = \{x_k : k \in \mathcal{I}_i\}$, $i = 1, 2$, is the solution to Problem **PARTITION**. $\square$

3.3   Problem 3

Problem 3 in the form of a decision problem is given below.

Problem 3A. *Given an N-element set* $\mathcal{Y} = \{y_1, \ldots, y_N\} \in \mathbb{R}^d$, *numbers* $A \in \mathbb{R}_+$ *and* $M \in \mathbb{N}$. *Question: are there non-empty disjoint subsets* $\mathcal{C}_1, \mathcal{C}_2 \subset \mathcal{Y}$ *such that*

$$\min(|\mathcal{C}_1|, |\mathcal{C}_2|) \geq M \tag{13}$$

and

$$F_3(\mathcal{C}_i) = F(\mathcal{C}_i, \bar{y}(\mathcal{C}_i)) \leq A, \ i = 1, 2. \tag{14}$$

Theorem 3. *Problem 3A is NP-complete even in the one-dimensional case.*

Proof. Given an arbitrary instance $\mathcal{X} = \{x_1, \ldots, x_{2K}\}$ of Problem **PARTITION**, we construct the following instance of Problem 3A (Fig. 5). In Problem 3A, we put $\mathcal{Y} = \{-B, -B, y_1, \ldots, y_{2K}\}$, $M = K + 1$, where

$$B = \frac{1}{2} \sum_{i=1}^{2K} y_i, \ y_i = x_i + D, \ D = \frac{1}{2(K+1)} \cdot \sum_{i=1}^{2K} x_i, \ A = \sum_{i=1}^{2K} y_i.$$

Next, we show that in the constructed instance of Problem 3A, multisubsets that satisfy inequalities (13) and (14) exist if and only if in Problem **PARTITION**, there exists a partition into multisubsets such that equality (6) holds.

$\Leftarrow$: Suppose that the required multisubsets exist in Problem **PARTITION**, i.e., there exist $\mathcal{S}_1$ and $\mathcal{S}_2$ such that $|\mathcal{S}_1| = |\mathcal{S}_2|$ and $\sum_{x \in \mathcal{S}_i} x = \frac{1}{2} \sum_{k=1}^{2K} x_k$, $i = 1, 2$.

For $\mathcal{S}_1$ and $\mathcal{S}_2$ there exist sets $\mathcal{I}_1, \mathcal{I}_2$ of indices such that

$$\mathcal{S}_i = \{x_k : k \in \mathcal{I}_i\}, \ i = 1, 2. \tag{15}$$

Define the multisubsets for Problem 3A as $\mathcal{C}_i = \{y_k : k \in \mathcal{I}_i\} \cup \{-B\}$, $k = 1, 2$, and prove that conditions (13) and (14) are satisfied.

Indeed, $\min(|\mathcal{C}_1|, |\mathcal{C}_2|) = \min(|\mathcal{I}_1|, |\mathcal{I}_2|) = K + 1 \geq M$; therefore, (13) is satisfied.

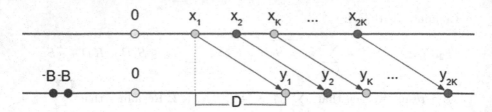

Fig. 5. Reduction for Problem 3A.

To check condition (14), we first compute the centroids $\bar{\mathcal{C}}_i := \bar{y}(\mathcal{C}_i)$:

$$(K+1)\bar{\mathcal{C}}_i = -B + \sum_{k\in\mathcal{I}_i} y_k = -\frac{1}{2}\sum_{k=1}^{2K} y_k + \sum_{k\in\mathcal{I}_i} y_k$$

$$= -\frac{1}{2}\sum_{k=1}^{2K} x_k - KD + \sum_{k\in\mathcal{I}_i} x_k + KD = 0, \; i = 1, 2. \qquad (16)$$

The former equation follows from (15) and definitions of y_i and B. From (16), it follows that the centroids of multisubsets $\mathcal{C}_1, \mathcal{C}_2$ are equal to zero. Then we estimate the left-hand side of (14):

$$F_3(\mathcal{C}_i) = \sum_{y\in\mathcal{C}_i} |y - \bar{\mathcal{C}}_i| = \sum_{y\in\mathcal{C}_i} |y| = B + \sum_{k\in\mathcal{I}_i} |y_k|$$

$$= \frac{1}{2}\sum_{k=1}^{2K} x_k + KD + \sum_{k\in\mathcal{I}_i}(x_k + D) = \sum_{k=1}^{2K} x_k + 2KD = A.$$

Therefore, condition (14) is also satisfied, and the required multisubsets in Problem 3A exist.

$\Leftarrow$: Now suppose that the multiset $\mathcal{Y}$ in Problem 3A has required multisubsets $\mathcal{C}_1, \mathcal{C}_2$ which satisfy conditions (13) and (14).

Firstly let's show that every $\mathcal{C}_i$ contains supplemented element $-B$. Assume that both elements $-B$ are contained in one multisubset $\mathcal{C}_i$. We may assume that $i = 1$. Then $\mathcal{C}_1$ has the following form:

$$\mathcal{C}_1 = \{-B, -B\} \cup \{y_i : i \in \mathcal{I}_1\}, \; |\mathcal{I}_1| = K - 1.$$

Now, if we recall the definition of $F_3(\mathcal{C}_1)$, we get that:

$$F_3(\mathcal{C}_1) = 2|B + \bar{\mathcal{C}}_1| + \sum_{k\in\mathcal{I}_1} |y_k - \bar{\mathcal{C}}_1|. \qquad (17)$$

Firstly find out the sign of the centroid $\bar{\mathcal{C}}_1$.

$$(K+1)\bar{\mathcal{C}}_1 = -2B + \sum_{k\in\mathcal{I}_1} y_k$$

$$= -S - 2KD + (K-1)D + \sum_{k\in\mathcal{I}_1} x_k < -S - (K+1)D + S = -(K+1)D \leq 0.$$

As we can see, $\bar{C}_1 \leq 0$ and therefore absolute values in the second sum of the right-hand side of (17) equal $y_k - \bar{C}_1$, because $y_k \geq 0$ by construction. This implies that (17) can be extended:

$$F_3(\mathcal{C}_1) = 2\left(\bar{C}_1 + B\right) + \sum_{k \in \mathcal{I}_1} \left(y_k - \bar{C}_1\right) = \bar{C}_1\left(3 - K\right) + 2B + \sum_{k \in \mathcal{I}_1} y_k$$

$$= \frac{3-K}{K+1}\left(-2B + \sum_{k \in \mathcal{I}_1} y_k\right) + 2B + \sum_{k \in \mathcal{I}_1} y_k = 2\frac{K-1}{K+1}\sum_{i=1}^{2K} y_i + \frac{4}{K+1}\sum_{k \in \mathcal{I}_1} y_k$$

$$= 2\left(1 - \frac{2}{K+1}\right)\sum_{i=1}^{2K} y_i + \frac{4}{K+1}\sum_{k \in \mathcal{I}_1} y_k > 2\left(1 - \frac{2}{K+1}\right)\sum_{i=1}^{2K} y_i. \quad (18)$$

Assuming without loss of generality that $K > 3$, we continue (18):

$$F_3(\mathcal{C}_1) > 2\left(1 - \frac{2}{3+1}\right)\sum_{i=1}^{2K} y_i = \sum_{i=1}^{2K} y_i = A.$$

Therefore, if both elements $-B$ are contained in $\mathcal{C}_1$, then $F_3(\mathcal{C}_1) > A$. This is a contradiction with the assumption that $\mathcal{C}_1$, $\mathcal{C}_2$ satisfy condition (14). Thus, every multisubset $\mathcal{C}_i$ contains one element $-B$.

Let $\mathcal{C}_i = \{-B\} \cup \{y_k : k \in \mathcal{I}_i\}$, $i = 1, 2$. We show that $\mathcal{S}_i = \{x_k : k \in \mathcal{I}_i\}$ is the solution to the initial Problem **PARTITION** instance. Suppose it isn't, i.e., the sum of the elements of one $\mathcal{S}_i$ is greater than the sum of the elements of the other. Without loss of generality, assume that $\sum\limits_{x \in \mathcal{S}_2} x > \frac{1}{2}\sum\limits_{i=1}^{2K} x_i$. We shall prove that $F_3(\mathcal{C}_2) > A$.

$$F_3(\mathcal{C}_2) = |B + \bar{C}_2| + \sum_{k \in \mathcal{I}_2} |y_k - \bar{C}_2|. \quad (19)$$

Let's estimate $\bar{C}_2$:

$$(K+1)\bar{C}_2 = -B + \sum_{k \in \mathcal{I}_2} y_k = \sum_{k \in \mathcal{I}_2} x_k - \frac{1}{2}S < S - \frac{1}{2}S = \frac{1}{2}S.$$

Therefore, $\bar{C}_2 < \frac{1}{2(K+1)}S = D$. Then $\forall k \in \{1, ..., 2K\}$: $y_k \geq \bar{C}_2$ due to the definition of y_k. It follows that in (19) $|y_k - \bar{C}_2| = y_k - \bar{C}_2$. Using this estimate for $\bar{C}_2$, (19) can be specified:

$$F_3\left(\mathcal{C}_2\right) = \left(B + \bar{C}_2\right) + \sum_{k \in \mathcal{I}_2} \left(y_k - \bar{C}_2\right) = (1 - K)\bar{C}_2 - B + \sum_{k \in \mathcal{I}_2} y_k$$

$$= -\frac{1-K}{1+K}B + \frac{1}{2}\sum_{i=1}^{2K} y_i + \frac{2}{1+K}\sum_{k \in \mathcal{I}_2} y_k = \frac{1}{2}\frac{2K}{K+1}\sum_{i=1}^{2K} y_i + \frac{2}{1+K}\sum_{k \in \mathcal{I}_2} y_k$$

$$> \frac{1}{2}\sum_{i=1}^{2K} y_i \frac{2K}{K+1} + \frac{1}{1+K}\sum_{i=1}^{2K} y_i = \sum_{i=1}^{2K} y_i = A.$$

This means that $F_3(\mathcal{C}_2) > A$; therefore, assumption $\sum\limits_{k\in\mathcal{I}_2} x_k > \frac{1}{2}\sum\limits_{i=1}^{2K} x_i$ is incorrect, i.e., $\sum\limits_{k\in\mathcal{I}_2} x_k \leq \frac{1}{2}\sum\limits_{i=1}^{2K} x_i$. Inequality $\sum\limits_{k\in\mathcal{I}_1} x_k \leq \frac{1}{2}\sum\limits_{i=1}^{2K} x_i$ can be proved similarly. Recalling that $\sum\limits_{k\in\mathcal{I}_1} x_k + \sum\limits_{k\in\mathcal{I}_2} x_k = \sum\limits_{i=1}^{2K} x_i$, we get that $\sum\limits_{k\in\mathcal{I}_1} x_k = \sum\limits_{k\in\mathcal{I}_2} x_k$.
Consequently, the required multisubsets $\mathcal{S}_1$ and $\mathcal{S}_2$ in Problem **PARTITION** exist.   □

Thus, NP-hardness in the one-dimensional case of Problems 1–3 is proved.

Corollary 1. *Problems 1–3 are NP-hard for every fixed dimension d of Euclidean space.*

Corollary 2. *Problems 1–3 are NP-hard when the dimension d is a part of the input.*

For such constraints that are considered in the paper, it would be natural to consider Problem 4, where the center of a cluster is its median.

Problem 4. *Given* an N-element set $\mathcal{Y} = \{y_1, \ldots, y_N\} \subset \mathbb{R}^d$ and a real number $A \in \mathbb{R}_+$.

Find non-empty disjoint subsets $\mathcal{C}_1, \mathcal{C}_2 \subset \mathcal{Y}$ such that (1) holds and $\mathcal{C}_1, \mathcal{C}_2$ satisfy

$$F_4(\mathcal{C}_i) = F(\mathcal{C}_i, m_i) := \sum_{y\in\mathcal{C}_i} \|y - m_i\| \leq A, \ i = 1, 2, \tag{20}$$

where m_i is a point minimizing $\sum\limits_{y\in\mathcal{C}_i} \|y - x\|$ over $x \in \mathbb{R}^d$.

In the one-dimensional case, this problem is equivalent to Problem 2. That follows from the fact that in this case, there is a point in $\mathcal{C}_i$ which minimizes $\sum\limits_{y\in\mathcal{C}_i} \|y - x\|$ over $x \in \mathbb{R}^d$ (similarly to the beginning of the proof of Proposition 3 in Theorem 2).

Also note that all statements in the article are valid if we replace the l_2 norm by any norm that is equivalent to l_2 in the one-dimensional case, for example by the l_1 (Manhattan) norm.

4   Conclusion

The paper shows NP-hardness of three previously unexplored maximin search problems of disjoint sets in a finite set of points in Euclidean space. It is shown that all three problems are NP-hard even in the one-dimensional case. An important direction of further research is the construction of efficient approximate algorithms and approximation schemes, as well as the search for subclasses of these problems that allow the construction of efficient exact algorithms.

Acknowledgments. The study presented was supported by the Russian Foundation for Basic Research, project 19-01-00308, 19-07-00397, and by the Russian Academy of Science (the Program of basic research), project 0314-2019-0015.

References

1. Ageev, A.A., Kel'manov, A.V., Pyatkin, A.V., Khamidullin, S.A., Shenmaier, V.V.: Approximation polynomial algorithm for the data editing and data cleaning problem. Pattern Recogn. Image Anal. **27**(3), 365–370 (2017). https://doi.org/10.1134/S1054661817030038
2. Aggarwal, C.C.: Data Mining: The Textbook. Springer, Switzerland (2015)
3. Aloise, D., Deshpande, A., Hansen, P., Popat, P.: NP-hardness of Euclidean sum-of-squares clustering. Mach. Learn. **75**(2), 245–248 (2009). https://doi.org/10.1007/s10994-009-5103-0
4. Bishop, C.M.: Pattern Recognition and Machine Learning. ISS, Springer, New York (2006)
5. Garey, M.R., Johnson, D.S.: Computers and Intractability: A Guide to the Theory of NP-completeness. Freeman, San Francisco (1979)
6. Hatami, B., Zarrabi-Zadeh, H.: A streaming algorithm for 2-center with outliers in high dimensions. Comput. Geom. **60**, 26–36 (2017). https://doi.org/10.1016/j.comgeo.2016.07.002
7. Kaufman, L., Rousseeuw, P.J.: Clustering by means of medoids. In: Dodge, Y. (ed.) Statistical Data Analysis based on the L_1 Norm, pp. 405–416. North-Holland, Amsterdam (1987)
8. Kel'manov, A., Khandeev, V., Pyatkin, A.: NP-hardness of some max-min clustering problems. In: Evtushenko, Y., Jaćimović, M., Khachay, M., Kochetov, Y., Malkova, V., Posypkin, M. (eds.) OPTIMA 2018. CCIS, vol. 974, pp. 144–154. Springer, Cham (2019). https://doi.org/10.1007/978-3-030-10934-9_11
9. Kel'manov, A.V., Romanchenko, S.M.: An FPTAS for a vector subset search problem. J. Appl. Ind. Math. **8**(3), 329–336 (2014). https://doi.org/10.1134/S1990478914030041
10. Kel'manov, A.V., Khandeev, V.I.: Polynomial-time solvability of the one-dimensional case of an NP-hard clustering problem. Comput. Math. Math. Phys. **59**(9), 1553–1561 (2019). https://doi.org/10.1134/S0965542519090112
11. Kel'manov, A.V., Ruzankin, P.S.: An accelerated exact algorithm for the one-dimensional M-variance problem. Pattern Recogn. Image Anal. **29**(4), 573–576 (2019). https://doi.org/10.1134/S1054661819040072
12. Masuyama, S., Ibaraki, T., Hasegawa, T.: The computational complexity of the m-center problems on the plane. IEICE Trans. **64**(2), 57–64 (1981)
13. Osborne, J.W.: Best Practices in Data Cleaning: A Complete Guide to Everything You Need to Do Before and After Collecting Your Data. SAGE Publication Inc, Los Angeles (2013)
14. Papadimitriou, C.H.: Worst-case and probabilistic analysis of a geometric location problem. SIAM J. Comput. **10**(3), 542–557 (1981)

Near-Optimal Decentralized Algorithms for Saddle Point Problems over Time-Varying Networks

Aleksandr Beznosikov[1,2]([✉]), Alexander Rogozin[1,2], Dmitry Kovalev[5], and Alexander Gasnikov[1,3,4]

[1] Moscow Institute of Physics and Technology, Moscow, Russia
beznosikov.an@phystech.edu
[2] Higher School of Economics, Moscow, Russia
[3] Institute for Information Transmission Problems RAS, Moscow, Russia
[4] Caucasus Mathematical Center, Adyghe State University, Maykop, Russia
[5] King Abdullah University of Science and Technology, Thuwal, Saudi Arabia

Abstract. Decentralized optimization methods have been in the focus of optimization community due to their scalability, increasing popularity of parallel algorithms and many applications. In this work, we study saddle point problems of sum type, where the summands are held by separate computational entities connected by a network. The network topology may change from time to time, which models real-world network malfunctions. We obtain lower complexity bounds for algorithms in this setup and develop near-optimal methods which meet the lower bounds.

Keywords: Saddle-point problem · Distributed optimization · Decentralized optimization · Time-varying network · Lower and upper bounds

1 Introduction

Distributed algorithms are an important part of solving many applied optimization problems [21,22,31]. They help to parallelize the computation process and make it faster. In this paper, we focus on the distributed methods for the saddle point problem:

$$\min_{x \in \mathcal{X}} \max_{y \in \mathcal{Y}} f(x,y) := \frac{1}{M} \sum_{m=1}^{M} f_m(x,y). \tag{1}$$

In this formulation of the problem, the original function f is divided into M parts, each of part f_m is stored on its own local device. Therefore, only the device with the number m knows information about f_m. Accordingly, in order

The research of A. Beznosikov, A. Rogozin and A. Gasnikov was supported by Russian Science Foundation (project No. 21-71-30005).

to obtain complete information about the function f, it is necessary to establish a communication process between devices. This process can be organized in two ways: centralized and decentralized. In a centralized approach, communication takes place via a central server, i.e. all devices can send some information about their local f_m function to the central server, the server collects information from the devices and does some additional calculations, and then can send new information or request to the devices. Then the process continues. With this approach, one can easily write centralized gradient descent for distributed sum minimization: $\min_x g(x) := \frac{1}{M} \sum_{m=1}^{M} g_m(x)$. All devices compute local gradients in the same current point and then send these gradients to the server, in turn, the server averages the gradients and makes a gradient descent step, thereby obtaining a new current point, which it sends to the devices. Centralized methods for (1) are discussed in detail, for example, in [5]. However, centralized approach has several problems, e.g. synchronization drawback or high requirements to the server. Possible approach to deal with these drawbacks is to use decentralized architecture [3]. In this case, there is no longer any server, and the devices are connected into a certain communication network and workers are able to communicate only with their neighbors and communications are simultaneous. The most popular and frequently used communication methods are the gossip protocol [7,13,24] and accelerated gossip protocol [30,32]. In the gossip protocol, nodes iteratively exchange data with their immediate neighbors using a communication matrix and in this way the information diffuses over the network. Decentralized algorithms are already widely developed for minimization problems, but not for saddle point problems. Meanwhile, saddle-point problems have a lot of applied applications, including those that require distributed computing. These are the already well-known and classic matrix game and Nash equilibrium [9,26], as well as modern problems in adversarial training [2,10], image deconvolution [8] and reinforcement, statistical learning [1,11].

This paper closes some of the open questions in decentralized saddle point problems.

1.1 Our Contribution

In particular, our contribution can be briefly described as follows

Lower Bounds. We present lower bounds for decentralized smooth strongly-convex-strongly-concave and convex-concave saddle-point problems on the time-varying networks. The lower bounds are derived under the assumption that the network is always a connected graph.

Near-Optimal Algorithm. The paper constructs a near-optimal algorithm that meets the lower bounds. The analysis of the algorithm is carried out for smooth strongly-convex-strongly-concave and convex-concave saddle-point problems
See our results in the column "time-varying" of Table 1.

1.2 Related Works

Our work is one of the first dedicated to decentralized saddle problems over time-varying networks. Among other works, we can highlight the following paper [4]. This work looks at a more general time-varying setting and suggests a new method. The upper bounds for their method are worse than for our method. We also mention papers on related topics:

Decentralized Saddle Point Problems
The next work is devoted to centralized and decentralized distributed saddle problems [5]. It carries out lower bounds and optimal algorithms in the case when the communication network is constant (non-time-varying). See Table 1 for comparison our results for time-varying topology and results from [5] for constant network.

Also note the following works devoted to decentralized min-max problems. In [19,27] one can find algorithms for saddle point problems on fixed network. In paper [27], it is near-optimal. Lower and upper bounds for decentralized min-max problems under data similarity condition are given in [6]. [18] studies the convergence of a decentralized methods for stochastic saddle point problems with homogeneous data on devices (all local functions f_m are the same).

Minimization on Time-Varying Networks
Decentralized methods are built upon combining iterations of classical first-order methods with communication steps. In the case of time-varying networks, a non-accelerated communication procedure is employed. Paper [24] can be named as an initial work on decentralized sub-gradient methods, and [23] proposed DIGing – the first-order minimization algorithm with linear convergence over time-varying networks. After that, PANDA, which is a dual method capable of working over time-varying graphs, was proposed in [20]. Analysis of DIGing and PANDA assumes that the underlying network is B-connected, that is, the union of B consequent networks is connected, while the network is allowed to be disconnected at some steps. Considering the time-varying graphs which stay connected at each iteration, decentralized Nesterov method [29] has an accelerated rate under the condition that graph changes happen rarely enough, ADOM [16] and ADOM+ [15] are first-order optimization methods which achieve lower complexity bounds [15]. APM-C [28], Acc-GT [17] are accelerated methods over time-varying graphs, as well. The mentioned results are devoted to minimization algorithms and can be generalized to saddle-point problems. In this paper we generalize lower bounds of [15] to min-max problems and obtain an algorithm which reaches them up to a logarithmic factor.

2 Preliminaries

We use $\langle z, u \rangle := \sum_{i=1}^{d} z_i u_i$ to denote standard inner product of $z, u \in \mathbb{R}^d$. It induces ℓ_2-norm in $\mathbb{R}^d$ in the following way $\|z\| := \sqrt{\langle z, z \rangle}$. We also introduce the following notation $\text{proj}_{\mathcal{Z}}(z) = \min_{u \in \mathcal{Z}} \|u - z\|$ – the Euclidean projection onto $\mathcal{Z}$.

Table 1. Lower and upper bounds for distributed smooth stochastic strongly-convex–strongly-concave (**sc**) or convex-concave (**c**) saddle-point problems in centralized and decentralized cases. Notation: L – smothness constant of f, μ – strongly-convex-strongly-concave constant, $R_0^2 = \|x_0 - x^*\|_2^2 + \|y_0 - y^*\|_2^2$, D – diameter of optimization set, χ – condition number of communication graph (in time-varying case maximum of all graphs), K – number of communication rounds. In the case of upper bounds in the convex-concave case, the convergence is in terms of the "saddle-point residual", in the rest – in terms of the (squared) distance to the solution.

	Time-varying network	Constant network [5]
	Lower	
sc	$\Omega\left(R_0^2 \exp\left(-\frac{\mu K}{256 L \chi}\right)\right)$	$\Omega\left(R_0^2 \exp\left(-\frac{\mu K}{128 L \sqrt{\chi}}\right)\right)$
c	$\Omega\left(\frac{L D^2 \chi}{K}\right)$	$\Omega\left(\frac{L D^2 \sqrt{\chi}}{K}\right)$
	Upper	
sc	$\tilde{O}\left(R_0^2 \exp\left(-\frac{\mu K}{8 L \chi}\right)\right)$	$\tilde{O}\left(R_0^2 \exp\left(-\frac{\mu K}{8 L \sqrt{\chi}}\right)\right)$
c	$\tilde{O}\left(\frac{L D^2 \chi}{K}\right)$	$\tilde{O}\left(\frac{L D^2 \sqrt{\chi}}{K}\right)$

We work with the problem (1), where the sets $\mathcal{X} \subseteq \mathbb{R}^{n_x}$ and $\mathcal{Y} \subseteq \mathbb{R}^{n_y}$ are convex sets. Additionally, we introduce the set $\mathcal{Z} = \mathcal{X} \times \mathcal{Y}$, $z = (x, y)$ and the operator F:

$$F_m(z) = F_m(x, y) = \begin{pmatrix} \nabla_x f_m(x, y) \\ -\nabla_y f_m(x, y) \end{pmatrix}. \tag{2}$$

This notation is needed for shortness.

Problem Setting. Next, we introduce the following assumptions:

Assumption 1(g). $f(x, y)$ is L - smooth, if for all $z_1, z_2 \subset \mathcal{Z}$

$$\|F(z_1) - F(z_2)\| \leq L\|z_1 - z_2\|. \tag{3}$$

Assumption 1(l). For all m, $f_m(x, y)$ is Lipschitz continuous with constant $L_{\max}$, it holds that for all $z_1, z_2 \in \mathcal{Z}$

$$\|F_m(z_1) - F_m(z_2)\| \leq L_{\max}\|z_1 - z_2\|. \tag{4}$$

Assumption 2(s). $f(x, y)$ is strongly-convex-strongly-concave with constant μ, if for all $z_1, z_2 \in \mathcal{Z}$

$$\langle F(z_1) - F(z_2), z_1 - z_2 \rangle \geq \mu\|z_1 - z_2\|^2. \tag{5}$$

Assumption 2(c). $f(x, y)$ is convex-concave, if $f(x, y)$ is strongly-convex-strongly-concave with 0.

Assumption 3. $\mathcal{Z}$ – compact bounded, i.e. for all $z, z' \in \mathcal{Z}$

$$\|z - z'\| \leq D. \tag{6}$$

All assumptions are standard in the literature.

Network Setting. In each moment of time (iteration) t, the communication network is modeled as a connected, undirected graph graph $\mathcal{G}(t) \triangleq (\mathcal{V}, \mathcal{E}(t))$, where $\mathcal{V} := \{1, \ldots, M\}$ denotes the vertex set–the set of devices (does not change in time) and $\mathcal{E}(t) := \{(i, j) \,|\, i, j \in \mathcal{V}\}$ represents the set of edges–the communication links at the moment t; $(i, j) \in \mathcal{E}(t)$ iff there exists a communication link between devices i and j in moment t.

As mentioned earlier, the gossip protocol is the most popular communication procedures in decentralized setting. This approach uses a certain matrix W. Local vectors during communications are "weighted" by multiplication of a vector with W. The convergence of decentralized algorithms is determined by the properties of this matrix. Therefore, we introduce the following assumption:

Assumption 4. We call a matrix $W(t)$ a gossip matrix at the moment t if it satisfies the following conditions: 1) $W(t)$ is an $M \times M$ symmetric, 2) $W(t)$ is positive semi-definite, 3) the kernel of $W(t)$ is the set of constant vectors, 4) $W(t)$ is defined on the edges of the network at the moment t: $W_{ij}(t) \neq 0$ only if $i = j$ or $(i, j) \in \mathcal{E}(t)$.

Let $\lambda_1(W(t)) \geq \ldots \geq \lambda_M(W(t)) = 0$ be the spectrum of $W(t)$, and define condition number $\chi = \max_t \chi(W(t)) = \max_t \frac{\lambda_1(W(t))}{\lambda_{M-1}(W(t))}$. Note that in practice we use not the matrix $W(t)$, but $\tilde{W}(t) = I - \frac{W(t)}{\lambda_1(W(t))}$, since this type of matrices are used in consensus algorithms [7]. To estimate the convergence speed, we introduce

$$
\rho = \max_t \lambda_2(\tilde{W}(t)) = \max_t \left[1 - \frac{\lambda_{M-1}(W(t))}{\lambda_1(W(t))} \right] = \max_t \left[1 - \frac{1}{\chi(W(t))} \right]
$$

$$
= 1 - \frac{1}{\max_t \chi(W(t))} = 1 - \frac{1}{\chi}.
$$

3   Main Part

We divide our contribution into two main parts, first we discuss lower bounds for decentralized saddle point problems over time-varying graphs. In the second part, we present an algorithm that achieves the lower bounds (up to logarithmic factors and numerical constants).

3.1   Lower Bounds

Before presenting lower bounds, we must restrict the class of algorithms for which our lower bounds are valid. For this we introduce the following black-box procedure.

Definition 1. *Each device m has its own local memories $\mathcal{M}_m^x$ and $\mathcal{M}_m^y$ for the x- and y-variables, respectively–with initialization $\mathcal{M}_m^x = \mathcal{M}_m^y = \{0\}$. $\mathcal{M}_m^x$ and $\mathcal{M}_m^x$ are updated as follows:*

- **Local computation:** *Each device* m *computes and adds to its* $\mathcal{M}_m^x$ *and* $\mathcal{M}_m^y$ *a finite number of points* x, y, *each satisfying*

$$x \in span\{x' , \nabla_x f_m(x'', y'')\}, \quad y \in span\{y' , \nabla_y f_m(x'', y'')\}, \quad (7)$$

for given $x', x'' \in \mathcal{M}_m^x$ *and* $y', y'' \in \mathcal{M}_m^y$.

- **Communication:** *Based upon communication round among neighbouring nodes at the moment* t, $\mathcal{M}_m^x$ *and* $\mathcal{M}_m^y$ *are updated according to*

$$\mathcal{M}_m^x := span\left\{ \bigcup_{(i,m)\in\mathcal{E}(t)} \mathcal{M}_i^x \right\}, \quad \mathcal{M}_m^y := span\left\{ \bigcup_{(i,m)\in\mathcal{E}(t)} \mathcal{M}_i^y \right\}. \quad (8)$$

- **Output:** *The final global output at the current moment of time is calculated as:*

$$x \in span\left\{ \bigcup_{m=1}^{M} \mathcal{M}_m^x \right\}, \quad y \in span\left\{ \bigcup_{m=1}^{M} \mathcal{M}_m^y \right\}.$$

This definition includes all algorithms capable of making local gradient updates, as well as exchanging information with neighbors. Notice that the proposed oracle builds on [30] for minimization problems over networks.

Theorem 1. *For any* L *and* μ , *there exists a saddle point problem in the form (1) with* $\mathcal{Z} = \mathcal{R}^{2d}$ *(where* d *is sufficiently large) and non-zero solution* y^*. *All local functions* f_m *of this problem are* L-*smooth,* μ-*strongly-convex-strongly-concave. Then, for any* $\chi \geq 1$, *there exists a sequence of gossip matrices* $W(t)$ *over the connected (at each moment) graph* $\mathcal{G}(t)$, *satisfying Assumption 4 with condition number* χ, *such that any decentralized algorithm satisfying Definition 1 and using the gossip matrices* $W(t)$ *produces the following estimate on the global output* $z = (x, y)$ *after* K *communication rounds:*

$$\|z^K - z^*\|^2 = \Omega\left(\exp\left(-\frac{256\mu}{L-\mu} \cdot \frac{K}{\chi} \right) \|y^*\|^2 \right).$$

The idea of finding lower bounds is to construct an example of "bad" functions and the "critical" location of these functions on the nodes. In papers [5,6], lower bounds for decentralized saddle point problems (but on fixed communication networks) were already investigated. Examples of "bad" functions and their analysis can be taken from these works. An example of "bad" time-varying topology of the node connection is a star with a changing center. Obtaining lower bounds using such varying networks for minimization problems was obtained in [15]. To prove Theorem 1 we need to combine results [5] and [15].

The following statement interprets Theorem 1 in terms of the number of local computations on each device and the number of communications between them.

Corollary 1. *In the setting of Theorem 1, the number of communication rounds required to obtain a ε-solution is lower bounded by*

$$\Omega\left(\chi\frac{L}{\mu}\cdot\log\left(\frac{\|y^*\|^2}{\varepsilon}\right)\right).$$

Additionally, we can get a lower bound for the number of local calculations on each of the devices:

$$\Omega\left(\frac{L}{\mu}\cdot\log\left(\frac{\|y^*\|^2}{\varepsilon}\right)\right).$$

Also we want to find lower bounds for the case of (non strongly) convex-concave problems, one can use regularization and consider the following objective function

$$f(x,y)+\frac{\varepsilon}{4D^2}\cdot\|x-x^0\|^2-\frac{\varepsilon}{4D^2}\cdot\|y-y^0\|^2,$$

which is strongly-convex-strongly-concave with constant $\mu=\frac{\varepsilon}{2D^2}$, where ε is a precision of the solution and D is the diameter of the sets $\mathcal{X}$ and $\mathcal{Y}$. The resulting new SPP problem is solved to $\varepsilon/2$-precision in order to guarantee an accuracy ε in computing the solution of the original problem. Therefore, we can easily deduce the lower bounds for convex-concave case

$$\Omega\left(\chi\frac{LD^2}{\varepsilon}\right)\ \text{communication rounds}\quad\text{and}\quad\Omega\left(\frac{LD^2}{\varepsilon}\right)\ \text{local computations}.$$

See Table 1 to compare with lower bounds for constant networks.

3.2   Near-Optimal Algorithm

In this part, we present an Algorithm that achieves lower bounds (up to logarithmic terms). Our Algorithm uses an auxiliary procedure for communication. This is a classic procedure - Gossip Algorithm.

Algorithm 1. Gossip Algorithm (`Gossip`)

Parameters: Vectors $z_1,...,z_M$, communic. rounds H.
Initialization: Construct matrix $\mathbf{z}$ with rows $z_1^T,...,z_M^T$.
Choose $\mathbf{z}^0=\mathbf{z}$.
for $h=0,1,2,\ldots,H$ **do**
    $\mathbf{z}^{h+1}=\tilde{W}(h)\cdot\mathbf{z}^h$
end for
Output: rows $z_1,...,z_M$ of $\mathbf{z}^{H+1}$.

The essence of the `Gossip` is very simple. Initially, there are vectors z_1 and z_M, which are stored on their devices. Our goal is to get a vector close to the $\bar{z}=\frac{1}{M}\sum_{m=1}^M z_m$ vector on all devices. At each iteration, each device exchange local vectors with its neighbors, and then modify its local vector by averaging local vector and vectors of neighbors with weights from the matrix $W(h)$.

We are now ready to present our main algorithm. It is based on the classical method for smooth saddle point problems - Extra Step Method (Mirror Prox) [12,25]. With the right choice of H, we can achieve averaging of all vectors with good accuracy. In particular, we can assume that $z_1^k \approx \ldots \approx z_M^k$. For more details about the choice of H and a detailed analysis of the algorithm (taking into account that in the general $z_1^k \neq \ldots \neq z_M^k$), see in the full version of the paper.

Algorithm 2. Time-Varying Decentralized Extra Step Method (TVDESM)

Parameters: Stepsize $\gamma \leq \frac{1}{4L}$, number of Gossip steps H.
Initialization: Choose $(x^0, y^0) = z^0 \in \mathcal{Z}$, $z_m^0 = z^0$.
for $k = 0, 1, 2, \ldots,$ do
    Each machine m computes $\hat{z}_m^{k+1/2} = z_m^k - \gamma \cdot F_m(z_m^k)$
    Communication: $\hat{z}_1^{k+1/2}, \ldots, \hat{z}_M^{k+1/2} = \text{Gossip}(\hat{z}_1^{k+1/2}, \ldots, \hat{z}_M^{k+1/2}, H)$
    Each machine m computes $z_m^{k+1/2} = \text{proj}_{\mathcal{Z}}(\hat{z}_m^{k+1/2})$,
    Each machine m computes $\hat{z}_m^{k+1} = z_m^k - \gamma \cdot F_m(z_m^{k+1/2})$
    Communication: $\hat{z}_1^{k+1}, \ldots, \hat{z}_M^{k+1} = \text{Gossip}(\hat{z}_1^{k+1}, \ldots, \hat{z}_M^{k+1}, H)$
    Each machine m computes $z_m^{k+1} = \text{proj}_{\mathcal{Z}}(\hat{z}_m^{k+1})$
end for

The analysis of Algorithm 2 is derived from the analysis of classical extrastep method. We study the convergence properties of sequence $\{\bar{z}^k\}_{k=0}^{\infty}$, where $\bar{z}^k = \frac{1}{M} \sum_{m=1}^{M} z_m^k$. Note that $\bar{z}^k$ is not held at any agent; instead, this quantity is only used in the analysis. Algorithm 2 employs gossip averaging after each extrastep. Therefore, the method uses an approximate value of $F(\bar{z}^k)$ when performing updates, and the approximation error is driven by the number of gossip iterations H. The analysis of Algorithm 2 comes down to studying extrastep method which uses inexact values of F at each iteration. Given a target accuracy ε, we choose the number of gossip iterations H proportional to ε. Since Gossip (Algorithm 1) is a linearly convergent method, H is proportional to $\log(1/\varepsilon)$. As a result, we have a $\log^2(1/\varepsilon)$ term in the number of communication rounds of Algorithm 2.

Theorem 2. *Let $\{z_m^k\}_{k \geq 0}^K$ denote the iterates of Algorithm 2 for solving problem (1) after K communication rounds. Let Assumptions 1(g,l) and 4 be satisfied. Then, if $\gamma \leq \frac{1}{4L}$, we have the following estimates in*
- *μ-strongly-convex-strongly-concave case (Assumption 2(s)):*

$$\|\bar{z}^{K+1} - z^*\|^2 = \tilde{\mathcal{O}}\left(\|z^0 - z^*\|^2 \exp\left(-\frac{\mu K}{8L\chi}\right)\right),$$

- *convex-concave case (Assumption 2 and 3):*

$$gap(\bar{z}_{avg}^{K+1}) = \tilde{\mathcal{O}}\left(\frac{L\Omega_z^2 \chi}{K}\right),$$

$$where \ \bar{z}^t = \frac{1}{M} \sum_{m=1}^{M} z_m^t, \ \bar{z}_{avg}^{k+1} = \frac{1}{M(k+1)} \sum_{t=0}^{k} \sum_{m=1}^{M} z_m^{t+1/2} \ and$$

$$gap(z) = \max_{y' \in \mathcal{Y}} f(x, y') - \min_{x' \in \mathcal{X}} f(x', y).$$

Corollary 2. *In the setting of Theorem 2, the number of communication rounds required for Algorithm 2 to obtain a ε-solution is upper bounded by*

$$\tilde{O}\left(\chi \frac{L}{\mu}\right)$$

in μ-strongly-convex–strongly-concave case and

$$\tilde{O}\left(\chi \frac{LD^2}{\varepsilon}\right)$$

in convex-concave case. Additionally, one can obtain upper bounds for the number of local calculations on each of the devices:

$$O\left(\frac{L}{\mu} \cdot \log\left(\frac{\|z^0 - z^*\|^2}{\varepsilon}\right)\right)$$

in μ-strongly-convex–strongly-concave case and

$$O\left(\frac{LD^2}{\varepsilon}\right)$$

in convex-concave case.

Corollary 2 illustrates that Algorithm 2 achieves lower bounds both for convex-concave and μ-strongly-convex-strongly-concave cases up to a logarithmic factor (the lower bounds are determined in Corollary 1). It can be observed that the complexity bounds are constituted of two factors: χ representing network connectivity and L/μ of LD^2/ε corresponding to the objective function. This effect is typical for decentralized optimization (see i.e. [30]). On the contrary to distributed minimization tasks, the dependence on function condition number L/μ is unimprovable for min-max problems (i.e. this factor cannot be enhanced to $\sqrt{L/\mu}$). Moreover, the dependence on χ cannot be improved to $\sqrt{\chi}$, since we focus on time-varying networks [15].

4   Conclusion

In conclusion, we briefly summarize the contributions of this paper and discuss the directions for future work. Our findings consist of two parts: lower bounds and optimal (up to a logarithmic factor) algorithms.

First, we derived the lower bounds for the classes of convex-concave and strongly-convex-strongly-concave min-max problems over time-varying graphs. The graph is assumed to be connected at each communication round. However, we studied only one class of time-varying networks. Other classes are connected to different assumptions on the network structure. In particular, in B-connected networks [23] the graph can be disconnected at some times, but the union of any B consequent graphs must be connected. Yet another possible assumption is the randomly changing graph with a contraction property of W in expectation [14]. Developing lower bounds for min-max problems for these two classes is an open question in decentralized optimization.

Second, we proposed a near-optimal algorithm with a gossip subroutine resulting in squared logarithmic factor. Developing an algorithm without an additional logarithmic factor would close the gap in theory and result in a more practical algorithm with less parameters to fine-tune. Possible directions for developing such an algorithm are generalizations of dual-based approaches for minimization [16,20] and gradient-tracking [20,23].

Finally, the comparison of our algorithm to existing works requires additional numerical experiments, which is left for future work.

References

1. Abadeh, S., Esfahani, P., Kuhn, D.: Distributionally robust logistic regression. In: Advances in Neural Information Processing Systems (NeurIPS), pp. 1576–1584 (2015)
2. Arjovsky, M., Chintala, S., Bottou, L.: Wasserstein generative adversarial networks. In: Proceedings of the 34th International Conference on Machine Learning (ICML), vol. 70, no. 1, pp. 214–223 (2017)
3. Bertsekas, D.P., Tsitsiklis, J.N.: Parallel and Distributed Computation: Numerical Methods, vol. 23. Prentice Hall Englewood Cliffs (1989)
4. Beznosikov, A., Dvurechensky, P., Koloskova, A., Samokhin, V., Stich, S.U., Gasnikov, A.: Decentralized local stochastic extra-gradient for variational inequalities. arXiv preprint arXiv:2106.08315 (2021)
5. Beznosikov, A., Samokhin, V., Gasnikov, A.: Local SGD for saddle-point problems. arXiv preprint arXiv:2010.13112 (2020)
6. Beznosikov, A., Scutari, G., Rogozin, A., Gasnikov, A.: Distributed saddle-point problems under similarity. arXiv preprint arXiv:2107.10706 (2021)
7. Boyd, S., Ghosh, A., Prabhakar, B., Shah, D.: Randomized gossip algorithms. IEEE Trans. Inf. Theory **52**(6), 2508–2530 (2006)
8. Chambolle, A., Pock, T.: A first-order primal-dual algorithm for convex problems with applications to imaging. J. Math. Imaging Vis. **40**(1), 120–145 (2011)
9. Facchinei, F., Pang, J.: Finite-Dimensional Variational Inequalities and Complementarity Problems. Springer Series in Operations Research and Financial Engineering, Springer, New York (2007). https://books.google.ru/books?id=lX_7Rce3_Q0C. https://doi.org/10.1007/b97543
10. Goodfellow, I.J., et al.: Generative adversarial networks (2014)
11. Jin, Y., Sidford, A.: Efficiently solving MDPs with stochastic mirror descent. In: III, H.D., Singh, A. (eds.) Proceedings of the 37th International Conference on Machine Learning. Proceedings of Machine Learning Research, vol. 119, pp. 4890–4900. PMLR, 13–18 July 2020
12. Juditsky, A., Nemirovskii, A.S., Tauvel, C.: Solving variational inequalities with Stochastic Mirror-Prox algorithm (2008)

13. Kempe, D., Dobra, A., Gehrke, J.: Gossip-based computation of aggregate information. In: 44th Annual IEEE Symposium on Foundations of Computer Science, 2003. Proceedings, pp. 482–491. IEEE (2003)
14. Koloskova, A., Loizou, N., Boreiri, S., Jaggi, M., Stich, S.U.: A unified theory of decentralized SGD with changing topology and local updates. arXiv preprint arXiv:2003.10422 (2020)
15. Kovalev, D., Gasanov, E., Richtárik, P., Gasnikov, A.: Lower bounds and optimal algorithms for smooth and strongly convex decentralized optimization over time-varying networks. arXiv preprint arXiv:2106.04469 (2021)
16. Kovalev, D., Shulgin, E., Richtárik, P., Rogozin, A., Gasnikov, A.: ADOM: accelerated decentralized optimization method for time-varying networks. arXiv preprint arXiv:2102.09234 (2021)
17. Li, H., Lin, Z.: Accelerated gradient tracking over time-varying graphs for decentralized optimization. arXiv preprint arXiv:2104.02596 (2021)
18. Liu, M., et al.: A decentralized parallel algorithm for training generative adversarial nets. arXiv preprint arXiv:1910.12999 (2019)
19. Liu, W., Mokhtari, A., Ozdaglar, A., Pattathil, S., Shen, Z., Zheng, N.: A decentralized proximal point-type method for saddle point problems. arXiv preprint arXiv:1910.14380 (2019)
20. Maros, M., Jaldén, J.: PANDA: a dual linearly converging method for distributed optimization over time-varying undirected graphs. In: 2018 IEEE Conference on Decision and Control (CDC), pp. 6520–6525 (2018)
21. McDonald, R., Hall, K., Mann, G.: Distributed training strategies for the structured perceptron. In: Human Language Technologies: The 2010 Annual Conference of the North American Chapter of the Association for Computational Linguistics, pp. 456–464 (2010)
22. McMahan, B., Moore, E., Ramage, D., Hampson, S., Arcas, B.A.: Communication-efficient learning of deep networks from decentralized data. In: Artificial Intelligence and Statistics, pp. 1273–1282. PMLR (2017)
23. Nedić, A., Olshevsky, A., Shi, W.: Achieving geometric convergence for distributed optimization over time-varying graphs. SIAM J. Optim. **27**(4), 2597–2633 (2017)
24. Nedic, A., Ozdaglar, A.: Distributed subgradient methods for multi-agent optimization. IEEE Trans. Autom. Control **54**(1), 48–61 (2009)
25. Nemirovski, A.: Prox-method with rate of convergence o(1/t) for variational inequalities with Lipschitz continuous monotone operators and smooth convex-concave saddle point problems. SIAM J. Optim. **15**, 229–251 (2004). https://doi.org/10.1137/S1052623403425629
26. von Neumann, J., Morgenstern, O., Kuhn, H.: Theory of Games and Economic Behavior (Commemorative Edition). Princeton University Press (2007)
27. Rogozin, A., Beznosikov, A., Dvinskikh, D., Kovalev, D., Dvurechensky, P., Gasnikov, A.: Decentralized distributed optimization for saddle point problems. arXiv preprint arXiv:2102.07758 (2021)
28. Rogozin, A., Gasnikov, A.: Projected gradient method for decentralized optimization over time-varying networks. arXiv preprint arXiv:1911.08527 (2019)
29. Rogozin, A., Uribe, C.A., Gasnikov, A.V., Malkovsky, N., Nedić, A.: Optimal distributed convex optimization on slowly time-varying graphs. IEEE Trans. Control Network Syst. **7**(2), 829–841 (2019)
30. Scaman, K., Bach, F., Bubeck, S., Lee, Y.T., Massoulié, L.: Optimal algorithms for smooth and strongly convex distributed optimization in networks. arXiv preprint arXiv:1702.08704 (2017)

31. Shalev-Shwartz, S., Ben-David, S.: Understanding Machine Learning: From Theory to Algorithms. Cambridge University Press (2014)
32. Ye, H., Luo, L., Zhou, Z., Zhang, T.: Multi-consensus decentralized accelerated gradient descent. arXiv preprint arXiv:2005.00797 (2020)

Towards Accelerated Rates for Distributed Optimization over Time-Varying Networks

Alexander Rogozin[1][(✉)], Vladislav Lukoshkin[2], Alexander Gasnikov[1], Dmitry Kovalev[3], and Egor Shulgin[3]

[1] Moscow Institute of Physics and Technology, Dolgoprudny, Russia
`aleksandr.rogozin@phystech.edu`
[2] Skolkovo Institute of Science and Technology, Moscow, Russia
[3] King Abdullah Institute of Science and Technology, Thuwal, Saudi Arabia

Abstract. We study the problem of decentralized optimization with strongly convex smooth cost functions. This paper investigates accelerated algorithms under time-varying network constraints. In our approach, nodes run a multi-step gossip procedure after taking each gradient update, thus ensuring approximate consensus at each iteration. The outer cycle is based on accelerated Nesterov scheme. Both computation and communication complexities of our method have an optimal dependence on global function condition number κ_g. In particular, the algorithm reaches an optimal computation complexity $O(\sqrt{\kappa_g} \log(1/\varepsilon))$.

Keywords: Distributed optimization · Time-varying network

1 Introduction

In this work, we study a sum-type minimization problem

$$f(x) = \frac{1}{n} \sum_{i=1}^{n} f_i(x) \rightarrow \min_{x \in \mathbb{R}^d} . \tag{1}$$

Convex functions f_i are stored separately by nodes in a communication network, which is represented by an undirected graph $\mathcal{G} = (V, E)$. This type of problems arise in distributed machine learning, drone or satellite networks, statistical inference [1] and power system control [2]. The computational agents over the network have access to their local f_i and can communicate only with their neighbors, but still aim to minimize the global objective in (1).

The research of A. Rogozin was partially supported by RFBR 19-31-51001 and was partially done in Sirius (Sochi). The research of A. Gasnikov was partially supported by the Ministry of Science and Higher Education of the Russian Federation (Goszadaniye) 075-00337-20-03, project no. 0714-2020-0005.

N. N. Olenev et al. (Eds.): OPTIMA 2021, LNCS 13078, pp. 258–272, 2021.
https://doi.org/10.1007/978-3-030-91059-4_19

The basic idea behind approach of this paper is to reformulate problem (1) as a problem with linear constraints. Let us assign each agent in the network a personal copy of parameter vector x_i and introduce

$$\mathbf{X} = (x_1 \ldots x_n)^\top \in \mathbb{R}^{n \times d}, \quad F(\mathbf{X}) = \sum_{i=1}^{n} f_i(x_i).$$

Now we equivalently rewrite problem (1) as

$$\min_{\mathbf{X}} F(\mathbf{X}) = \sum_{i=1}^{n} f_i(x_i) \text{ s.t. } x_1 = \ldots = x_n. \tag{2}$$

This reformulation increases the number of variables, but induces additional constraints at the same time. Problem (2) has the same optimal value as problem (1).

Let us denote the set of consensus constraints $\mathcal{C} = \{x_1 = \ldots = x_n\}$. Also, for each $\mathbf{X} \in \mathbb{R}^{n \times d}$ denote average of its columns $\overline{x} = \frac{1}{n} \sum_{i=1}^{n} x_i$ and introduce its projection onto the constraint set:

$$\overline{\mathbf{X}} = \frac{1}{n} \mathbf{1}_n \mathbf{1}_n^\top \mathbf{X} = \Pi_\mathcal{C}(\mathbf{X}) = (\overline{x} \ldots \overline{x})^\top,$$

where $\mathbf{1}_n = (1 \ldots 1)^\top$ is a vector consisting of n ones. Note that $\mathcal{C}$ is a linear subspace in $\mathbb{R}^{n \times d}$, and therefore projection operator $\Pi_\mathcal{C}(\cdot)$ is linear.

Decentralized optimization methods aim at minimizing the objective function and maintaining consensus accuracy between nodes. The optimization part is performed by using gradient steps. At the same time, keeping every agent's parameter vector close to average over the nodes is done via communication steps. Alternating gradient and communication updates allows both to minimize the objective and control consensus constraint violation.

In a centralized scenario, there exists a server that is able to communicate with every agent in the network. In particular, a common parameter vector is maintained at all of the nodes. However, in decentralized setting it is only possible to ensure that agent's vectors are approximately equal with desired accuracy. The algorithm studied in this paper runs a sequence of communication rounds after every optimization step. We refer to this series of communications as a *consensus subroutine*. Such information exchange allows to reach approximate consensus between nodes after each gradient update, while the accuracy is controlled by the number of communication rounds.

On the one hand, a method that employs a consensus subroutine after each gradient update mimics a centralized algorithm. The difference is that in presence of a master node all computational entities have an opportunity to hold strictly equal copies of the variable, while in decentralized case consensus constraints are satisfied only with nonzero accuracy. On the other hand, consensus subroutine may be interpreted as an inexact projection onto the constraint set $\mathcal{C}$, which is done in a number of iterations. Each of iterations represented by a communication round. Therefore, our approach fits the inexact oracle framework that has been studied in [3,4]. We note that a similar approach to decentralized optimization is studied in [5], but this paper studies only time-static graphs.

We aim at building a first-order method with trajectory lying in neighborhood of $\mathcal{C}$. A simple example would be GD with inexact projections.

$$\mathbf{X}^{k+1} \approx \Pi_{\mathcal{C}}(\mathbf{X}^k - \gamma\nabla F(\mathbf{X}^k)) = \overline{\mathbf{X}}^k - \gamma\overline{\nabla F}(\mathbf{X}^k), \tag{3}$$

where $\nabla F(\mathbf{X}^k) = (\nabla f_1(x_1^k) \ldots \nabla f_n(x_n^k))^\top$ denotes the gradient of F.

Algorithm with update rule 3 can be viewed as a gradient descent with an inexact oracle. If the oracle was exact, the update rule would write as

$$\overline{\mathbf{X}}^{k+1} = \overline{\mathbf{X}}^k - \gamma\overline{\nabla F}(\overline{\mathbf{X}}^k),$$

thus making the method trajectory stay precisely in $\mathcal{C}$. In this particular example, inexact gradient $\overline{\nabla F}(\mathbf{X}^k)$ approximates exact gradient $\overline{\nabla F}(\overline{\mathbf{X}}^k)$.

Throughout the paper, $\langle\cdot,\cdot\rangle$ denotes the inner product of vectors or matrices. Correspondingly, by $\|\cdot\|$ we denote a 2-norm for vectors or Frobenius norm for matrices.

1.1 Related Work

A decentralized algorithm makes two types of steps: local updates and information exchange. The complexity of such methods depends on objective condition number κ and a term χ representing graph connectivity (namely, the eigengap of a graph-associated communication matrix).

Local steps may use gradient [6–12] or sub-gradient [13] computations. In primal-only methods, the agents compute gradients of their local functions and alternate taking gradient steps and communication procedures. Under cheap communication costs, it may be beneficial to replace a single consensus iteration with a series of information exchange rounds. Such methods as MSDA [14], D-NC [15] and Mudag [11] employ multi-step gossip procedures.

Typically, non-accelerated methods need $O(\kappa\chi\log(1/\varepsilon))$ iterations to yield a solution with ε-accuracy [16]. Nesterov acceleration may be employed to improve dependence on κ or χ and obtain algorithms with $O(\sqrt{\kappa\chi}\log(1/\varepsilon))$ complexity. In order to achieve this, one may distribute accelerated methods directly [9,11,12,15,17] or use a Catalyst framework [18]. Accelerated methods meet the lower complexity bounds for decentralized optimization [14,19,20].

Consensus restrictions $x_1 = \ldots = x_n$ may be treated as linear constraints, thus allowing for a dual reformulation of problem (1). Dual-based methods include dual ascent and its accelerated variants [14,21–23]. Primal-dual approaches like ADMM [24,25] are also implementable in decentralized scenarios.

In [7], the authors developed algorithms for non-convex objectives and provided lower complexity bounds for non-convex case, as well.

Time-varying networks open a new venue in research. Changing topology requires new approaches to decentralized methods and a more complicated theoretical analysis. The first method with provable linear convergence was proposed in [6]. Such primal algorithms as Push-Pull Gradient Method [8] and DIGing [6] are robust to network changes and have theoretical guarantees of convergence over time-varying graphs. Recently, a dual method for time-varying architectures was introduced in [26].

1.2 Summary of Contributions

This paper focuses on smooth strongly convex objectives. Our analysis is bounded to the following

Assumption 1. *For every $i = 1, \ldots, n$, function f_i is differentiable, μ_i-strongly convex and L_i-smooth (μ_i, $L_i > 0$).*

Under this assumption it holds

- (local constants) $F(X)$ is μ_l-strongly convex and L_l-smooth on $\mathbb{R}^{n \times d}$, where $\mu_l = \min_i \mu_i$, $L_l = \max_i L_i$.
- (global constants) $F(X)$ is μ_g-strongly convex and L_g-smooth on $\mathcal{C}$, where $\mu_g = \frac{1}{n} \sum_{i=1}^n \mu_i$, $L_g = \frac{1}{n} \sum_{i=1}^n L_i$.

The global conditioning may be significantly better than local (see i.e. [14] for details). The analysis shows that performance of Algorithm 2 depends on global constants. Our approach uses multi-step gossip averaging, and the analysis is based on the inexact oracle framework.

The proposed algorithm (Algorithm 2) requires $O(\sqrt{\kappa_g} \chi \log^2(1/\varepsilon))$ communication rounds, where κ_g denotes the (global) condition number of f and χ is a term characterizing graph connectivity, which is defined later in the paper. For a static graph, $\chi = 1/\gamma$, where γ denotes the normalized eigengap of communication matrix associated with the network. Our result has an accelerated rate on function condition number (the number of iterations depends on $\sqrt{\kappa_g}$, not κ_g) and is derived for time-varying networks.

Our complexity estimate includes the condition number κ_g, i.e. *global* strong convexity and smoothness constants instead of *local* ones. If the data among the nodes is strongly heterogeneous, it is possible that $\kappa_l \gg \kappa_g$, where κ_l is the local condition number [14,19,27]. Therefore, the method with complexity depending on κ_g may perform significantly better. A recently proposed method Mudag [11] has a complexity depending on $\sqrt{\kappa_g}$, as well, but the method is designed for time-static graphs.

The lower bound for number of communications is $\Omega(\sqrt{\kappa_l \chi} \log(1/\varepsilon))$ [14]. Our result is obtained for time-varying graphs and has a worse dependence on χ. On the other hand, we derive a better dependence on condition number, i.e. we use κ_g instead of κ_l. However, this by no means breaks the lower bounds. First, our result includes $\log^2(1/\varepsilon)$ instead of $\log(1/\varepsilon)$. Second, a function in [14] on which the lower bounds are attained has $\kappa_g \sim \kappa_l$. Namely, for the bad function it holds $\kappa_g \geqslant \kappa_l/16$ (see Appendix A.1 in [14] for details).

2   Inexact Oracle Framework

In this section, we describe the inexact oracle construction for the objective function f.

2.1   Preliminaries

Initially we recall the definition of (δ, L, μ)-oracle from [4]. Let $h(x)$ be a convex function defined on a convex set $Q \subseteq \mathbb{R}^m$. We say that $(h_{\delta,L,\mu}(x), s_{\delta,L,\mu}(x))$ is a (δ, L, μ)-model of $h(x)$ at point $x \in Q$ if for all $y \in Q$ it holds

$$\frac{\mu}{2} \|y - x\|^2 \leqslant h(y) - (h_{\delta,L,\mu}(x) + \langle s_{\delta,L,\mu}(x), y - x \rangle) \leqslant \frac{L}{2} \|y - x\|^2 + \delta. \qquad (4)$$

2.2   Inexact Oracle for F

Consider $\overline{x}, \overline{y} \in \mathbb{R}^d$ and define $\overline{\mathbf{X}} = (\overline{x} \dots \overline{x})^\top$, $\overline{\mathbf{Y}} = (\overline{y} \dots \overline{y})^\top \in \mathcal{C}$. Let $\mathbf{X} \in \mathbb{R}^{n \times d}$ be such that $\Pi_{\mathcal{C}}(\mathbf{X}) = \overline{\mathbf{X}}$ and $\left\| \overline{\mathbf{X}} - \mathbf{X} \right\|^2 \leqslant \delta'$.

Lemma 1. *Define*

$$\delta = \frac{1}{2n} \left(\frac{L_l^2}{L_g} + \frac{2L_l^2}{\mu_g} + L_l - \mu_l \right) \delta', \qquad (5)$$

$$f_{\delta,L,\mu}(\overline{x}, \mathbf{X}) = \frac{1}{n} \left[F(\mathbf{X}) + \langle \nabla F(\mathbf{X}), \overline{\mathbf{X}} - \mathbf{X} \rangle + \frac{1}{2} \left(\mu_l - \frac{2L_l^2}{\mu_g} \right) \left\| \overline{\mathbf{X}} - \mathbf{X} \right\|^2 \right],$$

$$g_{\delta,L,\mu}(\overline{x}, \mathbf{X}) = \frac{1}{n} \sum_{i=1}^n \nabla f_i(x_i).$$

Then $(f_{\delta,L,\mu}(\overline{x}, \mathbf{X}), g_{\delta,L,\mu}(\overline{x}, \mathbf{X}))$ is a $(\delta, 2L_g, \mu_g/2)$-model of f at point $\overline{x}$, i.e.

$$\frac{\mu_g}{4} \|\overline{y} - \overline{x}\|^2 \leqslant f(\overline{y}) - f_{\delta,L,\mu}(\overline{x}, \mathbf{X}) - \langle g_{\delta,L,\mu}(\overline{x}, \mathbf{X}), \overline{y} - \overline{x} \rangle \leqslant L_g \|\overline{y} - \overline{x}\|^2 + \delta.$$

Proof. We aim at obtaining estimates for $F(\overline{\mathbf{Y}})$ similar to (4). First, we get a lower bound on $F(\overline{\mathbf{Y}})$.

$$F(\overline{\mathbf{Y}}) \geqslant F(\mathbf{X}) + \left[\langle \nabla F(\mathbf{X}), \overline{\mathbf{X}} - \mathbf{X} \rangle + \frac{\mu_l}{2} \|\mathbf{X} - \overline{\mathbf{X}}\|^2 \right] + \left[\langle \nabla F(\overline{\mathbf{X}}), \overline{\mathbf{Y}} - \overline{\mathbf{X}} \rangle + \frac{\mu_g}{2} \|\overline{\mathbf{Y}} - \overline{\mathbf{X}}\|^2 \right]$$

$$= \left[F(\mathbf{X}) + \langle \nabla F(\mathbf{X}), \overline{\mathbf{X}} - \mathbf{X} \rangle + \frac{\mu_l}{2} \|\mathbf{X} - \overline{\mathbf{X}}\|^2 \right] + \langle \nabla F(\overline{\mathbf{X}}), \overline{\mathbf{Y}} - \overline{\mathbf{X}} \rangle$$

$$+ \langle \nabla F(\overline{\mathbf{X}}) - \nabla F(\mathbf{X}), \overline{\mathbf{Y}} - \overline{\mathbf{X}} \rangle + \frac{\mu_g}{2} \|\overline{\mathbf{Y}} - \overline{\mathbf{X}}\|^2. \qquad (6)$$

Let us lower bound the term $\langle \nabla F(\overline{\mathbf{X}}) - \nabla F(\mathbf{X}), \overline{\mathbf{Y}} - \overline{\mathbf{X}} \rangle$ using Young inequality $\langle a, b \rangle \leqslant \frac{\|a\|^2}{2p} + \frac{p}{2} \|b\|^2$, $p > 0$.

$$\langle \nabla F(\overline{\mathbf{X}}) - \nabla F(\mathbf{X}), \overline{\mathbf{Y}} - \overline{\mathbf{X}} \rangle \geqslant -\frac{1}{2p} \left\| \nabla F(\overline{\mathbf{X}}) - \nabla F(\mathbf{X}) \right\|^2 - \frac{p}{2} \left\| \overline{\mathbf{Y}} - \overline{\mathbf{X}} \right\|^2$$

$$\geqslant -\frac{L_l^2}{2p} \left\| \overline{\mathbf{X}} - \mathbf{X} \right\|^2 - \frac{p}{2} \left\| \overline{\mathbf{Y}} - \overline{\mathbf{X}} \right\|^2.$$

Returning to (6) and noting that $\langle \overline{\nabla F}(\mathbf{X}), \overline{\mathbf{Y}} - \overline{\mathbf{X}} \rangle = \langle \nabla F(\mathbf{X}), \overline{\mathbf{Y}} - \overline{\mathbf{X}} \rangle$, we get

$$F(\overline{\mathbf{Y}}) \geqslant \left[F(\mathbf{X}) + \langle \nabla F(\mathbf{X}), \overline{\mathbf{X}} - \mathbf{X} \rangle + \frac{1}{2} \left(\mu_l - \frac{L_l^2}{p} \right) \|\mathbf{X} - \overline{\mathbf{X}}\|^2 \right]$$
$$+ \langle \nabla F(\mathbf{X}), \overline{\mathbf{Y}} - \overline{\mathbf{X}} \rangle + \frac{\mu_g - p}{2} \|\overline{\mathbf{Y}} - \overline{\mathbf{X}}\|^2 \tag{7}$$

Second, we get an upper estimate on $F(\overline{\mathbf{Y}})$.

$$F(\overline{\mathbf{Y}}) \leqslant \left[F(\mathbf{X}) + \langle \nabla F(\mathbf{X}), \overline{\mathbf{X}} - \mathbf{X} \rangle + \frac{L_l}{2} \|\overline{\mathbf{X}} - \mathbf{X}\|^2 \right] + \left[\langle \overline{\nabla F}(\overline{\mathbf{X}}), \overline{\mathbf{Y}} - \overline{\mathbf{X}} \rangle + \frac{L_g}{2} \|\overline{\mathbf{Y}} - \overline{\mathbf{X}}\|^2 \right]$$
$$= \left[F(\mathbf{X}) + \langle \nabla F(\mathbf{X}), \overline{\mathbf{X}} - \mathbf{X} \rangle + \frac{L_l}{2} \|\overline{\mathbf{X}} - \mathbf{X}\|^2 \right] + \langle \overline{\nabla F}(\mathbf{X}), \overline{\mathbf{Y}} - \overline{\mathbf{X}} \rangle$$
$$+ \langle \overline{\nabla F}(\overline{\mathbf{X}}) - \overline{\nabla F}(\mathbf{X}), \overline{\mathbf{Y}} - \overline{\mathbf{X}} \rangle + \frac{L_g}{2} \|\overline{\mathbf{Y}} - \overline{\mathbf{X}}\|^2 . \tag{8}$$

Analogously, we estimate the term $\langle \overline{\nabla F}(\overline{\mathbf{X}}) - \overline{\nabla F}(\mathbf{X}), \overline{\mathbf{Y}} - \overline{\mathbf{X}} \rangle$ with Young inequality.

$$\langle \overline{\nabla F}(\overline{\mathbf{X}}) - \overline{\nabla F}(\mathbf{X}), \overline{\mathbf{Y}} - \overline{\mathbf{X}} \rangle \leqslant \frac{1}{2q} \|\overline{\nabla F}(\overline{\mathbf{X}}) - \overline{\nabla F}(\mathbf{X})\|^2 + \frac{q}{2} \|\overline{\mathbf{Y}} - \overline{\mathbf{X}}\|^2$$
$$\leqslant \frac{L_l^2}{2q} \|\overline{\mathbf{X}} - \mathbf{X}\|^2 + \frac{q}{2} \|\overline{\mathbf{Y}} - \overline{\mathbf{X}}\|^2 , \quad q > 0.$$

Plugging it into (8) and once again using $\langle \overline{\nabla F}(\mathbf{X}), \overline{\mathbf{Y}} - \overline{\mathbf{X}} \rangle = \langle \nabla F(\mathbf{X}), \overline{\mathbf{Y}} - \overline{\mathbf{X}} \rangle$ yields

$$F(\overline{\mathbf{Y}}) \leqslant \left[F(\mathbf{X}) + \langle \nabla F(\mathbf{X}), \overline{\mathbf{X}} - \mathbf{X} \rangle + \frac{1}{2} \left(\mu_l - \frac{L_l^2}{p} \right) \|\mathbf{X} - \overline{\mathbf{X}}\|^2 \right]$$
$$+ \langle \nabla F(\mathbf{X}), \overline{\mathbf{Y}} - \overline{\mathbf{X}} \rangle + \frac{L_g + q}{2} \|\overline{\mathbf{Y}} - \overline{\mathbf{X}}\|^2$$
$$+ \frac{1}{2} \left(\frac{L_l^2}{q} + \frac{L_l^2}{p} + L_l - \mu_l \right) \|\mathbf{X} - \overline{\mathbf{X}}\|^2 \tag{9}$$

Consequently, smoothness and strong convexity constants for inexact oracle are $L = L_g + q$, $\mu = \mu_g - p$, respectively. Choosing q and p allows to control condition number L/μ. Letting $q = L_g$, $p = \mu_g/2$ leads to

$$L = 2L_g, \tag{10a}$$
$$\mu = \mu_g/2. \tag{10b}$$

Noting that

$$F(\overline{\mathbf{X}}) = nf(\overline{x}), \quad F(\overline{\mathbf{Y}}) = nf(\overline{y}),$$
$$\|\overline{\mathbf{Y}} - \overline{\mathbf{X}}\|^2 = n \|\overline{y} - \overline{x}\|^2,$$
$$\langle \overline{\nabla F}(\mathbf{X}), \overline{\mathbf{Y}} - \overline{\mathbf{X}} \rangle = n \langle g_{\delta, L, \mu}(\overline{x}), \overline{y} - \overline{x} \rangle$$

and combining (7) and (9) leads to

$$\frac{\mu}{2}\left\|\overline{y}-\overline{x}\right\|^2 \leqslant f(\overline{y}) - f_{\delta,L,\mu}(\overline{x},\mathbf{X}) - \langle g_{\delta,L,\mu}(\overline{x},\mathbf{X}),\overline{y}-\overline{x}\rangle \leqslant \frac{L}{2}\left\|\overline{y}-\overline{x}\right\|^2 + \delta,$$

which concludes the proof.

3 Algorithm and Results

We take Algorithm 2 from [28] as a basis for our method. The algorithm is designed for the inexact oracle model and achieves an accelerated rate.

Consensus Subroutine
We consider a sequence of non-directed communication graphs $\{\mathcal{G}^k = (V,E^k)\}_{k=0}^\infty$ and a sequence of corresponding mixing matrices $\{\mathbf{W}^k\}_{k=0}^\infty$ associated with it.

We impose the following

Assumption 2. *Mixing matrix sequence* $\{\mathbf{W}^k\}_{k=0}^\infty$ *satisfies the following properties.*

- *(Decentralized property) If* $(i,j) \notin E_k$, *then* $[\mathbf{W}^k]_{ij} = 0$.
- *(Double stochasticity)* $\mathbf{W}^k\mathbf{1}_n = \mathbf{1}_n$, $\mathbf{1}_n^\top\mathbf{W}^k = \mathbf{1}_n^\top$.
- *(Contraction property) There exist* $\tau \in \mathbb{Z}_{++}$ *and* $\lambda \in (0,1)$ *such that for every* $k \geqslant \tau - 1$ *it holds*

$$\left\|\mathbf{W}_\tau^k\mathbf{X} - \overline{\mathbf{X}}\right\| \leqslant (1-\lambda)\left\|\mathbf{X}-\overline{\mathbf{X}}\right\|,$$

where $\mathbf{W}_\tau^k = \mathbf{W}^k \dots \mathbf{W}^{k-\tau+1}$.

The contraction property in Assumption 2 generalizes several assumptions in the literature.

- Time-static connected graph: $\mathbf{W}^k = \mathbf{W}$. In this classical case we have $\lambda = 1 - \sigma_2(\mathbf{W})$, where $\sigma_2(\mathbf{W})$ denotes the second largest singular value of $\mathbf{W}$.
- Sequence of connected graphs: every $\mathcal{G}_k$ is connected. In this scenario $\lambda = 1 - \sup_{k \geqslant 0} \sigma_2(\mathbf{W}^k)$.
- τ-connected graph sequence (i.e. for every $k \geqslant 0$ graph $\mathcal{G}_\tau^k = (V, E^k \cup E^{k+1} \cup \dots \cup E^{k+\tau-1})$ is connected [6]). For τ-connected graph sequences it holds $1 - \lambda = \sup_{k \geqslant 0} \sigma_{\max}(\mathbf{W}_\tau^k - \frac{1}{n}\mathbf{1}_n\mathbf{1}_n^\top)$.

Algorithm 1. Consensus

Require: Initial $\mathbf{X}^0 \in \mathcal{C}$, number of iterations K.
for $k = 0, \dots, K-1$ **do**
  $\mathbf{X}^{k+1} = \mathbf{W}^k\mathbf{X}^k$
end for
return $\mathbf{X}^K$

A stochastic variant of this contraction property is also studied in [29].

During every communication round, the agents exchange information according to the rule

$$x_i^{k+1} = w_{ii}^k + \sum_{(i,j) \in E^k} w_{ij}^k x_j^k.$$

In matrix form, this update rule writes as $\mathbf{X}^{k+1} = \mathbf{W}^k \mathbf{X}^k$. The contraction property in Assumption 2 is needed to ensure linear convergence of Algorithm 1 to the average of nodes' initial vectors, i.e. to $\bar{x}^0$. In particular, the contraction property holds for τ-connected graphs with Metropolis weights choice for $\mathbf{W}^k$, i.e.

$$[\mathbf{W}^k]_{ij} = \begin{cases} 1/(1 + \max\{d_i^k, d_j^k\}) & \text{if } (i,j) \in E^k, \\ 0 & \text{if } (i,j) \notin E^k, \\ 1 - \displaystyle\sum_{(i,m) \in E^k} [\mathbf{W}^k]_{im} & \text{if } i = j, \end{cases}$$

where d_i^k denotes the degree of node i in graph $\mathcal{G}^k$.

Algorithm and Result. We are now in position to introduce the decentralized algorithm of interest.

Algorithm 2. Decentralized AGD with consensus subroutine

Require: Initial guess $\mathbf{X}^0 \in \mathcal{C}$, constants $L, \mu > 0$, $\mathbf{U}^0 = \mathbf{X}^0$, $\alpha^0 = A^0 = 0$

1: **for** $k = 0, 1, 2, \ldots$ **do**

2:     Find α^{k+1} as the greater root of
        $(A^k + \alpha^{k+1})(1 + A^k \mu) = L(\alpha^{k+1})^2$

3:     $A^{k+1} = A^k + \alpha^{k+1}$

4:     $\mathbf{Y}^{k+1} = \dfrac{\alpha^{k+1} \mathbf{U}^k + A^k \mathbf{X}^k}{A^{k+1}}$

5:     $\mathbf{V}^{k+1} = \dfrac{\mu \mathbf{Y}^{k+1} + (1 + A^k \mu) \mathbf{U}^k}{1 + A^k \mu + \mu} - \dfrac{\alpha^{k+1} \nabla F(\mathbf{Y}^{k+1})}{1 + A^k \mu + \mu}$

6:     $\mathbf{U}^{k+1} = \text{Consensus}(\mathbf{V}^{k+1}, T^k)$

7:     $\mathbf{X}^{k+1} = \dfrac{\alpha^{k+1} \mathbf{U}^{k+1} + A^k \mathbf{X}^k}{A^{k+1}}$

8: **end for**

In the next theorem, we provide computation and communication complexities of Algorithm 2.

Theorem 3. *Define*

$$D_1 = \frac{L_l}{L_g^{1/2} \mu_g} \left[8\sqrt{2} L_l \left\| \bar{u}^0 - x^* \right\| \left(\frac{L_g}{\mu_g} \right)^{3/4} + \frac{4\sqrt{2} \left\| \nabla F(\mathbf{X}^*) \right\|}{\sqrt{n}} \left(\frac{L_g}{\mu_g} \right)^{1/4} \right],$$

$$D_2 = \frac{L_l}{L_g^{1/2} \mu_g} \left[3\sqrt{\mu_g} + 4\sqrt{2n} \left(\frac{L_g}{\mu_g} \right)^{1/4} \right] \qquad (11)$$

.

For a fixed $\varepsilon > 0$ introduce

$$\delta' = \frac{n\varepsilon}{32} \frac{\mu_g^{3/2}}{L_g^{1/2} L_l^{2}}, \quad D = \delta' \left(\frac{D_1}{\sqrt{\varepsilon}} + D_2 \right)^2 \text{ and } T_k = T = \frac{\tau}{2\lambda} \log \frac{D}{\delta'}.$$

Then Algorithm 2 requires

$$N = 2\sqrt{\frac{L_g}{\mu_g}} \log \left(\frac{\left\| \bar{u}^0 - x^* \right\|^2}{2\varepsilon L_g} \right) \tag{12}$$

gradient computations at each node and

$$N_{tot} = N \cdot T = 2\sqrt{\frac{L_g}{\mu_g} \frac{\tau}{\lambda}} \cdot \log \left(\frac{2L_g \left\| \bar{u}^0 - x^* \right\|^2}{\varepsilon} \right) \log \left(\frac{D_1}{\sqrt{\varepsilon}} + D_2 \right) \tag{13}$$

communication steps to yield $\mathbf{X}^N$ such that

$$f(\bar{x}^N) - f(x^*) \leqslant \varepsilon, \quad \left\| \mathbf{X}^N - \overline{\mathbf{X}}^N \right\|^2 \leqslant \delta'.$$

We provide the proof of Theorem 3 in Appendix A.

The number of gradient computations in (12) reaches the lower bounds for non-distributed optimization up to a constant factor. Number of communication steps includes an additional multiplicator of τ/λ, which characterizes graph connectivity.

4   Numerical Experiments

We consider the logistic regression problem with L2 regularizer:

$$f(x) = \frac{1}{n} \sum_{i=1}^{n} \log \left(1 + \exp(-b_i \langle a_i, x \rangle) \right) + \frac{\theta}{2} \|x\|_2^2$$

Here $a_1, \ldots, a_n \in \mathbb{R}^d$ denote the data points, $b_1, \ldots, b_n \in \{-1, 1\}$ denote class labels and $\theta > 0$ is a penalty coefficient. We run experiments on a least-squares task:

$$f(x) = \frac{1}{2} \|\mathbf{A}x - b\|_2^2.$$

The blocks of data matrix $\mathbf{A}$ and vector b are distributed among the agents in the network.

The simulations are run on LIBSVM datasets [30]. Among benchmark methods are EXTRA [10], DIGing [6], Mudag [11] and APM-C [31].

Logistic regression is carried out on a9a data-set, inner iterations are set to $T = 5$ for Mudag and DAccGD. Generation of random geometric graph goes on 20 nodes.

Figures 1 and 2 show that DAccGD outperforms its competitors both in the number of gradient steps and communication steps. On the other hand, DAccGD has a relatively worse consensus accuracy.

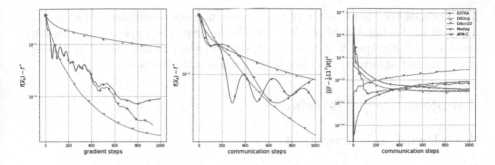

Fig. 1. a9a (logistic regression), 100 nodes

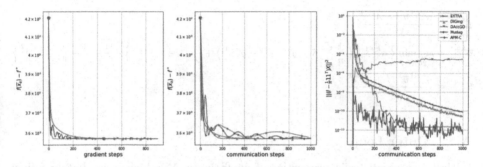

Fig. 2. Cadata (least squares), 20 nodes

5   Conclusion

This paper studies an inexact oracle-based approach to decentralized optimization. The paper focuses on a specific case of strongly convex smooth functions, but the inexact oracle framework introduced in [3] is also applicable to non-strongly convex functions. The development of this framework in [28] also enables to generalize the results of this article to composite optimization problems and distributed algorithms for saddle-point problems and variational inequalities.

Another interesting application of inexact oracle approach lies in stochastic decentralized algorithms. Consider a class of L-smooth μ-strongly convex objectives with gradient noise of each f_i being upper-bounded by σ^2. For this class of problems, lower complexity bounds write as [32]

$$O(\sqrt{L/\mu\chi}\log(1/\varepsilon))\qquad \text{stochastic oracle calls per node}$$

$$O\left(\max\left(\frac{\sigma^2}{m\mu\varepsilon}, \sqrt{L/\mu}\right)\log(1/\varepsilon)\right)\qquad \text{communication steps.}$$

At the moment, there exist methods that are optimal either in the number of oracle calls or in the number of communication steps. Several algorithms in the literature achieve lower bounds for stochastic oracle calls per node but do not meet the lower bounds for communication rounds. On the other hand, there are methods that achieve the optimal communication complexity but are sub-optimal in the number of oracle calls. We believe that approach of this paper combined with a specific batch-size choice described in [33] allows to develop a decentralized algorithm reaching both optimal complexities.

Supplementary Material

A   Proof of Theorem 3

First, note that Algorithm 2 comes down to the following iterative procedure in $\mathbb{R}^d$.

$$\overline{y}^{k+1} = \frac{\alpha^{k+1}\overline{u}^k + A^k\overline{x}^k}{A^{k+1}}$$

$$\overline{u}^{k+1} = \arg\min_{\overline{z}\in\mathbb{R}^d}\left\{\alpha^{k+1}\left(\left\langle\frac{1}{n}\sum_{i=1}^{n}\nabla f(y_i^{k+1}),\overline{z}-\overline{y}^{k+1}\right\rangle + \frac{\mu}{2}\left\|\overline{z}-\overline{y}^{k+1}\right\|^2\right) + \frac{1+A^k\mu}{2}\left\|\overline{z}-\overline{u}^k\right\|^2\right\}$$

$$\overline{x}^{k+1} = \frac{\alpha^{k+1}\overline{u}^{k+1} + A^k\overline{x}^k}{A^{k+1}}$$

A.1   Outer Loop

Initially we recall basic properties of coefficients A^k which immediately follow from Lemma 3.7 in [28] (for details see the full technical report of this paper [34]).

Lemma 2. *For coefficients A^k, it holds*

$$1.\ A^N \geqslant \frac{1}{L}\left(1 + \sqrt{\frac{\mu}{2L}}\right)^{2(N-1)} \qquad 2.\ \frac{\sum_{i=1}^{k}A^i}{A^k} \leqslant 1 + \sqrt{\frac{L}{\mu}}.$$

Lemma 3. *Provided that consensus accuracy is δ', i.e. $\left\|\mathbf{U}^j - \overline{\mathbf{U}}^j\right\|^2 \leqslant \delta'$ for $j = 1,\ldots,k$, we have*

$$f(\overline{x}^k) - f(x^*) \leqslant \frac{\left\|\overline{u}^0 - x^*\right\|^2}{2A^k} + \frac{2\sum_{j=1}^{k}A^j\delta}{A^k}$$

$$\left\|\overline{u}^k - x^*\right\|^2 \leqslant \frac{\left\|\overline{u}^0 - x^*\right\|^2}{1+A^k\mu} + \frac{4\sum_{j=1}^{k}A^j\delta}{1+A^k\mu}$$

where δ is given in (5).

Proof. First, assuming that $\left\|\mathbf{U}^j - \overline{\mathbf{U}}^j\right\|^2 \leqslant \delta'$, we show that $\mathbf{Y}^j, \mathbf{U}^j, \mathbf{X}^j$ lie in $\sqrt{\delta'}$-neighborhood of $\mathcal{C}$ by induction. At $j = 0$, we have $\left\|\mathbf{X}^0 - \overline{\mathbf{X}}^0\right\| = \left\|\mathbf{U}^0 - \overline{\mathbf{U}}^0\right\| = 0$. Using $A^{j+1} = A^j + \alpha^j$, we get an induction pass $j \to j+1$.

$$\left\| \mathbf{Y}^{j+1} - \overline{\mathbf{Y}}^{j+1} \right\| \leqslant \frac{\alpha^{j+1}}{A^{j+1}} \left\| \mathbf{U}^{j} - \overline{\mathbf{U}}^{j} \right\| + \frac{A^{j}}{A^{j+1}} \left\| \mathbf{X}^{j} - \overline{\mathbf{X}}^{j} \right\| \leqslant \sqrt{\delta'}$$

$$\left\| \mathbf{X}^{j+1} - \overline{\mathbf{X}}^{j+1} \right\| \leqslant \frac{\alpha^{j+1}}{A^{j+1}} \left\| \mathbf{U}^{j+1} - \overline{\mathbf{U}}^{j+1} \right\| + \frac{A^{j}}{A^{j+1}} \left\| \mathbf{X}^{j} - \overline{\mathbf{X}}^{j} \right\| \leqslant \sqrt{\delta'}$$

Therefore, $g(\overline{y}) = \frac{1}{n} \sum_{i=1}^{n} \nabla f(y_i)$ is a gradient from (δ, L, μ)-model of f, and the desired result directly follows from Theorem 3.4 in [28].

A.2   Consensus Subroutine Iterations

We specify the number of iteration required for reaching accuracy δ' in the following Lemma, which is proved in the extended version of this paper [34].

Lemma 4. *Let consensus accuracy be maintained at level δ', i.e. $\left\| \mathbf{U}^{j} - \overline{\mathbf{U}}^{j} \right\|^{2} \leqslant \delta'$ for $j = 1, \ldots, k$ and let Assumption 2 hold. Define*

$$\sqrt{D} := \left(\frac{2L_l}{\sqrt{L\mu}} + 1 \right) \sqrt{\delta'} + \frac{L_l}{\mu} \sqrt{n} \left(\left\| \overline{u}^{0} - x^{*} \right\|^{2} + \frac{8\delta'}{\sqrt{L\mu}} \right)^{1/2} + \frac{2 \left\| \nabla F(\mathbf{X}^{*}) \right\|}{\sqrt{L\mu}}$$

Then it is sufficient to make $T_k = T = \frac{\tau}{2\lambda} \log \frac{D}{\delta'}$ consensus iterations in order to ensure δ'-accuracy on step $k + 1$, i.e. $\left\| \mathbf{U}^{k+1} - \overline{\mathbf{U}}^{k+1} \right\|^{2} \leqslant \delta'$.

A.3   Putting the Proof Together

Let us show that choice of number of subroutine iterations $T_k = T$ yields

$$f(\overline{x}^{k}) - f(x^{*}) \leqslant \frac{\left\| \overline{u}^{0} - x^{*} \right\|^{2}}{2A^{k}} + \frac{2 \sum_{j=1}^{k} A^{j} \delta}{A^{k}}$$

by induction. At $k = 0$, we have $\left\| \mathbf{U}^{0} - \overline{\mathbf{U}}^{0} \right\| = 0$ and by Lemma 3 it holds

$$f(\overline{x}^{1}) - f(x^{*}) \leqslant \frac{\left\| \overline{u}^{0} - x^{*} \right\|^{2}}{2A^{1}} + \frac{2A^{1} \delta}{A^{1}}.$$

For induction pass, assume that $\left\| \mathbf{U}^{j} - \overline{\mathbf{U}}^{j} \right\|^{2} \leqslant \delta'$ for $j = 0, \ldots, k$. By Lemma 4, if we set $T_k = T$, then $\left\| \mathbf{U}^{k+1} - \overline{\mathbf{U}}^{k+1} \right\|^{2} \leqslant \delta'$. Applying Lemma 3 again, we get

$$f(\overline{x}^{k+1}) - f(x^{*}) \leqslant \frac{\left\| \overline{u}^{0} - x^{*} \right\|^{2}}{2A^{k+1}} + \frac{2 \sum_{j=1}^{k+1} A^{j} \delta}{A^{k+1}}$$

Recalling a bound on A^k from Lemma 2 gives

$$f(\overline{x}^{N}) - f(x^{*}) \leqslant \frac{L \left\| \overline{u}^{0} - x^{*} \right\|^{2}}{2} \left(1 + \sqrt{\frac{\mu}{2L}} \right)^{-2(N-1)} + 2 \left(1 + \sqrt{\frac{L}{\mu}} \right) \delta$$

$$\overset{\textcircled{1}}{=} L_g \left\| \overline{u}^{0} - x^{*} \right\|^{2} \left(1 + \frac{1}{2} \sqrt{\frac{\mu_g}{2L_g}} \right)^{-2(N-1)} + 2 \left(1 + 2 \sqrt{\frac{L_g}{\mu_g}} \right) \delta$$

Here in ① we used the definition of L, μ in (10): $L = 2L_g$, $\mu = \frac{\mu_g}{2}$. For ε-accuracy:

$$L_g \left\| \overline{u}^0 - x^* \right\|^2 \left(1 + \frac{1}{2}\sqrt{\frac{\mu_g}{2L_g}} \right)^{-2(N-1)} \leqslant \frac{\varepsilon}{2} \longrightarrow N \geqslant 1 + 2\sqrt{\frac{L_g}{2\mu_g}} \log\left(\frac{2L_g \left\| \overline{u}^0 - x^* \right\|^2}{\varepsilon} \right)$$

$$2\left(1 + 2\sqrt{\frac{L_g}{\mu_g}} \right) \delta \leqslant \frac{\varepsilon}{2} \longrightarrow \delta' \leqslant \frac{n\varepsilon}{2}\left(1 + 2\sqrt{\frac{L_g}{\mu_g}} \right)^{-1} \left(L_l^2\left(\frac{1}{L_g} + \frac{2}{\mu_g} \right) + L_l - \mu_l \right)^{-1}$$

It is sufficient to choose

$$N = 2\sqrt{\frac{L_g}{\mu_g}} \log\left(\frac{2L_g \left\| \overline{u}^0 - x^* \right\|^2}{\varepsilon} \right)$$

$$\delta' = \frac{n\varepsilon}{2} \cdot \frac{1}{2} \cdot \frac{1}{2}\sqrt{\frac{\mu_g}{L_g}} \cdot \left(4\frac{L_l^2}{\mu_g} \right)^{-1} = \frac{n\varepsilon}{32} \frac{\mu_g^{3/2}}{L_g^{1/2} L_l^2}$$

Let us estimate the term $\frac{D}{\delta'}$ under log.

$$\sqrt{\frac{D}{\delta'}} = \left(\frac{2L_l}{\sqrt{L\mu}} + 1 \right) + \frac{L_l}{\mu}\sqrt{n}\left(\frac{\left\| \overline{u}^0 - x^* \right\|^2}{\delta'} + \frac{8}{\sqrt{L\mu}} \right)^{1/2} + \frac{2\left\| \nabla F(X^*) \right\|}{\sqrt{L\mu}\sqrt{\delta'}}$$

$$\leqslant \frac{3L_l}{\sqrt{L_g\mu_g}} + \frac{2L_l\sqrt{n}}{\mu_g}\left(\sqrt{\frac{\left\| \overline{u}^0 - x^* \right\|^2}{n\varepsilon} \cdot \frac{32L_g^{1/2}L_l^2}{\mu_g^{3/2}}} + \sqrt{\frac{8}{\sqrt{L_g\mu_g}}} \right) + \frac{2\left\| \nabla F(X^*) \right\|}{\sqrt{L_g\mu_g}} \cdot \frac{\sqrt{32}\, L_g^{1/4}L_l}{\sqrt{n\varepsilon}\, \mu_g^{3/4}}$$

$$= \frac{3L_l}{\sqrt{L_g\mu_g}} + \frac{8\sqrt{2}L_l^2 L_g^{1/4}\left\| \overline{u}^0 - x^* \right\|}{\mu_g^{7/4}\sqrt{\varepsilon}} + \frac{4\sqrt{2}L_l\sqrt{n}}{L_g^{1/4}\mu_g^{5/4}} + \frac{4\sqrt{2}\left\| \nabla F(X^*) \right\| \cdot L_l}{L_g^{1/4}\mu_g^{5/4}\sqrt{n\varepsilon}}$$

$$= \frac{L_l}{L_g^{1/2}\mu_g}\left[3\sqrt{\mu_g} + \frac{8\sqrt{2}L_l\left\| \overline{u}^0 - x^* \right\|}{\sqrt{\varepsilon}}\left(\frac{L_g}{\mu_g} \right)^{3/4} + 4\sqrt{2n}\left(\frac{L_g}{\mu_g} \right)^{1/4} + \frac{4\sqrt{2}\left\| \nabla F(X^*) \right\|}{\sqrt{n\varepsilon}}\left(\frac{L_g}{\mu_g} \right)^{1/4} \right]$$

$$= \frac{D_1}{\sqrt{\varepsilon}} + D_2,$$

where D_1, D_2 are defined in 11. Finally, the total number of iterations is

$$N_{\text{tot}} = N \cdot T = 2\sqrt{\frac{L_g}{\mu_g}} \log\left(\frac{2L_g \left\| \overline{u}^0 - x^* \right\|^2}{\varepsilon} \right) \cdot \frac{\tau}{2\lambda} \cdot 2\log\sqrt{\frac{D}{\delta'}}$$

$$= 2\sqrt{\frac{L_g}{\mu_g}}\frac{\tau}{\lambda} \cdot \log\left(\frac{2L_g \left\| \overline{u}^0 - x^* \right\|^2}{\varepsilon} \right) \log\left(\frac{D_1}{\sqrt{\varepsilon}} + D_2 \right).$$

References

1. Nedić, A., Olshevsky, A., Uribe, C.A.: Fast convergence rates for distributed non-bayesian learning. IEEE Trans. Autom. Control **62**(11), 5538–5553 (2017)
2. Ram, S.S., Veeravalli, V.V., Nedic, A.: Distributed non-autonomous power control through distributed convex optimization. In: IEEE INFOCOM 2009, pp. 3001–3005. IEEE (2009)
3. Devolder, O., Glineur, F., Nesterov, Y.: First-order methods of smooth convex optimization with inexact oracle. Math. Program. 37–75 (2013). https://doi.org/10.1007/s10107-013-0677-5

4. Devolder, O., Glineur, F., Nesterov, Yu.: First-order methods with inexact oracle: the strongly convex case. CORE Discussion Papers 2013016:47 (2013)
5. Jakovetić, D., Xavier, J., Moura, J.M.F.: Fast distributed gradient methods. IEEE Trans. Autom. Control **59**(5), 1131–1146 (2014)
6. Nedić, A., Olshevsky, A., Shi, W.: Achieving geometric convergence for distributed optimization over time-varying graphs. SIAM J. Optim. **27**(4), 2597–2633 (2017)
7. Scaman, K., Bach, F., Bubeck, S., Lee, Y.T., Massoulié, L.: Optimal algorithms for non-smooth distributed optimization in networks. In: Advances in Neural Information Processing Systems, pp. 2740–2749 (2018)
8. Pu, S., Shi, W., Xu, J., Nedich, A.: A push-pull gradient method for distributed optimization in networks. In: 2018 IEEE Conference on Decision and Control (CDC), pp. 3385–3390 (2018)
9. Qu, G., Li, N.: Accelerated distributed Nesterov gradient descent. In: 2016 54th Annual Allerton Conference on Communication, Control, and Computing (2016)
10. Shi, W., Ling, Q., Gang, W., Yin, W.: Extra: an exact first-order algorithm for decentralized consensus optimization. SIAM J. Optim. **25**(2), 944–966 (2015)
11. Ye, H., Luo, L., Zhou, Z., Zhang, T.: Multi-consensus decentralized accelerated gradient descent. arXiv preprint arXiv:2005.00797 (2020)
12. Li, H., Fang, C., Yin, W., Lin, Z.: A sharp convergence rate analysis for distributed accelerated gradient methods. arXiv:1810.01053 (2018)
13. Nedic, A., Ozdaglar, A.: Distributed subgradient methods for multi-agent optimization. IEEE Trans. Autom. Control **54**(1), 48–61 (2009)
14. Scaman, K., Bach, F., Bubeck, S., Lee, Y.T., Massoulié, L.: Optimal algorithms for smooth and strongly convex distributed optimization in networks. In: International Conference on Machine Learning, pp. 3027–3036 (2017)
15. Jakovetic, D.: A unification and generalization of exact distributed first order methods. IEEE Trans. Signal Inf. Process. Netw. 31–46 (2019)
16. Rogozin, A., Gasnikov, A.: Projected gradient method for decentralized optimization over time-varying networks (2019). https://doi.org/10.1007/978-3-030-62867-3_18
17. Dvinskikh, D., Gasnikov, A.: Decentralized and parallelized primal and dual accelerated methods for stochastic convex programming problems (2019). https://doi.org/10.1515/jiip-2020-0068
18. Li, H., Lin, Z.: Revisiting extra for smooth distributed optimization (2020). https://doi.org/10.1137/18M122902X
19. Hendrikx, H., Bach, F., Massoulie, L.: An optimal algorithm for decentralized finite sum optimization. arXiv preprint arXiv:2005.10675 (2020)
20. Li, H., Lin, Z., Fang, Y.: Optimal accelerated variance reduced EXTRA and DIGing for strongly convex and smooth decentralized optimization. arXiv preprint arXiv:2009.04373 (2020)
21. Wu, X., Lu, J.: Fenchel dual gradient methods for distributed convex optimization over time-varying networks. In: 2017 IEEE 56th Annual Conference on Decision and Control (CDC), pp. 2894–2899, December 2017
22. Zhang, G., Heusdens, R.: Distributed optimization using the primal-dual method of multipliers. IEEE Trans. Signal Inf. Process. Netw. **4**(1), 173–187 (2018)
23. Uribe, C.A., Lee, S., Gasnikov, A., Nedić, A.: A dual approach for optimal algorithms in distributed optimization over networks. Optim. Methods Softw. 1–40 (2020)
24. Arjevani, Y., Bruna, J., Can, B., Gürbüzbalaban, M., Jegelka, S., Lin, H.: Ideal: inexact decentralized accelerated augmented Lagrangian method. arXiv preprint arXiv:2006.06733 (2020)

25. Wei, E., Ozdaglar, A.: Distributed alternating direction method of multipliers. In: 2012 IEEE 51st IEEE Conference on Decision and Control (CDC), pp. 5445–5450. IEEE (2012)

26. Maros, M., Jaldén, J.: PANDA: a dual linearly converging method for distributed optimization over time-varying undirected graphs. In: 2018 IEEE Conference on Decision and Control (CDC), pp. 6520–6525 (2018)

27. Tang, J., Egiazarian, K., Golbabaee, M., Davies, M.: The practicality of stochastic optimization in imaging inverse problems (2019). https://doi.org/10.1109/TCI.2020.3032101

28. Stonyakin, F., et al.: Inexact relative smoothness and strong convexity for optimization and variational inequalities by inexact model. arXiv:2001.09013 (2020)

29. Koloskova, A., Loizou, N., Boreiri, S., Jaggi, M., Stich, S.U.: A unified theory of decentralized SGD with changing topology and local updates (2020). http://proceedings.mlr.press/v119/koloskova20a.html

30. Chang, C.-C., Lin, C.-J.: LIBSVM: a library for support vector machines. ACM Trans. Intell. Syst. Technol. (TIST) **2**(3), 27 (2011)

31. Li, H., Fang, C., Yin, W., Lin, Z.: Decentralized accelerated gradient methods with increasing penalty parameters. IEEE Trans. Signal Process. **68**, 4855–4870 (2020)

32. Arjevani, Y., Shamir, O.: Communication complexity of distributed convex learning and optimization. Adv. Neural Inf. Process. Syst. **28**, 1756–1764 (2015)

33. Dvinskikh, D.M., Turin, A.I., Gasnikov, A.V., Omelchenko, S.S.: Accelerated and non accelerated stochastic gradient descent in model generality. Matematicheskie Zametki **108**(4), 515–528 (2020)

34. Rogozin, A., Lukoshkin, V., Gasnikov, A., Kovalev, D., Shulgin, E.: Towards accelerated rates for distributed optimization over time-varying networks. arXiv preprint arXiv:2009.11069 (2020)

Game Theory and Mathematical Economics

Stackelberg and Nash Equilibria in Games with Linear-Quadratic Payoff Functions as Models of Public Goods

Victor Gorelik[1,2] and Tatiana Zolotova[3]

[1] FRC CSC RAS, Vavilova Street 40, 119333 Moscow, Russia
[2] Moscow Pedagogical State University, M. Pirogovskaya Street 1/1, 119991 Moscow, Russia
[3] Financial University under the Government of RF, Leningradsky Prospekt 49, 125993 Moscow, Russia

Abstract. The paper proposes a game model with an additive convolution of two criteria, describing public and personal interests. The first (general) criterion depends on strategies of all players and represents losses from the intensity of their activity. The second (particular) criterion for each player is a function of his strategy and reflects the income from his activities. The negative definite quadratic form is taken as a general criterion. The particular criterion of each player is linear, which is quite natural for the formalization of the income function. It turns out that the resulting game with linear-quadratic payoff functions has good properties, in particular, the independence of the leader's strategy in the Stackelberg equilibrium from the parameters of the follower's linear functions (in contrast to the Nash equilibrium). This property means that the leader does not need accurate information about the follower's objective function, and his strategy has the property of robustness.

Keywords: Stackelberg equilibrium · Nash equilibrium · Public goods · Linear-quadratic payoff function

1 Introduction

The Stackelberg equilibrium [1,2], proposed by the author for the analysis of duopoly, subsequently found wide application in various branches of mathematical economics and control theory. Among the most famous works we can mention the theory of contracts [3–5], which was awarded the Nobel Prize. In Russian studies, Stackelberg's idea was developed in the information theory of hierarchical systems [6–8] and the theory of active systems [9]. These theories consider control problems without feedback (direct Stackelberg problem) and with feedback (inverse Stackelberg problem), dynamic games, hierarchical games with an

---

The work was carried out within the framework of the project No. AAAA-A20-120122190034-9.

N. N. Olenev et al. (Eds.): OPTIMA 2021, LNCS 13078, pp. 275–287, 2021.
https://doi.org/10.1007/978-3-030-91059-4_20

ambiguous lower-level response and incomplete information, general methods of two-level optimization, etc.

Stackelberg equilibrium methods are computationally very complex. Most of the work in the literature has been based on simple structure or specific models. The inverse Stackelberg game is difficult to solve even for one leader - one follower [10]. Full characterization of the problem with multiple leaders and followers includes two-level optimization, MPEC or EPEC [11–14]. Therefore, research related to the construction of classes of models that have a practical interpretation, for which solutions can be found in a constructive form, seem promising.

One of the important areas of application of game theory and, in particular, the principle of hierarchical management, are the problems of public goods, the most urgent of which are currently the problems of ecology [15–17].

The public goods game is a classic model in economics. The pervasive and profound "tragedy of the commons" [18] is that social well-being will be optimal if all players cooperate, which, however, cannot be achieved in the Nash equilibrium, since the individual player will benefit more from the free rider role. Such problems occur widely, for example, overgrazing of common land, overfishing in the ocean, carbon emissions [15,16]. These game issues have received a lot of attention from researchers from various fields, such as economics, biology, politics, management technology, etc. (see, for example, [17]). Various mechanisms are introduced, such as reward and punishment, evolution, spatial structure, threshold, etc. (see, for example, [19–23]). In [24,25] a hierarchical structure is presented and its influence on social welfare (cooperation) in a discrete public goods game and the Prisoner's Dilemma Game (public goods game for two persons) is investigated. The architectural structure is naturally determined by the number of levels. In [26] the equilibrium and the optimal structure in a continuous game of public goods with common hierarchical structures are formulated by the inverse Stackelberg game. The author studies the influence of hierarchical structures on outcome, especially the social welfare of the public goods game. The reverse Stackelberg game with general hierarchical structures is investigated analytically for a specific game model.

Related works include a study of the problem of pricing Internet services with one leader and one follower [27], on the problem of tolls and the energy market [28,29], on the optimal affine function of a leader and an algorithm of optimal nonlinear functions with one leader - one follower [30,31].

In game-theoretic models of this type of problems, the payoff functions of the players are a convolution of two criteria, describing public and personal interests. In this paper, a game model is proposed that uses an additive convolution of two criteria. The first criterion depends on the strategy of all players and represents losses from the intensity of their activity. The second criterion for each player is a function of his strategy and reflects the income from his activities. It is assumed, that the first (general) criterion has a single global maximum, and without loss of generality, we can assume that this maximum is located at the origin with the criterion value equal to zero (there is no loss in the absence of activity). In the model, considered below, the negative definite quadratic form is taken as a general criterion. This, of course, significantly limits the generality of consideration, however, a certain basis is the mathematical fact that a twice differentiable function in the vicinity of the global maximum is well approximated by a negative definite quadratic form. The second (particular) criterion of each player is linear, which is quite natural for the formalization of the income function. The Stackelberg equilibrium is considered as a solution to the emerging game. It turns out, that the resulting game with linear-quadratic payoff functions has a number of good properties, the most interesting of which is the independence of the leader's strategy in the Stackelberg equilibrium from the parameters of the follower's linear function (in contrast to the Nash equilibrium). This property means that the leader does not need accurate information about the follower's objective function, which in reality is difficult to obtain, and his strategy has the property of robustness.

2   The Two-Persons Game Model

So, the objective functions of the players are of the form

$$F(x, y) = \frac{1}{2} \langle x, Ax \rangle + \langle x, By \rangle + \frac{1}{2} \langle y, Cy \rangle + \langle p, x \rangle,$$

$$G(x, y) = \frac{1}{2} \langle x, Ax \rangle + \langle x, By \rangle + \frac{1}{2} \langle y, Cy \rangle + \langle q, y \rangle,$$

where n-dimensional vector x is the strategy of the player with the payoff function F, the m-dimensional vector y is the strategy of the player with the payoff function G, $D = \begin{pmatrix} A & B \\ B^T & C \end{pmatrix}$ is a symmetric negative definite matrix of size $(n + m) \times (n + m)$. The functions F and G are strictly concave by the set of variables, so they have unique global maxima, respectively, (x^1, y^1) and (x^2, y^2).

Theorem 1. *The optimal leader strategy in the game with payoff functions $F(x, y)$ and $G(x, y)$ is the choice of controlled variables of the global maximum of its function, i.e. $x^F = x^1$, $y^G = y^2$. Wherein, at equilibrium points and points of absolute maxima of the payoff functions the equality holds*

$$F\left(x^F, y^F\right) + G\left(x^F, y^F\right) = F\left(x^1, y^1\right) + G\left(x^1, y^1\right)$$

$$= F\left(x^G, y^G\right) + G\left(x^G, y^G\right) = F\left(x^2, y^2\right) + G\left(x^2, y^2\right).$$

Proof. If the leader chooses the strategy x and communicates it to the follower, the latter maximizes its payoff function and from the condition that the gradient G'_y of the function G by variables y is equal to zero we have $B^T x + Cy + q = 0$. The matrix C is obviously negative definite, therefore it is not degenerate and the follower's optimal response is

$$y(x) = -C^{-1}(B^T x + q).$$

Then

$$F(x, y(x)) = \frac{1}{2}\langle x, Ax \rangle - \langle B^T x, C^{-1}(B^T x + q)\rangle$$

$$+ \frac{1}{2}\langle B^T x + q, C^{-1}(B^T x + q)\rangle + \langle p, x \rangle$$

$$= \frac{1}{2}\langle x, Ax \rangle - \frac{1}{2}\langle B^T x - q, C^{-1}(B^T x + q)\rangle + \langle p, x \rangle$$

$$= \frac{1}{2}\langle x, Ax \rangle - \frac{1}{2}\langle B^T x, C^{-1}B^T x\rangle + \frac{1}{2}\langle q, C^{-1}q\rangle + \langle p, x \rangle.$$

Equating gradient $F(x, y(x))$ as a complex function from x to zero, we have

$$Ax + p - BC^{-1}B^T x = 0. \tag{1}$$

Conditions for the global maximum of the function F are $F'_x = F'_y = 0$, so we have

$$Ax + p + By = 0, \quad B^T x + Cy = 0. \tag{2}$$

Since the matrix D is not degenerate, from the system (2) we have

$$\begin{pmatrix} x^1 \\ y^1 \end{pmatrix} = D^{-1}\begin{pmatrix} -p \\ 0 \end{pmatrix}, \tag{3}$$

that is, the function F has a unique global maximum. On the other hand, expressing y from the second equation of the system (2) $y = -C^{-1}B^T x$ and substituting into the first, we obtain the same Eq. (1) for x. So the matrix $E_1 = A - BC^{-1}B^T$ is not degenerate (however, this is a well-known fact for negative definite block matrices) and

$$x^F = x^1 = -E_1^{-1}p. \tag{4}$$

wherein $y^1 = -C^{-1}B^T x^1$ and

$$y^F = -C^{-1}(B^T x^F + q) = -C^{-1}(B^T x^1 + q) = y^1 - C^{-1}q.$$

Conditions for the global maximum of the function G are $G'_x = G'_y = 0$, so we have

$$Ax + By = 0, \quad B^T x + Cy + q = 0. \tag{5}$$

From the system (5) we have

$$\begin{pmatrix} x^2 \\ y^2 \end{pmatrix} = D^{-1}\begin{pmatrix} 0 \\ -q \end{pmatrix}, \tag{6}$$

i.e., the function G has a unique global maximum, and similarly to the previous

$$y^G = y^2 = -E_2^{-1}q, \tag{7}$$

where $E_2 = C - B^T A^{-1} B$.

Further, obviously, we have the equality

$$F(x,y) + G(x,y) = \langle x, Ax \rangle + \langle x, By \rangle + \langle p, x \rangle$$
$$+ \langle B^T x, y \rangle + \langle y, Cy \rangle + \langle q, y \rangle = \langle F_x'(x,y), x \rangle + \langle G_y'(x,y), y \rangle. \tag{8}$$

As $F_x'\left(x^1, y^1\right) = 0$, and $G_y'\left(x^1, y^1\right) = F_y'\left(x^1, y^1\right) + q = q$, then $F\left(x^1, y^1\right) + G\left(x^1, y^1\right) = \langle q, y^1 \rangle$. As $G_y'\left(x^F, y^F\right) = 0$, then $F\left(x^F, y^F\right) + G\left(x^F, y^F\right) = \langle F_x'(x^F, y^F), x^F \rangle$. Considering that $x^F = x^1$, $y^F = y^1 - C^{-1}q$, we have

$$F_x'\left(x^F, y^F\right) = Ax^F + p + By^F = Ax^1 + p + B\left(y^1 - C^{-1}q\right) = -BC^{-1}q.$$

Then $F\left(x^F, y^F\right) + G\left(x^F, y^F\right) = -\langle BC^{-1}q, x^1 \rangle = -\langle q, C^{-1}B^T x^1 \rangle = \langle q, y^1 \rangle$. Thus, we have proved the first equality of the statement of the theorem:

$$F\left(x^F, y^F\right) + G\left(x^F, y^F\right) = F\left(x^1, y^1\right) + G\left(x^1, y^1\right) = \langle q, y^1 \rangle.$$

It can be proved similarly that

$$F\left(x^G, y^G\right) + G\left(x^G, y^G\right) = F\left(x^2, y^2\right) + G\left(x^2, y^2\right) = \langle p, x^2 \rangle.$$

Now we will show, that $\langle p, x^2 \rangle = \langle q, y^1 \rangle$.

Using (3), (6) and the expression for the inverse matrix

$$D^{-1} = \frac{1}{|D|} \begin{pmatrix} D_{11} & D_{21} & \cdots & D_{n+m\ 1} \\ D_{12} & D_{22} & \cdots & D_{n+m\ 2} \\ \cdots & \cdots & \cdots & \cdots \\ D_{1\ n+m} & D_{2\ n+m} & \cdots & D_{n+m\ n+m} \end{pmatrix},$$

where D_{ij} is the cofactor of the corresponding element of the matrix D, we have

$$\langle q, y^1 \rangle = -\frac{1}{|D|} \sum_{j=1}^{m} q_j \sum_{i=1}^{n} p_i D_{i\ n+j},$$

$$\langle p, x^2 \rangle = -\frac{1}{|D|} \sum_{i=1}^{n} p_i \sum_{j=1}^{m} q_j D_{n+j\ i}.$$

Whence, taking into account the symmetry of the matrix D we have $\langle p, x^2 \rangle = \langle q, y^1 \rangle$. The theorem is proved.

From the first result of Theorem 1 it follows, that the leader and the follower does not need information about the individual parameters p and q in the payoff function of the other player, while choosing their optimal strategies. This is an important property from a practical point of view, since such information is practically unavailable, and an attempt to clarify it faces a possible bluff.

From the second result of Theorem 1 it follows that if one player prefers the equilibrium (x^F, y^F), then the other prefers (x^G, y^G). Therefore, depending on the parameters of the payoff functions, either a struggle for leadership or for the opportunity to be a follower can arise, and both cases can take place. To determine the benefits of leadership, we define the difference between the values of the payoff function F in points (x^F, y^F) and (x^G, y^G).

Obviously, the following representation of the players payoff functions takes place

$$
\begin{aligned}
F(x,y) &= \tfrac{1}{2}\left[\left\langle F_x'(x,y),x\right\rangle + \left\langle F_y'(x,y),y\right\rangle + \langle p,x\rangle\right] \\
&= \tfrac{1}{2}\left[\left\langle F_x'(x,y),x\right\rangle + \left\langle G_y'(x,y),y\right\rangle - \langle q,y\rangle + \langle p,x\rangle\right],
\end{aligned}
$$

$$
\begin{aligned}
G(x,y) &= \tfrac{1}{2}\left[\left\langle G_x'(x,y),x\right\rangle + \left\langle G_y'(x,y),y\right\rangle + \langle q,y\rangle\right] \\
&= \tfrac{1}{2}\left[\left\langle F_x'(x,y),x\right\rangle + \left\langle G_y'(x,y),y\right\rangle - \langle p,x\rangle + \langle q,y\rangle\right],
\end{aligned}
\tag{9}
$$

whence, when substituting the values of the arguments, we obtain the value of the specified difference

$$
F\left(x^F,y^F\right) - F\left(x^G,y^G\right)
$$
$$
= \frac{1}{2}\left[\langle\left(A^{-1} - E_1^{-1}\right)p,p\rangle + \langle\left(C^{-1} - E_2^{-1}\right)q,q\rangle\right] - \left\langle C^{-1}B^T E_1^{-1}p,q\right\rangle.
$$

Further, we will make sure with examples that it can be both positive and negative.

Now let's find the Nash equilibrium in this game. It satisfies the system of equations

$$
Ax + By + p = 0, \quad B^T x + Cy + q = 0 . \tag{10}
$$

From the system (10) we have, that the Nash equilibrium exists, is unique and is equal to

$$
\begin{pmatrix} x^N \\ y^N \end{pmatrix} = D^{-1}\begin{pmatrix} -p \\ -q \end{pmatrix},
$$

and from (8) it follows that $F\left(x^N,y^N\right) + G\left(x^N,y^N\right) = 0$.

Since the Nash equilibrium is strict, the leader's payoff is obviously not less than his result at the Nash equilibrium. The follower can receive both more and less than in the Nash equilibrium. As the examples show, both can take place for the follower.

3   The n-Persons Game Model

Let us generalize the considered model for a game with an arbitrary number of players. We will consider the strategies of the players as scalar quantities. The strategy of the k-th player will be denoted by x_k. The objective functions of the players are of the form

$$F_k(x) = \frac{1}{2} \langle x, Ax \rangle + p_k x_k, \ \ k = 1, \ldots, n.$$

Here $x = (x_1, \ldots, x_k, \ldots, x_n)$, A is a symmetric negative definite matrix.

Let the players choose their strategies sequentially in numerical order and k-th player by the time of its turn knows the choice of strategies by the players with numbers $1, \ldots, k-1$ and the sequence, in which the players with numbers $k+1, \ldots, n$ will make their moves. It is also assumed, that when choosing a strategy, all players are guided by the general principle of behavior: to maximize their payoff function, taking into account the corresponding reaction of the players of the next levels.

Theorem 2. *The optimal strategy of the k-th player in the game with a sequence of moves is to choose the controlled variable x_k in the maximum point of its payoff function $F_k(x)$ by the arguments, remaining free at the time of his move $x_k, x_{k+1}, \ldots, x_n$, for known fixed values of the arguments $x_1, \ldots, x_{k-1}$.*

Proof. Let us carry out the proof by induction on the number of players. For $n = 1$ the statement of the theorem means simple optimization, and for $n = 2$ was proved above. Suppose the theorem is true for the number of players $n-1$ and will proof for n. If the first player chooses the strategy x_1, then we have the game with $n-1$ players, in which, according to the inductive assumption, the players choose their strategies $x_2^0, x_3^0, \ldots, x_n^0$ from the conditions for the maximum of their payoff functions with respect to free arguments. So these strategies satisfy the following systems of linear equations

$$\begin{cases} a_{21}x_1 + a_{22}x_2^0 + a_{23}x_3^{(2)} + \cdots + a_{2n}x_n^{(2)} + p_2 = 0, \\ a_{i1}x_1 + a_{i2}x_2^0 + a_{i3}x_3^{(2)} + \cdots + a_{in}x_n^{(2)} = 0, \ \ i = 3, \ldots, n. \end{cases}$$

$$\begin{cases} a_{31}x_1 + a_{32}x_2^0 + a_{33}x_3^0 + a_{34}x_4^{(3)} + \cdots + a_{3n}x_n^{(3)} + p_3 = 0, \\ a_{i1}x_1 + a_{i2}x_2^0 + a_{i3}x_3^0 + a_{i4}x_4^{(3)} + \cdots + a_{in}x_n^{(3)} = 0, \ \ i = 4, \ldots, n. \end{cases}$$

$$\cdots\cdots\cdots$$

$$\begin{cases} a_{n-1\,1}x_1 + a_{n-1\,2}x_2^0 + \cdots + a_{n-1\,n-1}\,x_{n-1}^0 + a_{n-1\,n}\,x_n^{(n-1)} + p_{n-1} = 0, \\ a_{n1}x_1 + a_{n2}x_2^0 + \cdots + a_{n\,n-1}x_{n-1}^0 + a_{nn}x_n^{(n-1)} = 0. \end{cases}$$

$$a_{n1}x_1 + a_{n2}x_2^0 + \cdots + a_{nn}x_n^0 + p_n = 0.$$

Since the matrix A is negative definite, the determinants of these systems, which are corner minors of the matrix A, are nonzero.

Therefore, the strategies $x_2^0, x_3^0, \ldots, x_n^0$ are ultimately single-valued functions of the argument x_1.

Let us calculate the derivatives of these functions. From the first system we have $\frac{dx_2^0(x_1)}{dx_1} = \frac{A_{12}}{A_{11}}$, where A_{ij} is the cofactor of the corresponding element a_{ij} in the determinant $|A|$. Let us suppose, that

$$\frac{dx_i^0(x_1)}{dx_1} = \frac{A_{1i}}{A_{11}}, \quad i = 2, 3, \ldots, k.$$

Let us denote by M_k the minor of the matrix A, obtained by deleting the first k rows and the first k columns, and by $\tilde{M}_k$ – determinant derived from minor M_k replacing each element of its first column $a_{i\,k+1}$ by expression

$$a_{i1}A_{11} + a_{i2}A_{12} + \cdots + a_{ik}A_{1k}, \quad i = k+1, \ldots, n.$$

Then we have from $(k+1)$-th system

$$\frac{dx_{k+1}^0(x_1)}{dx_1} = -\frac{1}{A_{11}} \cdot \frac{\tilde{M}_k}{M_k}.$$

Using the well-known property of determinants, we have for $i = k+1, \ldots, n$ equality

$$a_{i1}A_{11} + a_{i2}A_{12} + \cdots + a_{ik}A_{1k} = -(a_{i\,k+1}A_{1\,k+1} + a_{i\,k+2}A_{1\,k+2} + \cdots + a_{in}A_{1n}).$$

Substitute the right-hand side of this equality in $\tilde{M}_k$. Taking into account that determinants with proportional columns are equal to zero, we have

$$\tilde{M}_k = -A_{1\,k+1}\,M_k, \quad \frac{dx_{k+1}^0(x_1)}{dx_1} = \frac{A_{1\,k+1}}{A_{11}}.$$

Thus, we have proved by induction that

$$\frac{dx_i^0(x_1)}{dx_1} = \frac{A_{1i}}{A_{11}}, \quad i = 2, 3, \ldots, n.$$

The first player chooses his strategy x_1 as to achieve the maximum of the function

$$f(x_1) = F_1\left(x_1, x_2^0(x_1), \ldots, x_n^0(x_1)\right).$$

We calculate the total derivative of the function $f(x_1)$:

$$\frac{df(x_1)}{dx_1} = \frac{\partial F_1\left(x_1, x_2^0(x_1), \ldots, x_n^0(x_1)\right)}{\partial x_1}$$

$$+ \sum_{i=2}^{n} \frac{\partial F_1\left(x_1, x_2^0(x_1), \ldots, x_n^0(x_1)\right)}{\partial x_i} \cdot \frac{dx_i^0(x_1)}{dx_1}$$

$$= a_{11}x_1 + \sum_{i=2}^{n} a_{i1}x_i^0(x_1) + p_1 + \sum_{i=2}^{n} \frac{A_{1i}}{A_{11}}(a_{1i}x_1 + \sum_{j=2}^{n} a_{ji}x_j^0(x_1))$$

$$= \frac{1}{A_{11}} \cdot \left[x_1 \sum_{i=1}^{n} a_{1i}A_{1i} + p_1 A_{11} + \sum_{j=2}^{n} \left(x_j^0(x_1) \sum_{i=1}^{n} a_{ji}A_{1i} \right) \right]$$

$$= \frac{1}{A_{11}} \cdot (x_1 |A| + p_1 A_{11}).$$

Equating this derivative to zero, we have $x_1 = -\frac{p_1 A_{11}}{|A|}$.

Since the necessary and sufficient condition for the negative definiteness of a matrix is the alternation of the signs of its major minors, then $\frac{d^2 f(x_1)}{dx_1^2} = \frac{|A|}{A_{11}} < 0$, i.e. $f(x_1)$ is strictly concave quadratic function.

Hence $f(x_1)$ reaches its global maximum at the point $x_1^0 = -\frac{p_1 A_{11}}{|A|}$, i.e. x_1^0 is the optimal strategy of the first player. It is easy to make sure that x_1^0 satisfies the system of equations

$$\begin{cases} a_{11}x_1^0 + a_{12}x_2^{(1)} + a_{13}x_3^{(1)} + \cdots + a_{1n}x_n^{(1)} + p_1 = 0, \\ a_{i1}x_1^0 + a_{i2}x_2^{(1)} + a_{i3}x_3^{(1)} + \cdots + a_{in}x_n^{(1)} = 0, \quad i = 2, \ldots, n, \end{cases}$$

so x_1^0 is the first coordinate of the point of the global maximum of the function $F_1(x_1, x_2, \ldots, x_n)$, QED.

The Nash equilibrium in the given n-persons game also exists, is unique and is determined from the system of linear equations

$$Ax = -p, \quad p = (p_1, \ldots, p_n).$$

The solution of this system has the form $x^N = -A^{-1}p$. The values of the players payoff functions in the Nash equilibrium are

$$F_k(x^N) = \frac{1}{2}\langle A^{-1}p, p \rangle + p_k x_k^N, \quad k = 1, \ldots, n,$$

and the total value

$$\sum_{k=1}^{n} F_k(x^N) = \left(\frac{n}{2} - 1 \right) \langle A^{-1}p, p \rangle.$$

Since the matrix A^{-1} is negatively definite, the total value is equal to zero for $n = 2$ (as already shown above) and negative for $n \geq 3$.

Naturally, finding the Nash equilibrium requires all players to know the parameters of the payment functions of other players.

4   Some Special Cases

Let us illustrate the results obtained for $n = 2$ in the case of one-dimensional strategies (Example 1) and for $n = 3$ (Example 2).

Example 1. Consider a practical situation, where two fishing companies operate in the same water basin. With an abundance of fish resources, their incomes linearly depend on the volume of production. When resources are depleted, their costs begin to increase superlinearly.
Let's the payoff functions are

$$F(x, y) = \frac{1}{2}ax^2 + bxy + \frac{1}{2}cy^2 + px,$$

$$G(x, y) = \frac{1}{2}ax^2 + bxy + \frac{1}{2}cy^2 + qx.$$

For quadratic form $\frac{1}{2}ax^2 + bxy + \frac{1}{2}cy^2$ necessary and sufficient conditions for negative definiteness have the form $a < 0$, $c < 0$, $ac - b^2 > 0$.
The coordinates of the equilibrium points according to Stackelberg, according to Nash and global maxima are equal:

$$x^F = x^1 = -\frac{cp}{ac - b^2}, \quad y^F = y^1 - \frac{q}{c} = \frac{bp}{ac - b^2} - \frac{q}{c},$$

$$x^G = x^2 - \frac{p}{a} = \frac{bq}{ac - b^2} - \frac{p}{a}, \quad y^G = y^2 = -\frac{aq}{ac - b^2},$$

$$x^N = \frac{bq - cp}{ac - b^2}, \quad y^N = \frac{bp - aq}{ac - b^2}.$$

Using formulas (9), we obtain the values of the payoff functions at the Stackelberg equilibriums:

$$F(x^F, y^F) = \frac{q^2}{2c} - \frac{cp^2}{2(ac - b^2)}, \quad G(x^F, y^F) = \frac{cp^2 + 2bpq}{2(ac - b^2)} - \frac{q^2}{2c},$$

$$F(x^G, y^G) = \frac{aq^2 + 2bpq}{2(ac - b^2)} - \frac{p^2}{2a}, \quad G(x^G, y^G) = \frac{p^2}{2a} - \frac{cp^2}{2(ac - b^2)},$$

and

$$F(x^F, y^F) - F(x^G, y^G) = G(x^G, y^G) - G(x^F, y^F)$$

$$= \frac{|a|b^2q^2 - 2acbpq + |c|b^2p^2}{2ac(ac - b^2)}.$$

As mentioned above, these differences can be either positive or negative. For example, for $a = c = -4$, $b = 3$, $p = q = 1$ we have

$$|a| \, b^2 q^2 - 2acbpq + |c| \, b^2 p^2 = 36 - 2 \cdot 48 + 36 < 0,$$

so it is advantageous to be the follower (to receive information).

If $a = c = -4$, $b = 3$, $p = 1, q = 4$ we have $|a| \, b^2 q^2 - 2acbpq + |c| \, b^2 p^2 = 576 - 2 \cdot 48 \cdot 4 + 36 > 0$, so it is beneficial to be the leader.

The values of the payoff functions at the Nash equilibrium are:

$$F\left(x^N, y^N\right) = \frac{aq^2 - cp^2}{2\left(ac - b^2\right)}, \quad G\left(x^N, y^N\right) = \frac{cp^2 - aq^2}{2\left(ac - b^2\right)}.$$

Further,

$$F\left(x^F, y^F\right) = F\left(x^N, y^N\right) + \frac{b^2 q^2}{2|c|(ac - b^2)} = F\left(x^1, y^1\right) - \frac{q^2}{|c|},$$

$$G\left(x^G, y^G\right) = G\left(x^N, y^N\right) + \frac{b^2 p^2}{2|a|(ac - b^2)} = G\left(x^2, y^2\right) - \frac{p^2}{|a|},$$

so, the value of the leader's payoff function is strictly greater than at the Nash equilibrium (for $b \neq 0$), and strictly less than the global maximum.

For the follower, $F\left(x^G, y^G\right) \geq F\left(x^N, y^N\right)$, if $2|a|bpq \geq b^2 p^2$, $G\left(x^F, y^F\right) \geq G\left(x^N, y^N\right)$, if $2|c|bpq \geq b^2 q^2$. Besides,

$$F\left(x^F, y^F\right) + G\left(x^F, y^F\right) = F\left(x^G, y^G\right) + G\left(x^G, y^G\right) = \frac{bpq}{ac - b^2},$$

that is, the "public good" in the Stackelberg equilibrium can be either more or less than in the Nash equilibrium.

Example 2. Consider a similar example with three companies. Let the payoff functions have the form

$$F_k\left(x\right) = \frac{1}{2}\left\langle x, Ax\right\rangle + p_k x_k, \quad k = 1, 2, 3,$$

where $A = \begin{pmatrix} -2 & -1 & 1 \\ -1 & -2 & 1 \\ 1 & 1 & -2 \end{pmatrix}$, $p = (2, 1, 1)$.

We have $A^{-1} = \begin{pmatrix} -\frac{3}{4} & \frac{1}{4} & -\frac{1}{4} \\ \frac{1}{4} & -\frac{3}{4} & -\frac{1}{4} \\ -\frac{1}{4} & -\frac{1}{4} & -\frac{3}{4} \end{pmatrix}$, $A^{-1}p = (-\frac{3}{2}, -\frac{1}{2}, -\frac{3}{2})$, $\left\langle A^{-1}p, p\right\rangle = -5$.

Thus, $x^N = -A^{-1}p = (\frac{3}{2}, \frac{1}{2}, \frac{3}{2})$, $\sum_{k=1}^{3} F_k\left(x^N\right) = (\frac{3}{2} - 1)\left\langle A^{-1}p, p\right\rangle = -\frac{5}{2}$. Wherein $F_1\left(x^N\right) = \frac{1}{2}$, $F_2\left(x^N\right) = -2$, $F_3\left(x^N\right) = -1$.

Depending on the order of moves, we have six variants of the Stackelberg equilibrium. Consider one of them for the order of moves $1 \to 2 \to 3$. Solving the corresponding systems of linear equations, we have for the given order the Stackelberg equilibrium $x^S = (\frac{3}{2}, \frac{1}{6}, \frac{4}{3})$, $F_1\left(x^S\right) = \frac{11}{12}$, $F_2\left(x^S\right) = -\frac{23}{12}$, $F_3\left(x^S\right) = -\frac{9}{12}$. In this case, for all participants, the Stackelberg equilibrium is preferable to the Nash equilibrium. Moreover, the total result is also negative: $\sum_{k=1}^{3} F_k (x^S) = -\frac{7}{4}$.

For comparison, we find the maximum of the total function $\sum_{k=1}^{3} F_k (x)$. It is equal to $\frac{5}{6}$ and is reached at the point $x^{max} = (\frac{1}{2}, \frac{1}{6}, \frac{1}{2})$. However, the achievement of such result is possible only in a cooperative version of the game, when "grand coalition" is formed and catch quotas are established.

5   Conclusion

In our opinion, the proposed game model with linear-quadratic payment functions, despite its specific form, can be considered as a meaningful description of the problem of providing public goods. First of all, these can be environmental problems, such as pollution of common water bodies, emissions of harmful substances in the region, depletion of public resources, etc. Both the Nash equilibrium and the Stackelberg equilibrium can be considered as a solution. At first glance, the Nash equilibrium is more natural and easier to implement, but it requires a general knowledge of all the parameters of the model and is not protected from bluffs. A remarkable property of the considered model is that the Stackelberg equilibrium is protected from bluffs, since in determining the optimal strategy by each player, one does not need to know the individual parameters of the payment functions of the other players. At the same time, the results obtained make it possible to reduce the complex problem of multilevel optimization to solving ordinary extreme problems.

References

1. Von Stackelberg, H.: Market Structure and Equilibrium (English). Springer, Heidelberg (Transl. Bazin, D., Urch, L., Hill, R. (2011). https://doi.org/10.1007/978-3-642-12586-7. Marktform und Gleichgewicht, Vienna, 1934
2. Stackleberg, H.V.: The Theory of the Market Economy. William Hodge, London (1952)
3. Hart, O.: Incomplete contracts and control. Am. Econ. Rev. **107**(7), 1731–1752 (2017)
4. Hart, O., Zingales, L.: Liquidity and inefficient investment. J. Eur. Econ. Assoc. **13**(5), 737–769 (2015)
5. Holmstrom, B., Milgrom P.: Multitask principal-agent analyses: incentive contracts, asset ownership, and job design. In: The Economic Nature of the Firm, pp. 232–244. Cambridge University Press, Cambridge (2009)
6. Germeier, Y.B.: Games with Nonconflicting Interests. Nauka, Moscow (1976)
7. Gorelik, V.A., Gorelov, M.A., Kononenko, A.F.: Analysis of Conflict Situations in Control Systems. Radio and Communication, Moscow (1991)
8. Moiseev, N.N.: Mathematical Problems of System Analysis. Nauka, Moscow (1981)
9. Burkov, V.N.: Fundamentals of the Mathematical Theory of Active Systems. Nauka, Moscow (1977)
10. Olsder, G.J.: Phenomena in inverse Stackleberg games, Part I: static problems. J. Optim. Theory Appl. **143**(3), 589–600 (2009)

11. Pang, J.S., Fukushima, M.: Quasi-variational inequalities, generalized Nash equilibria, and multi-leader-follower games. Comput. Manag. Sci. **2**(1), 21–56 (2005)
12. Luo, Z.Q., Pang, J.S., Ralph, D.: Mathematical Programs with Equilibrium Constraints. Cambridge University Press, Cambridge (1996)
13. Ye, J., Zhu, D.: New necessary optimality conditions for bilevel programs by combining the MPEC and value function approaches. SIAM J. Optim. **20**(4), 1885–1905 (2010)
14. Su, C.L.: Equilibrium Problems with Equilibrium Constraints: Stationarities, Algorithms and Applications. Ph.D. thesis. Stanford University, Stanford (2005)
15. Dixit, A.K., Nalebuff, B.J.: The Art of Strategy: A Game Theorist's Guide to Success in Business and Life. W.W. Norton Company, New York (2010)
16. Baliga, S., Maskin, E.: Mechanism design for the environment. In: Handbook of Environmental Economics, vol. 1, 305–324. Elsevier, Amsterdam (2003)
17. Ostrom, E.: Governing the Commons. Cambridge University Press, Cambridge (1990)
18. Hardin, G.: The tragedy of the commons. Science **162**(3859), 1243–1248 (1968)
19. Sefton, M., Shupp, R., Walker, J.M.: The effect of rewards and sanctions in provision of Public Good. Econ. Inquiry **45**(4), 671–690 (2007)
20. Fehr, E., Gachter, S.: Cooperation and punishment in public goods experiments. Am. Econ. Rev. **90**(4), 980–994 (2000)
21. Hauert, C., Holmes, M., Doebeli, M.: Evolutionary games and population dynamics: maintenance of cooperation in public goods games. Proc. R. Soc. Lond. B Biol. Sci. **273** (1600), pp. 2565–2571 (2006)
22. Zhang, J., Zhang, C., Cao, M.: How insurance affects altruistic provision in threshold public goods games. Sci. Rep. **5**, Article number: 9098 (2015)
23. Gorelik, V.A., Zolotova, T.V.: Models of hierarchial control in ecological-economic systems. J. Math. Sci. **2**(5), 612–626 (2016)
24. Mu, Y., Guo, L.: How cooperation arises from rational players? Sci. China Inf. Sci. **56**(11), 1–9 (2013)
25. Mu, Y.: Inverse Stackelberg Public Goods Game with multiple hierarchies under global and local information structures. J. Optim. Theory Appl. **163**(1), 332–350 (2014)
26. Mu, Y.: Stackelberg-Nash equilibrium, social welfare and optimal structure in hierarchical continuous Public Goods game. J. Syst. Control Lett. **112**(1), 1–8 (2018)
27. Shen, H., Basar, T.: Incentive-based pricing for network games with complete and incomplete information. Adv. Dyn. Game Theory **9**, 431–458 (2007)
28. Staňková, K, Olsder, G.J., Bliemer, M.C.J.: Bi-level optimal toll design problem solved by the inverse Stackelberg games approach. WIT Trans. Built Environ. **89** (2006)
29. Staňková, K., Olsder, G.J., De Schutter, B.: On European electricity market liberalization: a game-theoretic approach. Inf. Syst. Oper. Res. **48**(4), 267–280 (2010)
30. Groot, N., Schutter, B.D., Hellendoorn, H.: On systematic computation of optimal nonlinear solutions for the reverse Stackelberg game. IEEE Trans. Syst. Man Cybern. Syst. **44**(10), 1315–1327 (2014)
31. Groot, N., Schutter, B.D., Hellendoorn, H.: Optimal affine leader functions in reverse Stackelberg games. J. Optim. Theory Appl. **168**(1), 348–374 (2016)

Equilibrium in the Piece-Wise Constant Pricing

Igor Bykadorov$^{(\boxtimes)}$ (iD)

Sobolev Institute of Mathematics, 4 Koptyug Avenue, 630090 Novosibirsk, Russia

Abstract. We consider a stylized distribution channel in the structure "manufacturer-retailer-consumer" under the assumption that the whole-sales and retail discounts are piece-wise constant. Earlier, we demonstrate the strict concavity of the profit of the manufacturer w.r.t. wholesale discount levels and strict concavity of the profit of retailer w.r.t. retail discount levels. This allows studying the equilibrium. The main result of the paper is: for the case of constant wholesale discount and piece-wise constant retail discount, the Stackelberg equilibrium under the leader-ship of the manufacturer can be calculated in closed form.

Keywords: Retailer · Piece-wise constant prices · Wholesale discount · Retail discount · Equilibrium

1 Introduction

Typically, economic agents stimulate production and sales through communications, as well as various types of influence on pricing. Moreover, in the structure of "producer - retailer - consumer," various types of discounts are often used.

One of the first work in this direction can be considered the paper [1], see also [2]. Among many works on this subject, let us note [3–8].

In the presented paper, we study dynamic marketing model. At every moment, the manufacturer stimulates retailer by wholesale discount as the manufacturer's control, while the retailer stimulates sales by retail discount as the retailer's control.

In [9], the maximization of manufacturer's profit with respect to the wholesale discount under constant and fixed retail discount is studied. Analogously, in [10], the maximization of retailer's profit with respect to the retailer discount under constant and fixed retail discount is studied. In the both cases, the result optimal control are continuous. Thus, the corresponding prices (wholesale and retail) are continuous. This results are elegant mathematically, but they seem inadequate economically. In practice, the prices are piece-wise constant.

We assume that these discounts are piece-wise constant; moreover, time switches of discount levels are fixed and known[1].

In [11,12] we demonstrate (under rather realistic conditions) the strict concavity of the profit of the manufacturer with respect to wholesale discount levels

---

[1] It seems realistic, cf. [11].

N. N. Olenev et al. (Eds.): OPTIMA 2021, LNCS 13078, pp. 288–302, 2021.
https://doi.org/10.1007/978-3-030-91059-4_21

and strict concavity of the profit of retailer with respect to retail discount levels. This allows to move on to the next stage of research: to study the equilibrium (Nash and Stackelberg) in the structure "manufacturer-retailer-consumer".

In the presented paper, we start to study the equilibrium. The first question is very relevant and non-trivial: is it possible to calculate the equilibrium explicitly, i.e., in closed form?

It seems that in the general case the answer to this question is "no". It is all the more interesting to identify cases when the answer to this question is "yes".

We study the case when the wholesale discount is constant while the retail discount is piece-wise constant. This simplification seems promising, since it keeps the wholesale price constant, while only the retail price turns out to be piece-wise constant. At the same time, this situation seems realistic in the case when the manufacturer is a "large" firm, and retailers are relatively "small".

The main result of the paper is (see Proposition 5 and (30), (31)):

For the case of constant wholesale discount and piece-wise constant retail discount, the Stackelberg equilibrium under the leadership of the manufacturer can be calculated in closed form.

The paper is organized as follows. In Sect. 2.1 repeat the material of Sect. 2 of [12]: we formulate the model and consider the case of piece-wise constant wholesale and retail discounts. In Section we study the case of constant wholesale discount and piece-wise constant retail discount. More precisely, we study the Stackelberg equilibrium under the leadership of manufacturer. Here we get the form of equilibrium retail discount as the function of wholesale discount (see Lemma 1). Also we get the main result: the equilibrium can be calculated in closed form (see Proposition 5). In Sect. 4 we discuss the obtained results. In Sect. 5 we present the example for the case of one switch. Section 6 contains the proofs of Lemma 1 and Proposition 5. Section 7 concludes.

2   Model

In this section, we remind briefly the model [9–12].

Let us consider a vertical distribution channel.

On the market, there are manufacturer ("firm"), retailer and consumer. The firm produces and sells a single product. To increase its profits, the firm uses the services of a retailer.

Let

- $[t_1, t_2]$ be the sales period[2],
- p be *the unit price* in a situation where the firm sells the product *directly* (i.e., bypassing the retailer) to the consumer, $p > 0$;
- c_0 be *the unit production cost*.

---

[2] We assume that the product is "seasonal" (cf. [9]), thus the sales period is rather small.

2.1 Basic Model

Wholesale Discount and Wholesale Price. To stimulate the retailer to sell the goods, the firm provides *the wholesale discount* $\alpha(t) \in [\mathcal{A}_1, \mathcal{A}_2] \subset [0,1]$. Thus, *the wholesale price* of the goods is

$$p_w(t) = (1 - \alpha(t))p. \tag{1}$$

Retail Discount, Retail Price, and Retailer's Profit per Unit. In turn, the retailer directs the *retail discount* ("pass-through"), i.e., a part $\beta(t) \in [\mathcal{B}_1, \mathcal{B}_2] \subset [0,1]$ of the discount $\alpha(t)$ to reduce the market price of the commodity. Thus, *the retail price* of the goods is equal to

$$(1 - \beta(t)\alpha(t))p. \tag{2}$$

Then *the retailer's profit per unit* from the sale is the difference between retail price and wholesale price, i.e., $\alpha(t)(1 - \beta(t))p$.

Accumulated Sales, Retailer's Motivation. Let

– $x(t)$ be the accumulated sales during the period $[t_1, t]$,
– $M(t)$ be the motivation of the retailer.

Motion Equations. We assume that accumulated sales $x(t)$ and the motivation of the retailer $M(t)$, satisfy the differential equations

$$\dot{M}(t) = \gamma \dot{x}(t) + \varepsilon\left(\alpha(t) - \overline{\alpha}\right),$$

$$\dot{x}(t) = -\theta x(t) + \delta M(t) + \eta \alpha(t)\beta(t),$$

where $\gamma > 0$, $\varepsilon > 0$, $\theta > 0$, $\delta >$, $\eta > 0$; see [11] for details[3].

Profits. At the end of the selling period, the total profit of the firm is[4]

$$\Pi_m = \int_{t_1}^{t_2} (p_w(t) - c_0)\,\dot{x}(t)dt = \int_{t_1}^{t_2} (q - \alpha(t)p)\,\dot{x}(t)dt,$$

where $q = p - c_0$. The total profit of the retailer is

$$\Pi_r = p \int_{t_1}^{t_2} \dot{x}(t)\alpha(t)(1 - \beta(t))dt.$$

---

[3] Parameter $\overline{\alpha} \in [\mathcal{A}_1, \mathcal{A}_2]$ takes into account the fact that the retailer has some expectations about the wholesale discount: the motivation is reduced if the retailer is dissatisfied with the wholesale discount, i.e., if $\alpha(t) < \overline{\alpha}$; on the contrary, the motivation increases if $\alpha(t) > \overline{\alpha}$.

[4] Since the sales period is rather small ("seasonal" products).

Model. Let $\overline{M} > 0$ be the initial motivation of the retailer.

Thus, the *Manufacturer-Retailer Problem* is

$$\Pi_m \longrightarrow \max_\alpha$$
$$\Pi_r \longrightarrow \max_\beta$$
$$\dot{x}(t) = -\theta x(t) + \delta M(t) + \eta\alpha(t)\beta(t),$$
$$\dot{M}(t) = \gamma\dot{x}(t) + \varepsilon\left(\alpha(t) - \overline{\alpha}\right),$$
$$x(t_1) = 0,\ M(t_1) = \overline{M},$$
$$\alpha(t) \in [\mathcal{A}_1, \mathcal{A}_2] \subset [0, 1],$$
$$\beta(t) \in [\mathcal{B}_1, \mathcal{B}_2] \subset [0, 1].$$

The above problem is the optimal control problem, where

- $x(t)$ and $M(t)$ are *the state variables,*
- $\alpha(t)$ and $\beta(t)$ are *the controls.*

In [9,10], the partial cases of this problem were studied:

- in [9], the maximization of Π_m w.r.t. wholesale discount under constant and fixed retail discount is studied;
- in [10], the maximization of Π_r w.r.t. retail discount under constant and fixed wholesale discount is studied.

In the both cases, the result optimal control are continuous. Thus, the corresponding prices (wholesale and retail) are continuous. This results are elegant mathematically, but they seem inadequate economically. In practice, the prices are piece-wise constant.

2.2 The Case: Wholesale Discount and Pass-Through Are Piece-Wise Constant

Let $I = \{1, \ldots n + 1\}$ and for some $t_1 = \tau_0 < \tau_1 < \ldots < \tau_n < \tau_{n+1} = t_2$

$$\alpha(t) = \alpha_i,\ \beta(t) = \beta_i,\ t \in (\tau_{i-1}, \tau_i),\ i \in I,$$

where α_i, β_i, $i \in I$, are *the discount levels.* Then, due to continuity of state variables,

$$x(t) = x_i(t),\ M(t) = M_i(t),\ t \in [\tau_{i-1}, \tau_i],\ i \in I,$$

where $x_i(t)$ and $M_i(t)$ are the solutions of the systems[5]

$$\dot{x}_i(t) = -\theta x_i(t) + \delta M_i(t) + \eta\alpha_i\beta_i,$$
$$\dot{M}_i(t) = \gamma\dot{x}_i(t) + \varepsilon\left(\alpha_i - \overline{\alpha}\right),$$
$$x_i(\tau_i) = x_{i-1}(\tau_i),$$
$$M_i(\tau_i) = M_{i-1}(\tau_i),$$
$$t \in [\tau_{i-1}, \tau_i], i \in I.$$

---

[5] Note that $x_1(\tau_0) = 0$ while $M_1(\tau_0) = \overline{M}$..

We get[6]

$$\Pi_r = p \cdot \sum_{i \in I} (1 - \beta_i) \, \alpha_i \, (x_i \, (\tau_i) - x_i \, (\tau_{i-1})), \qquad (3)$$

$$\Pi_m = p \cdot \sum_{i \in I \setminus \{n+1\}} (\alpha_{i+1} - \alpha_i) \, x_i \, (\tau_i) + (q - \alpha_{n+1}p) \, x \, (t_2). \qquad (4)$$

Therefore, we need the expressions for $x_i \, (\tau_i)$. Let $a = \theta - \gamma \delta$. Under rather natural condition (the concavity of cumulative sales for constant wholesale price, see details in [9,10]), we assume

$$a > 0. \qquad (5)$$

Let us define the functions[7]

$$K(t) = \frac{\delta \overline{M}}{a} \cdot \left(1 - e^{a(t_1 - t)} \right) + \frac{\overline{\alpha} \delta \varepsilon}{a^2} \cdot \left(1 - e^{a(t_1 - t)} + a \, (t_1 - t) \right), \qquad (6)$$

$$H_i(t) = \frac{\eta}{a} \cdot \left(1 - e^{a(\tau_{i-1} - t)} \right), \; t \geq \tau_{i-1}, \; i \in I, \qquad (7)$$

$$L_i(t) = -\frac{\delta \varepsilon}{a^2} \cdot \left(1 - e^{a(\tau_{i-1} - t)} + a \, (\tau_{i-1} - t) \right), \; t \geq \tau_{i-1}, \; i \in I. \qquad (8)$$

Proposition 1. *(See [11]) For $t \in [\tau_{i-1}, \tau_i]$, $i \in I$,*

$$x_i(t) = K(t) + (H_i(t) \beta_i + L_i(t)) \alpha_i$$

$$+ \sum_{j=1}^{i-1} ((H_j(t) - H_{j+1}(t)) \beta_j + L_j(t) - L_{j+1}(t)) \alpha_j .$$

Further, let us recall the result about the strict concavity of retailer's profit Π_r with respect to the retail discount levels.

Proposition 2. *(See [11]) The retailer's profit Π_r is strictly concave with respect to the retail discount levels β_i, $i \in I$.*

As to the manufacturer's profit Π_m, in [12] we demonstrate the strict concavity of Π_m, with respect to the wholesale discount levels, but only when the retail discount is constant.

---

[6] Now we study not the optimal control problem, but the optimization problem with the variables α_i, β_i, $i \in I$.

[7] Due to (5), these functions are well defined.

3   The Case: Wholesale Discount Is Constant, Pass-Through Is Piece-Wise Constant

Let
$$\alpha(t) = \alpha, \ t \in [t_1, t_2],$$
and for some $t_1 = \tau_0 < \tau_1 < \ldots < \tau_n < \tau_{n+1} = t_2$
$$\beta(t) = \beta_i, \ t \in (\tau_{i-1}, \tau_i), \ i \in I.$$

Then, due to continuity of space variables,
$$x(t) = x_i(t), \ M(t) = M_i(t), \ t \in [\tau_{i-1}, \tau_i], \ i \in I,$$

where $x_i(t)$ and $M_i(t)$ are the solutions of the systems[8]
$$
\begin{aligned}
&\dot{x}_i(t) = -\theta x_i(t) + \delta M_i(t) + \eta\alpha\beta_i, \\
&\dot{M}_i(t) = \gamma\dot{x}_i(t) + \varepsilon(\alpha - \overline{\alpha}), \\
&x_i(\tau_i) = x_{i-1}(\tau_i), \\
&M_i(\tau_i) = M_{i-1}(\tau_i), \\
&t \in [\tau_{i-1}, \tau_i], i \in I.
\end{aligned}
$$

We get
$$\Pi_r = p\alpha \cdot \sum_{i \in I} (1 - \beta_i)(x_i(\tau_i) - x_i(\tau_{i-1})), \tag{9}$$

$$\Pi_m = (q - \alpha p) x(t_2). \tag{10}$$

Now, Proposition 1 has the form (the definition of $K(t), H_i(t), L_i(t)$ see in (6)–(8))

Proposition 3. *For* $t \in [\tau_{i-1}, \tau_i], \ i \in I$

$$x_i(t) = K(t) + \left(H_i(t)\beta_i + L_1(t) + \sum_{j=1}^{i-1} (H_j(t) - H_{j+1}(t))\beta_j \right) \cdot \alpha.$$

Moreover, since

$$\frac{\partial^2 \Pi_m}{\partial \alpha^2} = -2p \cdot \left(H_i(t)\beta_i + L_1(t) + \sum_{j=1}^{i-1} \left(H_j(\tau_j) \cdot \prod_{l=j+1}^{n+1} e^{T_l} \right) \cdot \beta_j \right) < 0,$$

where

$$T_i = (\tau_{i-1} - \tau_i)a < 0, \ i \in I, \tag{11}$$

we get the strict concavity of retailer's profit (10) with respect to the wholesale discount, i.e., the following Proposition holds.

Proposition 4. *The manufacturer's profit (10) is strictly concave with respect to the wholesale discount level* α.

Proposition and Proposition allow us to study the equilibrium.

---

[8] Note that $x_1(\tau_0) = 0$ while $M_1(\tau_0) = \overline{M}$.

3.1   Stackelberg Equilibrium Under the Leadership of Manufacturer

Let us consider the manufacturer and the retailer as the two players of a Stack-elberg game. Moreover, let us consider the manufacturer as the channel leader: in this case we assume that she can only choose a constant trade discount during the whole sales period. This way we formulate the following Stackelberg game:

   Game ML: maximize (10) where, for each fixed $\alpha \in [\mathcal{A}_1, \mathcal{A}_2]$, *the values* $\beta_i, i \in I$, *are the optimal solution of the problem: maximize (9) subject to* $\beta_i \in [\mathcal{B}_1, \mathcal{B}_2]$.

   One has for every $i \in I$

$$\frac{\partial \Pi_r}{\partial \beta_i} = p\alpha \cdot \left(x_i\left(\tau_{i-1}\right) - x_i\left(\tau_i\right) + \sum_{k \in I} \frac{\partial}{\partial \beta_i}\left(x_k\left(\tau_k\right) - x_k\left(\tau_{k-1}\right)\right)\left(1 - \beta_k\right) \right). \quad (12)$$

   Let us consider the system

$$\frac{\partial \Pi_r}{\partial \beta_i} = 0, i \in I, \quad (13)$$

as the system of linear algebraic equations with respect to $\beta_i, i \in I$. Then the solution of this system depends on α. Let us denote the solution of (13) as

$$\beta_1\left(\alpha\right), \dots, \beta_{n+1}\left(\alpha\right). \quad (14)$$

Lemma 1. *The solution (14) can be expressed via* α *as*

$$\beta_i\left(\alpha\right) = \frac{A_i}{\alpha} + B_i \, , \, i \in I, \quad (15)$$

where A_i *and* B_i *are some constants.*

Proof. See Sect. 6.1.

   Lemma 1 allows to get the main result of the paper.

Proposition 5. *For the case of constant wholesale discount and piece-wise constant retail discount, the Stackelberg equilibrium under the leadership of the manufacturer can be calculated in closed form.*

Proof. See Sect. 6.2.

4   Discussion

The derived closed form description of the Stackelberg equilibrium looks very appealing as it leads to instant computational time, which is very good. Moreover, simple solutions are always nice from analytic point of view – those can be better understood and explained to even non-professionals.

   But the natural question arises: why we study the leadership of the manufacturer? What is "economically adequate": the leadership of the manufacturer or the leadership of the retailer? Moreover, it seems that "in reality" the leadership of the retailer is more adequate.

In [13] we studied the case of the constant discounts (both wholesale and retail) and get that if the expression

$$K = \frac{\delta \overline{M}}{a} \cdot \left(1 - e^{(t_1 - t_2)a}\right) + \frac{\overline{a}\delta\varepsilon}{a^2} \cdot \left(1 - e^{(t_1 - t_2)a} + (t_1 - t_2)a\right)$$

(cf. (6)) is negative, then both manufacturer and retailer want to be leader (**the struggle to be leader**); while if K is positive then both manufacturer and retailer want to be follower (**the struggle to be follower**). May be, the similar situation can be in the case of piece-wise constant discounts? In any case case, we plan to study the case of the retailer's leadership in the future.

5 Example: The Case $n = 1$

Let $n = 1, (t_1 - \tau_1)a = (\tau_1 - t_2)a = T$. Then system (13) is

$$\begin{cases} -\dfrac{2}{1-r} \cdot \beta_1 + \beta_2 = C_1(\alpha) \\ \beta_1 - \dfrac{2}{1-r} \cdot \beta_2 = C_2(\alpha) \end{cases}$$

where

$$r = e^T,$$

$$C_1(\alpha) = \frac{\dfrac{K(\tau_1)}{\alpha} + L_1(\tau_1)}{(1-r)H_1(\tau_1)} + 1,$$

$$C_2(\alpha) = \frac{\dfrac{K(\tau_2) - K(\tau_1)}{\alpha} + L_1(\tau_2) - L_1(\tau_1)}{(1-r)H_1(\tau_1)}.$$

We get

$$\beta_1(\alpha) = -\frac{1-r}{(1+r)(3-r)} \cdot (2 \cdot C_1(\alpha) + (1-r) \cdot C_2(\alpha)), \qquad (16)$$

$$\beta_2(\alpha) = -\frac{1-r}{(1+r)(3-r)} \cdot ((1-r) \cdot C_1(\alpha) + 2 \cdot C_2(\alpha)). \qquad (17)$$

Since

$$x_2(\tau_2) = K(\tau_2) + ((H_1(\tau_2) - H_2(\tau_2))\beta_1 + H_2(\tau_2)\beta_2 + L_1(\tau_2)) \cdot \alpha$$

$$= K(\tau_2) + (H_1(\tau_1)(r\beta_1 + \beta_2) + L_1(\tau_2)) \cdot \alpha,$$

we get due to (16) and (17)

$$\tilde{x}_2(\tau_2) = K(\tau_2) + (H_1(\tau_1)(r\beta_1(\alpha) + \beta_2(\alpha)) + L_1(\tau_2)) \cdot \alpha$$

$$= K\left(\tau_2\right) + \left(-\frac{1-r}{(3-r)} \cdot H_1\left(\tau_1\right)\left(C_1\left(\alpha\right) + (2-r)\cdot C_2\left(\alpha\right)\right) + L_1\left(\tau_2\right)\right)\cdot\alpha$$

$$= \frac{(1-r)\cdot K\left(\tau_1\right) + K\left(\tau_2\right) + \left((1-r)\cdot\left(L_1\left(\tau_1\right) - H_1\left(\tau_1\right)\right) + L_1\left(\tau_2\right)\right)\cdot\alpha}{3-r}.$$

Further,

$$\tilde{\Pi}_m = (q - p\alpha)\,\tilde{x}_2\left(\tau_2\right).$$

Hence

$$\frac{\partial\tilde{\Pi}_m}{\partial\alpha} = -p\tilde{x}_2\left(\tau_2\right) + (q - p\alpha)\cdot\frac{\partial\tilde{x}_2\left(\tau_2\right)}{\partial\alpha}$$

$$= -p\cdot\frac{(1-r)\cdot K\left(\tau_1\right) + K\left(\tau_2\right)}{3-r}$$

$$+ (q - 2p\alpha)\cdot\frac{(1-r)\cdot\left(L_1\left(\tau_1\right) - H_1\left(\tau_1\right)\right) + L_1\left(\tau_2\right)}{3-r}.$$

So

$$\frac{\partial\tilde{\Pi}_m}{\partial\alpha} = 0 \iff \alpha = \frac{q}{2p} - \frac{1}{2}\cdot\frac{(1-r)\cdot K\left(\tau_1\right) + K\left(\tau_2\right)}{(1-r)\cdot\left(L_1\left(\tau_1\right) - H_1\left(\tau_1\right)\right) + L_1\left(\tau_2\right)}.$$

and the Stackelberg equilibrium under the leadership of manufacturer (ML) is[9]

$$\alpha^{ML} = \frac{q}{2p} - \frac{1}{2}\cdot\frac{(1-r)\cdot K\left(\tau_1\right) + K\left(\tau_2\right)}{(1-r)\cdot\left(L_1\left(\tau_1\right) - H_1\left(\tau_1\right)\right) + L_1\left(\tau_2\right)}$$

$$\beta_1^{ML} = \beta_1\left(\alpha^{ML}\right) = -\frac{1-r}{(1+r)(3-r)}\cdot\left(2\cdot C_1\left(\alpha^{ML}\right) + (1-r)\cdot C_2\left(\alpha^{ML}\right)\right)$$

$$\beta_2^{ML} = \beta_2\left(\alpha^{ML}\right) = -\frac{1-r}{(1+r)(3-r)}\cdot\left((1-r)\cdot C_1\left(\alpha^{ML}\right) + 2\cdot C_2\left(\alpha^{ML}\right)\right)$$

Further,

$$\tilde{x}_2\left(\tau_2\right) = \frac{p\left((1-r)\cdot K\left(\tau_1\right) + K\left(\tau_2\right)\right) + q\left((1-r)\cdot\left(L_1\left(\tau_1\right) - H_1\left(\tau_1\right)\right) + L_1\left(\tau_2\right)\right)}{2p(3-r)}$$

$$q - p\alpha^{ML} = \frac{p\left((1-r)\cdot K\left(\tau_1\right) + K\left(\tau_2\right)\right) + q\left((1-r)\cdot\left(L_1\left(\tau_1\right) - H_1\left(\tau_1\right)\right) + L_1\left(\tau_2\right)\right)}{2\left((1-r)\cdot\left(L_1\left(\tau_1\right) - H_1\left(\tau_1\right)\right) + L_1\left(\tau_2\right)\right)}.$$

$$\tilde{\Pi}_m = \frac{\left(p\left((1-r)\cdot K\left(\tau_1\right) + K\left(\tau_2\right)\right) + q\left((1-r)\cdot\left(L_1\left(\tau_1\right) - H_1\left(\tau_1\right)\right) + L_1\left(\tau_2\right)\right)\right)^2}{4p(3-r)\left((1-r)\cdot\left(L_1\left(\tau_1\right) - H_1\left(\tau_1\right)\right) + L_1\left(\tau_2\right)\right)}.$$

---

[9] Of course, it is equilibrium only if $\alpha^{ML}\in[\mathcal{A}_1,\mathcal{A}_2], \beta_1^{ML}\in[\mathcal{B}_1,\mathcal{B}_2], \beta_2^{ML}\in[\mathcal{B}_1,\mathcal{B}_2]$. Otherwise, we need to consider the "boundary conditions".

6   Proofs

6.1   Proof of Lemma 1

One has that

$$\frac{\partial \Pi_r}{\partial \beta_i} = 0, \ i \in I, \tag{18}$$

if and only if the conditions

$$x_i(\tau_i) - x_i(\tau_{i-1}) = \sum_{k \in I} \frac{\partial}{\partial \beta_i} (x_k(\tau_k) - x_k(\tau_{k-1})) (1 - \beta_k), \ i \in I, \tag{19}$$

hold.

We get

$$x_i(\tau_i) = K(\tau_i) + \left(H_i(\tau_i) \beta_i + L_1(\tau_i) + \sum_{j=1}^{i-1} \left(\prod_{l=j+1}^{i} r_l \right) H_j(\tau_i) \cdot \beta_j \right) \cdot \alpha \tag{20}$$

and

$$x_i(\tau_{i-1}) = K(\tau_{i-1}) + \left(L_1(\tau_{i-1}) + \sum_{j=1}^{i-1} \left(\prod_{l=j+1}^{i-1} r_l \right) H_j(\tau_{i-1}) \cdot \beta_j \right) \cdot \alpha, \tag{21}$$

where

$$r_i = e^{T_i}, \ i \in I, \tag{22}$$

while $T_i, \ i \in I$, are defined in (11).

Hence

$$x_i(\tau_i) - x_i(\tau_{i-1}) = K(\tau_i) - K(\tau_{i-1})$$

$$+ \left(H_i(\tau_i) \left(\beta_i - \sum_{j=1}^{i-1} \left(\prod_{l=j+1}^{i-1} r_l \right) (1 - r_j) \cdot \beta_j \right) + L_1(\tau_i) - L_1(\tau_{i-1}) \right) \cdot \alpha$$

and

$$\frac{\partial}{\partial \beta_i} (x_k(\tau_k) - x_k(\tau_{k-1})) = \begin{cases} H_i(\tau_i) \cdot \alpha, & k = i \\ H_k(\tau_k)(r_i - 1) \displaystyle\prod_{l=i+1}^{k-1} r_l \cdot \alpha, & k \geq i+1 \end{cases} \tag{23}$$

Now, let us substitute (20), (21) and (23) in (19).

We get that the conditions (18) hold if and only if the conditions

$$K\left(\tau_i\right) - K\left(\tau_{i-1}\right)$$

$$+ \left(H_i\left(\tau_i\right) \left(\beta_i - \sum_{j=1}^{i-1} \left(\prod_{l=j+1}^{i-1} r_l \right) \left(1 - r_j\right) \cdot \beta_j \right) + L_1\left(\tau_i\right) - L_1\left(\tau_{i-1}\right) \right) \cdot \alpha$$

$$= \left(H_i\left(\tau_i\right) \left(1 - \beta_i\right) - \left(1 - r_i\right) \sum_{k=i+1}^{n+1} \left(\prod_{l=i+1}^{k-1} r_l \right) H_k\left(\tau_k\right) \left(1 - \beta_k\right) \right) \cdot \alpha, \ i \in I,$$

hold. Let us rewrite these conditions as

$$\sum_{j=1}^{i-1} \left(\left(1 - r_j\right) \prod_{l=j+1}^{i-1} r_l \right) \cdot \beta_j - 2\beta_i + \sum_{j=i+1}^{n+1} \left(\left(1 - r_j\right) \prod_{l=i+1}^{j-1} r_l \right) \cdot \beta_j$$

$$= \frac{K\left(\tau_i\right) - K\left(\tau_{i-1}\right)}{H_i\left(\tau_i\right)} \cdot \frac{1}{\alpha} + \frac{L_1\left(\tau_i\right) - L_1\left(\tau_{i-1}\right)}{H_i\left(\tau_i\right)} + \sum_{j=i+1}^{n+1} \left(\left(1 - r_j\right) \prod_{l=i+1}^{j-1} r_l \right), \ i \in I.$$

In what follows, let us assume that[10]

$$\tau_0 - \tau_1 = \tau_1 - \tau_2 = \ldots = \tau_{n-1} - \tau_n = \tau_n - \tau_{n+1},$$

i.e., that[11]

$$T_i = T, \ i \in I.$$

and

$$r_i = e^T = r, \ i \in I,$$

Then

$$H_i\left(\tau_i\right) = \frac{\eta}{a} \cdot \left(1 - e^T\right) = H_1\left(\tau_1\right).$$

Thus, the conditions (18) are equivalent to the conditions

$$\sum_{j=1}^{i-1} r^{(i-j-1)} \cdot \beta_j - \frac{2}{(1-r)} \cdot \beta_i + \sum_{j=i+1}^{n+1} r^{(j-i-1)} \cdot \beta_j = C_i\left(\alpha\right), \ i \in I, \quad (24)$$

where

$$C_i\left(\alpha\right) = \frac{K\left(\tau_i\right) - K\left(\tau_{i-1}\right)}{(1-r) H_1\left(\tau_1\right)} \cdot \frac{1}{\alpha} + \frac{L_1\left(\tau_i\right) - L_1\left(\tau_{i-1}\right)}{(1-r) H_1\left(\tau_1\right)} + \sum_{j=i}^{n} r^{j-i}, \ i \in I. \quad (25)$$

---

[10] The general case can be considered analogously, but with more complicated technical calculations.

[11] The constants T_i, $i \in I$, and r_i, $i \in I$, are defined in (11) and (22) respectively.

Let us consider (24) as the system of linear algebraic equations of β_i, $i \in I$. The matrix of the system is

$$
D = \begin{pmatrix}
-\dfrac{2}{1-r} & 1 & r & \cdots & r^{n-1} \\
1 & \ddots & \ddots & \ddots & \vdots \\
r & & \ddots & \ddots & \ddots & r \\
\vdots & & & \ddots & \ddots & 1 \\
r^{n-1} & \cdots & r & 1 & 1-\dfrac{2}{1-r}
\end{pmatrix}.
\tag{26}
$$

One has

$$
D = \frac{1+r}{1-r} \cdot F \cdot G \cdot F^T,
\tag{27}
$$

where

$$
F = \begin{pmatrix}
1 & -r & 0 & \cdots & 0 \\
0 & \ddots & \ddots & \ddots & \vdots \\
\vdots & \ddots & \ddots & \ddots & 0 \\
\vdots & & \ddots & \ddots & -r \\
0 & \cdots & \cdots & 0 & 1
\end{pmatrix}, \quad
G = \begin{pmatrix}
-\dfrac{2}{1+r} & 1 & 0 & \cdots & 0 \\
1 & -2 & \ddots & \ddots & \vdots \\
0 & \ddots & \ddots & \ddots & 0 \\
\vdots & \ddots & \ddots & \ddots & 1 \\
0 & \cdots & 0 & 1 & -2
\end{pmatrix}.
$$

Moreover

$$
G = J \cdot S \cdot J^T,
\tag{28}
$$

where

$$
J = \begin{pmatrix}
1 & 0 & \cdots & \cdots & \cdots & 0 \\
\dfrac{n}{n+1} & 1 & \ddots & & & \vdots \\
\dfrac{n-1}{n+1} & \dfrac{n-1}{n} & 1 & \ddots & & \vdots \\
\vdots & \vdots & & \ddots & \ddots & \vdots \\
\vdots & \vdots & & \ddots & 1 & 0 \\
\dfrac{1}{n+1} & \dfrac{1}{n} & \cdots & \cdots & \dfrac{1}{2} & 1
\end{pmatrix},
$$

$$
S = \begin{pmatrix}
-\dfrac{2+n\cdot(1-r)}{1+r}\cdot\dfrac{1}{n+1} & 0 & \cdots & \cdots & 0 \\
0 & -\dfrac{n+1}{n} & \ddots & \ddots & \vdots \\
\vdots & & \ddots & \ddots & \vdots \\
\vdots & & & -\dfrac{3}{2} & 0 \\
0 & & \cdots & 0 & -2
\end{pmatrix}.
$$

Since

$$\det F = \det J = 1,$$

we get due to (27) and (28)

$$\det D = \left(\frac{1+r}{1-r}\right)^{n+1} \cdot \det S,$$

i.e.,

$$\det D = (-1)^{n+1} \left(\frac{1+r}{1-r}\right)^{n+1} \cdot \frac{2 + n \cdot (1-r)}{1+r} \neq 0. \tag{29}$$

Due to (29), system (24) has unique solution (14).

To finish the proof, it is sufficient

− to note that matrix (26) does not depend on α,
− to take into consideration the form of (25).

6.2   Proof of Proposition 5

One has, see (10), $\Pi_m = (q - p\alpha) x_{n+1}(t_2)$. Hence

$$\tilde{\Pi}_m = (q - p\alpha) \tilde{x}_{n+1}(t_2),$$

where[12]

$$\tilde{x}_{n+1}(t_2) = K(t_2) + \left(\sum_{j=1}^{n+1} (H_j(t_2) - H_{j+1}(t_2)) \beta_j(\alpha) + L_1(t_2)\right) \cdot \alpha,$$

$$H_{n+2}(t_2) = 0.$$

Due to Lemma 1, see (15),

$$\tilde{x}_{n+1}(t_2) = K(t_2)$$

$$+ \sum_{j=1}^{n+1} (H_j(t_2) - H_{j+1}(t_2)) A_j + \left(\sum_{j=1}^{n+1} (H_j(t_2) - H_{j+1}(t_2)) B_j + L_1(t_2)\right) \alpha.$$

We get

$$\frac{\partial \tilde{\Pi}_m}{\partial \alpha} = -p \cdot K(t_2) + \sum_{j=1}^{n+1} (qB_j - pA_j)(H_j(t_2) - H_{j+1}(t_2)) + q \cdot L_1(t_2)$$

$$-2p \cdot \left(\sum_{j=1}^{n+1} (H_j(t_2) - H_{j+1}(t_2)) B_j + L_1(t_2)\right) \alpha.$$

---

[12] $\beta_j(\alpha)$, $i \in I$, are as in (15).

Hence

$$\frac{\partial \tilde{\Pi}_m}{\partial \alpha} = 0 \iff$$

$$\iff \alpha = \frac{-p \cdot K(t_2) + \sum_{j=1}^{n+1} (qB_j - pA_j)(H_j(t_2) - H_{j+1}(t_2)) + q \cdot L_1(t_2)}{2p \cdot \left(\sum_{j=1}^{n+1} (H_j(t_2) - H_{j+1}(t_2)) B_j + L_1(t_2) \right)}.$$

Thus,

$$\alpha^{ML} = \frac{q}{2p} - \frac{1}{2} \cdot \frac{K(t_2) + \sum_{j=1}^{n+1} (H_j(t_2) - H_{j+1}(t_2)) A_j}{L_1(t_2) + \sum_{j=1}^{n+1} (H_j(t_2) - H_{j+1}(t_2)) B_j} \qquad (30)$$

and due to (15)

$$\beta_i^{ML} = \beta_i \left(\alpha^{ML} \right) = \frac{A_i}{\alpha^{ML}} + B_i, \ i \in I,$$

i.e., see (30)

$$\beta_i^{ML} = \frac{A_i}{\frac{q}{2p} - \frac{1}{2} \cdot \frac{K(t_2) + \sum_{j=1}^{n+1} (H_j(t_2) - H_{j+1}(t_2)) A_j}{L_1(t_2) + \sum_{j=1}^{n+1} (H_j(t_2) - H_{j+1}(t_2)) B_j}} + B_i, \ i \in I. \quad (31)$$

If $\alpha^{ML} \in [\mathcal{A}_1, \mathcal{A}_2]$ and $\beta_i^{ML} \in [\mathcal{B}_1, \mathcal{B}_2]$, $i \in I$, then α^{ML}, β_i^{ML}, $i \in I$, is the Stackelberg equilibrium under the leadership of manufacturer.

Otherwise, we need to consider the "boundary conditions": in these cases, $\alpha^{ML} \in \{\mathcal{A}_1, \mathcal{A}_2\}$ and/or for some $i \in I$, $\beta_i^{ML} \in \{\mathcal{B}_1, \mathcal{B}_2\}$.

7   Conclusion

In this paper, we study a stylized vertical control distribution channel in the structure "manufacturer-retailer-consumer". More precisely, we consider the situation when the wholesale discount and pass-through are piece-wise constant. The switching times are assumed to be known and fixed.

The main result of the paper is the following: for the case of constant wholesale discount and piece-wise constant retail discount, the Stackelberg equilibrium under the leadership of the manufacturer can be calculated in closed form.

As for the topics of further research, we plan to study the possibility of calculating in closed form:

- the Nash equilibrium;
- the Stackelberg equilibrium under the leadership of the retailer;
- the equilibrium when not only the retail discount is piece-wise constant, but also the manufacturer discount is piece-wise constant.

Acknowledgments. The study was carried out within the framework of the state contract of the Sobolev Institute of Mathematics (project no. 0314-2019-0018). The work was supported in part by the Russian Foundation for Basic Research, project 19-010-00910. Besides, many thanks to the anonymous referees for their useful comments and suggestions.

References

1. Nerlove, M., Arrow, K.J.: Optimal advertising policy under dynamic conditions. Economica **29**(144), 129–142 (1962)
2. Vidale, M.L., Wolfe, H.B.: An operations-research study of sales response to advertising. In: Lecture Notes in Economics and Mathematical Systems (Operations Research), vol. 132, pp. 223–225 (1976). https://doi.org/10.1007/978-3-642-51565-1_72
3. Bykadorov, I.A., Ellero, A., Moretti, E.: Minimization of communication expenditure for seasonal products. RAIRO Oper. Res. **36**(2), 109–127 (2002)
4. Mosca, S., Viscolani, B.: Optimal goodwill path to introduce a new product. J. Optim. Theory Appl. **123**(1), 149–162 (2004)
5. Giri, B.C., Bardhan, S.: Coordinating a two-echelon supply chain with price and inventory level dependent demand, time dependent holding cost, and partial backlogging. Int. J. Math. Oper. Res. **8**(4), 406–423 (2016)
6. Printezis, A., Burnetas, A.: The effect of discounts on optimal pricing under limited capacity. Int. J. Oper. Res. **10**(2), 160–179 (2011)
7. Lu, L., Gou, Q., Tang, W., Zhang, J.: Joint pricing and advertising strategy with reference price effect. Int. J. Prod. Res. **54**(17), 5250–5270 (2016)
8. Zhang, S., Zhang, J., Shen, J., Tang, W.: A joint dynamic pricing and production model with asymmetric reference price effect. J. Ind. Manag. Optim. **15**(2), 667–688 (2019)
9. Bykadorov, I., Ellero, A., Moretti, E., Vianello, S.: The role of retailer's performance in optimal wholesale price discount policies. Euro. J. Oper. Res. **194**(2), 538–550 (2009)
10. Bykadorov, I.: Dynamic marketing model: optimization of retailer's role. Commun. Comput. Inf. Sci. **974**, 399–414 (2019)
11. Bykadorov, I.: Dynamic marketing model: the case of piece-wise constant pricing. Commun. Comput. Inf. Sci. **1145**, 150–163 (2020)
12. Bykadorov, I.: Pricing in dynamic marketing: the cases of piece-wise constant sale and retail discounts. In: Olenev, N., Evtushenko, Y., Khachay, M., Malkova, V. (eds.) OPTIMA 2020. LNCS, vol. 12422, pp. 27–39. Springer, Cham (2020). https://doi.org/10.1007/978-3-030-62867-3_3
13. Bykadorov, I.A., Ellero, A., Moretti, E.: Trade discount policies in the differential games framework. Int. J. Biomed. Soft Comput. Hum. Sci. **18**(1), 15–20 (2013)

Analyzing Stability of Extreme Portfolios

Yury Nikulin[1]([✉]) [iD] and Vladimir Emelichev[2]

[1] Faculty of Science, University of Turku, 20014 Turku, Finland
`yurnik@utu.fi`
[2] Faculty of Mechanics and Mathematics, Belarusian State University,
220030 Minsk, Belarus

Abstract. On the basis of the portfolio theory, a multicriteria investment Boolean problem of minimizing lost profits is formulated. The problem considered is to find a set of all extreme portfolios. The quality of such portfolios is assessed by examining stability of the set of extreme portfolios to perturbations of Savage's minimax risk criterion parameters. The lower and upper bounds on the radius of the strong stability are obtained under the assumption that arbitrary Hölder's norms are specified in the three spaces of the problem's initial data. The case of the investment problem with linear criteria is considered separately. For this case, the attainability of the bounds is proven.

Keywords: Multicriteria optimization · Savage's risk criteria · Set of extreme portfolios · Strong stability radius · Hölder's metric · Investment problem

1 Introduction

As mentioned in [17], assessing the quality of decisions while selecting project portfolios becomes an inherent part of the decision-making process when the project parameters are inaccurate or uncertain. Many business and management decisions are made in an uncertain and risky environment caused by the influence of various factors, such as an inadequacy of the mathematical models used by real processes, measurements errors or rounding, etc.

Many problems of making multipurpose decisions in management, planning, and design can be formulated as multicriteria (multiobjective) problems of continuous and/or discrete optimization. While classical portfolio optimization models deal with continuous variables [3,24], we propose a discrete analogue based on binary decision variables which encode decision maker's choices about what assets should or should not be included in optimal portfolios. The usage of Savage's criterion [30] as optimization objective decreases the investment risk in the worst market situation.

The term stability is commonly used for the phase of an algorithm at which a solution (or a set of solutions) of the problem has been already found, and additional calculations are performed in order to investigate how this solution depends on changes in the problem's data. In 1923, Jacques Hadamard recognized the

N. N. Olenev et al. (Eds.): OPTIMA 2021, LNCS 13078, pp. 303–317, 2021.
https://doi.org/10.1007/978-3-030-91059-4_22

problem of stability as one of the central in mathematical research. He postulated that in order to be well-posed, a mathematical problem should satisfy three properties: solution existence, uniqueness as well as solution stability, i.e. continuous dependency on the data [4]. Problems that are not well-posed in the sense of Hadamard are usually termed ill-posed. Despite the fact that multiple objective discrete optimization problems may not formally satisfy the property of solution uniqueness, carefully considering stability is believed necessary and important.

Despite the existence of numerous approaches to stability analysis of optimization problems, two major directions can be pointed out: quantitative and qualitative. In the first direction, many authors focus on studying different types of stability and deriving optimality conditions for them (see e.g. [8, 10, 11, 14, 20, 21]). The second direction is focused on obtaining quantitative estimates of permissible changes in the problem's initial data that retains some invariance property of optimal solutions, and on the development of computational algorithms for large classes of single objective [1, 23] and multiobjective [13, 18, 19] discrete optimization problems. The key concept here is the stability radius that determines numerical bounds and conditions when an optimal solution retains its optimal properties. For example, [7] contains a survey of the results describing analytical expressions and bounds for calculating the stability radii of the multicriteria integer programming model with linear criteria. Sometimes, instead of stability radius, other measures of stability are scrutinized when the perturbations in the initial data exceed the level determined by the stability radius, and the perturbed initial data are outside the stability region. In this case, the concept of the stability and accuracy function is used to evaluate the quality of the chosen solution. Being originally proposed in [22] for multicriteria linear combinatorial optimization problem, later some of the results were generalized under game theoretic framework [26], and for scheduling problems [27].

A relatively new direction in research is to analyze stability for multicriteria investment problems. In [9], some bounds on the stability radii of one Pareto optimal portfolio were obtained in the cases where the three-dimensional space of the problem parameters is equipped with different combinations of linear and Chebyshev norms. The case of general l_p norm is analyzed in [16]. Most recently, the performances of the stability function and the optimality threshold are shown in the case study using global risk assessments for projects participating in the Belt and Road Initiative. The computation results demonstrate the ability through the stability function to evaluate the quality and optimal properties of feasible project portfolios [17].

Our current work continues research towards a similar direction, with focus on a different optimality principle, namely, the so-called extreme solutions and their stability are investigated. The paper is organized as follows. In Sect. 2, we introduce basic concepts and formulate the problem. In this section we also introduce a type of stability named strong stability, i.e. a situation when for any admissible perturbation there exists at least one extreme portfolio preserving its own optimality. Section 3 contains auxiliary technical statements required for the proof of the main result. As a result of the parametric analysis, in Sect. 4 the lower and upper bounds on strong stability radius are obtained in the case with arbitrary Hölder's norms are specified in the three spaces of the problem's initial data. Separately, a particular case where all the criteria are linear is considered in Sect. 5.

2 Problem Formulation and Basic Definitions

Consider a multicriteria discrete variant of the investment optimization problem with the following parameters specified below. Let

$N_n = \{1, 2, \ldots, n\}$ be a variety of alternatives (investment assets);

N_m be a set of possible financial market states (market situations, scenarios);

N_s be a set of possible risks;

r_{ijk} be a numerical measure of economic risk of type $k \in N_s$ if investor chooses project $j \in N_n$ given the market state $i \in N_m$;

$R = [r_{ijk}] \in \mathbf{R}^{m \times n \times s}$ be a matrix specifying risks;

$x = (x_1, x_2, \ldots, x_n)^T \in \mathbf{E}^n$ be an investment portfolio, where $\mathbf{E} = \{0, 1\}$,

$$x_j = \begin{cases} 1, \text{ if investor chooses project } j, \\ 0, \text{ otherwise;} \end{cases}$$

$X \subset \mathbf{E}^n$ be a set of all admissible investment portfolios, i.e. those whose realization provides the investor with the expected income and does not exceed his/her initial capital;

$\mathbf{R}^m$ be a financial market state space;

$\mathbf{R}^n$ be a portfolio space;

$\mathbf{R}^s$ be a risk space.

In our model, we assume that the risk measure is addictive, i.e. the total risk of one portfolio is a sum of risks of the projects included in the portfolio. The risk of each project can be measured, for instance, by means of the associated implementation cost.

The risk factor is an essential attribute of the functioning of the financial market. Extensive literature is devoted to methods of quantitative assessment of economic risks, their classification and characterization. Recently, experts propose quantifying risks through the prism of five R: Robustness, Redundancy, Resourcefulness, Response and Recovery [5]. It leads to the necessity of multicriteria decision making tools.

Assume that the efficiency of a chosen portfolio (Boolean vector) $x \in X$, $|X| \geq 2$, is evaluated by a vector objective function

$$f(x, R) = (f_1(x, R_1), f_2(x, R_2), \ldots, f_s(x, R_s)),$$

each partial objective represents minimax Savage's risk criterion [30].

$$f_k(x, R_k) = \max_{i \in N_m} r_{ik} x = \max_{i \in N_m} \sum_{j \in N_n} r_{ijk} x_j \rightarrow \min_{x \in X}, \qquad k \in N_s,$$

where $R_k \in \mathbf{R}^{m \times n}$ is the k-th cut $R = [r_{ijk}] \in \mathbf{R}^{m \times n \times s}$ with rows

$$r_{ik} = (r_{i1k}, \; r_{i2k}, \ldots \; r_{ink}) \in \mathbf{R}^n, \; i \in N_m.$$

Following the criterion of a bottleneck [2], an investor in the conditions of economic instability and uncertainty of the market state is extremely cautious, optimizing the total risk of the portfolio in the most unfavorable situation, namely when the risk is maximum. Such caution is appropriate because any investment is the exchange of a certain current value for a possibly uncertain future income. Obviously, this approach is dictated by the safest and most protective rule prescribing to assume the worst.

The problem of finding extreme portfolios is referred to as the multicriteria investment Boolean problem with Savage's risk criteria and denoted $Z_s^m(R)$, $s, m \in \mathbf{N}$. The set of extreme portfolios is defined as follows:

$$E_s^m(R) = \{x \in X \; : \; \exists k \in N_s \; \; \forall x' \in X \; \; (f_k(x, R_k) \le f_k(x', R_k))\}.$$

This set can equivalently be written as follows:

$$E_s^m(R) = \{x \in X \; : \; \exists k \in N_s \; \; (E_k(x, C_k) = \emptyset)\},$$

where

$$E_k(x, C_k) = \{x' \in X \; : \; f_k(x, R_k) > f_k(x', R_k)\}, \; k \in N_s.$$

Thus, the choice of extreme portfolios can be interpreted as finding best solutions for each of s criteria, and then combining them into one set. The vector composed of optimal objective values constitutes the ideal vector that is of great importance in theory and methodology of multiobjective optimization [25, 31]. This also justifies our particular interest in studying some properties of extreme solutions. Obviously, $E_1^m(R)$, $R \in \mathbf{R}^{m \times n}$ is the set of optimal solutions for scalar problem $Z_1^m(C)$.

Given the assumption that X is finite, the following statements are true for any $R \in \mathbf{R}^{m \times n \times s}$:

$$E_s^m(R) = S_s^m(R) \setminus (P_s^m(R) \setminus L_s^m(R)) = L_s^m(R) \cup (S_s^m(R) \setminus P_s^m(R)),$$

$$E_s^m(R) \cap P_s^m(R) = L_s^m(R), \; L_s^m(R) \subseteq E_s^m(R) \subseteq S_s^m(R), \; L_s^m(R) \subseteq P_s^m(R) \subseteq S_s^m(R),$$

where $P_s^m(R)$ is the Pareto set [29], $S_s^m(R)$ is the S set [28], $L_s^m(R)$ is lexicographic set [25, 28]. These sets are defined as follows:

$$P_s^m(R) = \{x \in X \; : \; \nexists x^0 \in X \; \; (f(x, R) \ge f(x^0, R) \; \& \; f(x, R) \ne f(x^0, R))\},$$

$$S_s^m(R) = \{x \in X \; : \; \nexists x^0 \in X \; \; \forall k \in N_s \; \; (f_k(x, R_k) > f_k(x^0, R_k))\},$$

$$L_s^m(R) = \bigcup_{\pi \in \Pi_s} L(R, \pi).$$

Here

$$L(R, \pi) = \left\{ x \in X \; : \; \forall x' \in X \; \; \left(f(x, R) \underset{\pi}{\le} f(x', R) \right) \right\},$$

Π_s is the set of all $s!$ permutations of numbers $1, 2, ..., s$; $\pi = \{\pi_1, \pi_2, ..., \pi_s\} \in \Pi_s$, is a binary relation $\underset{\pi}{\leq}$ between two vectors $y = (y_1, y_2, ..., y_s)$ and $y' = (y'_1, y'_2, ..., y'_s)$ belonging to $\mathbf{R}^s$ and defined as:

$$y \underset{\pi}{\leq} y' \Leftrightarrow (y = y') \vee \left(\exists u \in N_s \quad \forall l \in N_{u-1} \quad (y_{\pi_u} < y'_{\pi_u} \ \& \ y_{\pi_l} = y'_{\pi_l}) \right),$$

where $N_0 = \emptyset$. Evidently, all the sets above are nonempty for any $R \in \mathbf{R}^{m \times n \times s}$.

We will perturb the elements of matrix R by adding elements of the perturbing matrix $R' \in \mathbf{R}^{m \times n \times s}$. Thus the perturbed problem $Z_s^m(R + R')$ of finding extreme solutions has the following form:

$$f(x, R + R') \to \min_{x \in X}.$$

The set of extreme portfolios in the perturbed problem is denoted by $E_s^m(R+R')$.

Recall that Hölder's norm l_p (also known as p-norm) in vector space $\mathbf{R}^n$ is the number

$$\|a\|_p = \begin{cases} \left(\sum_{j \in N_n} |a_j|^p \right)^{1/p} & \text{if } 1 \leq p < \infty, \\ \max\{|a_j| \ : \ j \in N_n\} & \text{if } p = \infty, \end{cases}$$

where $a = (a_1, a_2, ..., a_n)^T \in \mathbf{R}^n$.

In the spaces $\mathbf{R}^n, \mathbf{R}^m$ and $\mathbf{R}^s$ we define three Hölder's norms l_p, l_q and l_t, where $p, q, t \in [1, \infty]$. So, the norm of matrix $R \in \mathbf{R}^{m \times n \times s}$ is the following number

$$\|R\|_{pqt} = \|(\|R_1\|_{pq}, \|R_2\|_{pq}, ..., \|R_s\|_{pq})\|_t,$$

with cuts

$$\|R_k\|_{pq} = \|(\|r_{1k}\|_p, \|r_{2k}\|_p, ..., \|r_{mk}\|_p)\|_q, \quad k \in N_s.$$

For any numbers $p, q, t \in [1, \infty]$ the following inequalities are valid:

$$\|r_{ik}\|_p \leq \|R_k\|_{pq} \leq \|R\|_{pqt}, \quad i \in N_m, \ k \in N_s. \tag{1}$$

Following [7], the strong stability (in terminology [8] T_1-stability) radius of $Z_s^m(R)$, $s, m \in \mathbf{N}$, with Hölder's norms l_p, l_q and l_t in spaces $\mathbf{R}^n, \mathbf{R}^m$ and $\mathbf{R}^s$, respectively, is defined as:

$$\rho = \rho_s^m(p, q, t) = \begin{cases} \sup \Xi_{pqt} & \text{if } \Xi_{pqt} \neq \emptyset, \\ 0 & \text{if } \Xi_{pqt} = \emptyset, \end{cases}$$

$\Xi_{pqt} = \{\varepsilon > 0 \ : \ \forall R' \in \Omega_{pqt}(\varepsilon) \quad (E_s^m(R) \cap E_s^m(R + R') \neq \emptyset)\}$, and $\Omega_{pqt}(\varepsilon) = \{R' \in \mathbf{R}^{m \times n \times s} \ : \ \|R'\|_{pqt} < \varepsilon\}$ is the set of perturbing matrices R' with cuts $R'_k \in \mathbf{R}^{m \times n}$, $k \in N_s$;

$E_s^m(R+R')$ is the set of extreme solutions for the perturbed problem $Z_s^m(R + R')$;

$\|R'\|_{pqt}$ is the norm of matrix $R' = [r'_{ijk}]$.

Thus the strong stability radius of the problem $Z_s^m(R)$ is an extreme level of independent perturbations of elements of matrix $R \in \mathbf{R}^{m \times n \times s}$, such that for each such perturbation there exists a portfolio that is optimal both in $Z_s^m(R+R')$ and $Z_s^m(R)$.

Obviously, if $E_s^m(R) = X$, then $E_s^m(R) \cap E_s^m(R+R') \neq \emptyset$ for any perturbing matrix $R' \in \Omega_{pqt}(\varepsilon)$, where $\varepsilon > 0$. So, the strong stability radius is not bounded in this case above. For this reason, the problem with $\overline{E_s^m}(R) = X \backslash E_s^m(R) \neq \emptyset$, is called *non-trivial*.

3 Auxiliary Statements and Lemmas

Let v be any of the above-numbers $p, q.t$. For the number v, let v^* be a number conjugated to v and defined as:

$$1/v + 1/v^* = 1, \quad 1 < v < \infty. \tag{2}$$

We also set $v^* = 1$ if $v = \infty$, and $v^* = \infty$ otherwise.

We assume that v and v^* are taken from $[1, \infty]$, and they satisfy 2. In addition to the above, we assume that $1/v = 0$ and $v = \infty$.

Further we will use the well-know Hölder's inequality [15]:

$$|a^T b| \leq \|a\|_v \|b\|_{v^*} \tag{3}$$

that is true for any two vectors a and b of the same dimension.

It is easy to see that for any $a = (a_1, a_2, ..., a_n)^T \in \mathbf{R}^n$ with

$$|a_j| = \alpha, \quad j \in N_n,$$

the following equality holds

$$\|a\|_v = \alpha n^{1/v} \tag{4}$$

for any $v \in [1, \infty]$,

The following two lemmas can easily be proven.

Lemma 1. *Given two portfolios $x, x^0 \in X$, two market states $i, i' \in N_m$ and a fixed risk $k \in N_s$, the following statement is true for any $p, q \in [1, \infty]$:*

$$r_{ik}x - r_{i'k}x^0 \geq -\|R_k\|_{pq}\|(\|x\|_{p^*}, \|x^0\|_{p^*})\|_\nu,$$

where $R_k \in \mathbf{R}^{m \times n}$ is the k-th cut of matrix $R \in \mathbf{R}^{m \times n \times s}$ with rows $r_{1k}, r_{2k}, ..., r_{mk}$, $\nu = \min\{p^*, q^*\}$.

Proof. Let $i \neq i'$. Then, using Hölder's inequality (3), we get

$$r_{ik}x - r_{i'k}x^0 \geq -(\|r_{ik}\|_p\|x\|_{p^*} + \|r_{i'k}\|_p\|x\|_{p^*})$$

$$\geq \|(\|r_{ik}\|_p, \|r_{i'k}\|_p)\|_q \|(\|x\|_{p^*}, \|x^0\|_{p^*})\|_{q^*}$$

$$\geq -\|R_k\|_{pq} \|(\|x\|_{p^*}, \|x^0\|_{p^*})\|_{q^*} \geq -\|R_k\|_{pq} \|(\|x\|_{p^*}, \|x^0\|_{p^*})\|_\nu.$$

For $i = i'$, using (1), and Hölder's inequality (3) we deduce

$$r_{ik}x - r_{i'k}x^0 \geq -\|r_{ik}\|_p \|x - x^0\|_{p^*} \geq -\|R_k\|_{pq} \|x - x^0\|_{p^*}$$

$$\geq -\|R_k\|_{pq} \|(\|x\|_{p^*}, \|x^0\|_{p^*})\|_{q^*} \geq -\|R_k\|_{pq} \|(\|x\|_{p^*}, \|x^0\|_{p^*})\|_\nu.$$

Lemma 2. *Given cardinal numbers s and m, optimal portfolio $x^0 \in E_s^m(R)$ and perturbing matrix $R' \in \mathbf{R}^{m \times n \times s}$ with cuts $R_k' \in \mathbf{R}^{m \times n}$, assume that for some index $l \in N_s$ the following equality holds:*

$$\overline{E_s^m}(R) \cap E_l(x^0, R_l + R_l') = \emptyset. \tag{5}$$

Then we have

$$E_s^m(R) \cap E_s^m(R + R') \neq \emptyset. \tag{6}$$

Proof. if $x^0 \in E_s^m(R + R')$, then the lemma is trivial. If $x^0 \notin E_s^m(R + R')$. Then there exists a portfolio $x^* \in E_s^m(R + R')$ such that $x^* \in E_l(x^0, R_l + R_l')$. Then due to (5), we conclude that $x^* \in E_s^m(R)$. This implies the validity of (6). $\qquad\square$

4   Strong Stability Radius Bounds

For non-trivial problem $Z_s^m(R)$, we introduce the following notation

$$\varphi = \varphi_s^m(p, q) = \min_{x \notin E_s^m(R)} \min_{k \in N_s} \max_{x' \in X \setminus \{x\}} \frac{g_k(x.x', R_k)}{\|(\|x\|_{p^*}, \|x'\|_{p^*})\|_\nu},$$

$$\psi = \psi_s^m(p, q) = \max_{x' \in E_s^m(R)} \max_{k \in N_s} \min_{x \notin E_s^m(R)} \frac{g_k(x, x', R_k)}{\|(\|x\|_{p^*}, \|x'\|_{p^*})\|_\nu},$$

$$\chi = \chi_s^m(p, q, t) = n^{1/p} m^{1/q} s^{1/t} \min_{x \notin E_s^m(R)} \max_{k \in N_s} \max_{x' \in E_s^m(R)} \frac{g_k(x, x', R_k)}{\|x - x'\|_1},$$

where

$$g_k(x, x', R_k) = f_k(x, R_k) - f_k(x', R_k), \quad k \in N_s,$$

$$\nu = \min\{p^*, q^*\}.$$

Theorem 1. *Given $s, m \in \mathbf{N}$ and $p, q, t \in [1, \infty]$, for the strong stability radius $\rho_s^m(p, q, t)$ of s-criteria non-trivial problem $Z_s^m(R)$, the following bounds are valid:*

$$0 < \max\{\varphi, \ \psi\} \leq \rho_s^m(p, q, t) \leq \min\{\chi, \ \|R\|_{pqt}\}.$$

Proof. Since,

$$\forall x \notin E_s^m(R) \quad \forall k \in N_s \quad \exists x^0 \in X \quad (f_k(x, R_k) > f_k(x^0, R_k)),$$

the following inequality is evident

$$\varphi > 0,$$

i.e. the lower bound and the radius itself $\rho_s^m(p, q, t)$ are positive numbers.
    Now we show that

$$\rho = \rho_s^m(p, q, t) \geq \varphi^s(p, q) = \varphi. \tag{7}$$

Let the perturbing matrix $R' = [r'_{ijk}] \in \mathbf{R}^{m \times n \times s}$ with cuts R'_k, $k \in N_s$, be taken from the set $\Omega_{pqt}(\varphi)$. According to the definition of the number φ, and due to inequality (1), we obtain

$$\forall x \notin E_s^m(R) \quad \forall k \in N_s \quad \exists x^0 \in X \backslash \{x\}$$

$$\left(\frac{g_k(x, x^0, R_k)}{\|(\|x\|_{p^*}, \|x^0\|_{p^*})\|_\nu} \geq \varphi > \|R'\|_{pqt} \geq \|R'_k\|_{pq} \right).$$

Thus, due to Lemma 1, for any criterion $k \in N_s$ there exists a portfolio $x^0 \neq x$ such that

$$g_k(x, x^0, R_k + R'_k) = f_k(x, R_k + R'_k) - f_k(x^0, R_k + R'_k)$$

$$= \max_{i \in N_m}(r_{ik} + r'_{ik})x - \max_{i \in N_m}(r_{ik} + r'_{ik})x^0$$

$$= \min_{i \in N_m} \max_{i' \in N_m} (r_{ik}x + r'_{ik}x - r_{i'k}x^0 - r'_{i'k}x^0) \tag{8}$$

$$\geq f_k(x, R_k) - f_k(x^0, R_k) - \|R'_k\|_{pq} \|(\|x\|_{p^*}, \|x^0\|_{p^*})\|_\nu$$

$$= g_k(x, x^0, R_k) - \|R'_k\|_{pq} \|(\|x\|_{p^*}, \|x^0\|_{p^*})\|_\nu > 0,$$

where r'_{ik} is the i-th row of the k-th cut R'_k of the matrix R'. This implies $x \notin E_s^m(R + R')$ if $x \notin E_s^m(R)$.

Summarizing, we conclude that any non-optimal portfolio in $Z_s^m(R)$ remains non-optimal in $Z_s^m(R + R')$, i.e. the following inclusion holds

$$\emptyset \neq E_s^m(R + R') \subseteq E_s^m(R).$$

Hence, $E_s^m(R) \cap E_s^m(R + R') \neq \emptyset$ for any perturbing matrix $R' \in \Omega_{pqt}(\varphi)$, i.e. inequality (7) is true.

Further, we prove the lower bound

$$\rho = \rho_s^m(p, q, t) \geq \psi_s^m(p, q, r) = \psi. \tag{9}$$

Since the problem $Z_s^m(R)$ is non-trivial, we have

$$\exists x' \in E_s^m(R) \quad \exists k \in N_s \quad \forall x \notin E_s^m(R) \quad (f_k(x, R_k) \geq f_k(x', R_k)).$$

Therefore $\psi > 0$.

As in the case considered above, let $R' = [r'_{ijk}] \in \mathbf{R}^{m \times n \times s}$ be a perturbing matrix taken from the set of perturbing matrices $\Omega_{pqt}(\psi)$. Then according to the definition of ψ, there exist $x^0 \in E_s^m(R)$ and $l \in N_s$ such that for any portfolio $x \notin E_s^m(R)$ due to (1) the following inequalities hold

$$\frac{q_l(x, x^0, R_l)}{\|(\|x\|_{p^*}, \|x^0\|_{p^*})\|_\nu} \geq \psi > \|R'\|_{pqt} \geq \|R'_l\|_{pt}.$$

Then using (8) (for $k = l$), we obtain that for any $x \notin E_s^m(R)$ and any $R' \in \Omega_{pqt}(\psi)$ the following inequalities are valid

$$q_l(x, x^0, R_l + R_l') \geq q_l(x, x^0, R_l) - \|R_l'\|_{pq}\|(\|x\|_{p^*}, \|x^0\|_{p^*})\|_\nu > 0.$$

Therefore, we have

$$\overline{E_s^m}(R) \cap E_l(x^0, R_l + R_l') = \emptyset.$$

Thus, due to Lemma 2, the inequality (6) is true for any perturbing matrix $R' \in \Omega_{pqt}(\psi)$, i.e. inequality (9) holds.

Further, we prove the upper bound

$$\rho = \rho_s^m(p, q, t) \leq \chi_s^m(p, q, t) = \chi. \tag{10}$$

According to the definition of χ and due to assumption about problem's non-triviality, we have

$$\exists x^0 = (x_1^0, x_2^0, ..., x_n^0)^T \notin E_s^m(R) \quad \forall k \in N_s \quad \forall x \in E_s^m(R)$$

$$\left(\chi\|x^0 - x\|_1 \geq n^{1/p} m^{1/q} s^{1/t} g_k(x^0, x, R_k)\right). \tag{11}$$

Let $\varepsilon > \chi$, and let the elements of perturbing matrix $R^0 = [r_{ijk}^0] \in \mathbf{R}^{m \times n \times s}$ with cuts R_k^0, $k \in N_s$, be defined as:

$$r_{ijk}^0 = \begin{cases} -\delta & \text{if } i \in N_m, \ x_j^0 = 1, \ k \in N_s, \\ \delta & \text{if } i \in N_m, \ x_j^0 = 0, \ k \in N_s, \end{cases}$$

where δ satisfies

$$\chi < \delta n^{1/p} m^{1/q} s^{1/t} < \varepsilon. \tag{12}$$

From the above according to (4), we get

$$\|r_{ik}^0\|_p = \delta n^{1/p}, \quad i \in N_m, \ k \in N_s,$$

$$\|R_k^0\|_{pq} = \delta n^{1/p} m^{1/q}, \quad k \in N_s,$$

$$\|R^0\|_{pqt} = \delta n^{1/p} m^{1/q} s^{1/t},$$

$$R^0 \in \Omega_{pqt}(\varepsilon).$$

In addition, all the rows r_{ik}^0, $i \in N_m$, of any k-th cut R_k^0, $k \in N_s$, are constructed identically and composed of δ and $-\delta$. So, setting $c = r_{ik}^0$, $i \in N_m$, $k \in N_s$, we deduce

$$c(x^0 - x) = -\delta\|x^0 - x\|_1 < 0$$

that is true for any portfolio $x \neq x^0$. Using (11) and (12), we conclude that for any portfolio $x \in E_s^m(R)$ and any risk $k \in N_s$, the following statements are true:

$$g_k(x^0, x, R_k + R_k^0) = f_k(x^0, R_k + R_k^0) - f_k(x, R_k + R_k^0)$$

$$= \max_{i \in N_m}(r_{ik} + c)x^0 - \max_{i \in N_m}(r_{ik} + c)x = \max_{i \in N_m} r_{ik}x^0 - \max_{i \in N_m} r_{ik}x + c(x^0 - x)$$

$$= g_k(x^0, x, R_k) + c(x^0 - x) \leq \left(\chi(n^{1/p}m^{1/q}s^{1/t})^{-1} - \delta\right)\|x^0 - x\|_1 < 0.$$

This implies $x \notin E_s^m(R + R^0)$ for $x \in E_s^m(R)$. Thus, for any $\varepsilon > \chi$ there exists a perturbing matrix $R^0 \in \Omega_{pqt}(\varepsilon)$ such that $E_s^m(R) \cap E_s^m(R + R^0) = \emptyset$, i.e. $\rho < \varepsilon$ for any $\varepsilon > \chi$. Hence, inequality (10) is true.

Finally, we show

$$\rho_s^m(p, q, t) \leq \|R\|_{pqt}. \tag{13}$$

Let $x^0 = (x_1^0, x_2^0, ..., x_n^0)^T \notin E_s^m(R)$ and $\varepsilon > \|R\|_{pqt}$. Let fix δ satisfying condition

$$0 < \delta n^{1/p}m^{1/q}s^{1/t} < \varepsilon - \|R\|_{pqt}. \tag{14}$$

We introduce an auxiliary matrix $V = [v_{ijk}] \in \mathbf{R}^{m \times n \times s}$ with cuts V_k, $k \in N_s$, defined as follows:

$$v_{ijk} = \begin{cases} -\delta & \text{if} \quad i \in N_m, \quad x_j^0 = 1, \quad k \in N_s, \\ \delta & \text{if} \quad i \in N_m, \quad x_j^0 = 0, \quad k \in N_s. \end{cases}$$

Using (4), we obtain

$$\|V_k\|_{pq} = \delta n^{1/p}m^{1/q}, \; k \in N_s,$$

$$\|V\|_{pqt} = \delta n^{1/p}m^{1/q}s^{1/t}. \tag{15}$$

It is easy to see that all rows of V_k, $k \in N_s$, are identical and composed of δ and $-\delta$. So, we get

$$f_k(x^0, V_k) - f_k(x, V_k) = -\delta\|x^0 - x\|_1 < 0, \; k \in N_s, \tag{16}$$

for any $x \neq x^0$, and in particular for $x \in E_s^m(R)$.

Further, let $R^0 \in \mathbf{R}^{m \times n \times s}$ be a perturbing matrix with cuts R_k^0, $k \in N_s$, defined as:

$$R_k^0 = V_k - R_k, \; k \in N_s, \tag{17}$$

i.e. $R^0 = V - R$. Using (14) and (15), we deduce

$$\|R^0\|_{pqt} \leq \|V\|_{pqt} + \|R\|_{pqt} = \delta n^{1/p}m^{1/q}s^{1/t} + \|R\|_{pqt} < \varepsilon,$$

i.e. $R^0 \in \Omega_{pqt}(\varepsilon)$.

Additionally, using (16) and (17) for any index $k \in N_s$, we have

$$g_k(x^0, x, R_k + R_k^0) = f_k(x^0, R_k + R_k^0) - f_k(x, R_k + R_k^0)$$

$$= f_k(x^0, V_k) - f_k(x, V_k) = -\delta\|x^0 - x\|_1 < 0,$$

i.e. $x \notin E_s^m(R + R^0)$ for $x \in E_s^m(R)$. Summarizing, we get

$$\forall \varepsilon > \|R\|_{pqt} \quad \exists R^0 \in \Omega_{pqt}(\varepsilon) \quad \left(E_s^m(R) \cap E_s^m(R + R^0) = \emptyset\right).$$

The last implies (13). □

From Theorem 1 we obtain the following result specifying the lower and upper bounds on the strong stability radius for the case of same Hölder's norm l_p is used in all the three spaces.

Corollary 1. *For $s, m \in \mathbf{N}$ and $p \in [1, \infty]$, the strong stability radius $\rho_s^m(p, p, p)$ of s-criteria non-trivial problem $Z_s^m(R)$ has the following valid lower and upper bounds:*

$$0 < \varphi_s^m(p, p) = \min_{x \notin E_s^m(R)} \min_{k \in N_s} \max_{x' \in X \setminus \{x\}} \frac{g_k(x, x', R_k)}{\|x + x'\|_1^{1/p^*}} \le \rho_s^m(p, p, p)$$

$$\le (nms)^{1/p} \min_{x \notin E_s^m(R)} \max_{k \in N_s} \max_{x' \in E_s^m(R)} \frac{g_k(x, x', R_k)}{\|x - x'\|_1} = \chi_s^m(p, p, p).$$

5   The Case of Linear Criteria

For $m = 1$, our problem transforms into s-criteria problem of Boolean linear programming. It is convenient to write it in the following form:

$$Z_s^1(R): \quad Rx = (R_1 x, R_2 x, ..., R_s x)^T \to \min_{x \in X},$$

for $x = (x_1, x_2, ..., x_n)^T \in X \subseteq \mathbf{E}^n$. Here $R = [r_{ijk}] \in \mathbf{R}^{s \times n}$ is a matrix with rows $R_k \in \mathbf{R}^n$, $k \in N_s$.

The case $(m = 1)$ can be interpreted as a situation in which the financial market is stable and there is no concern for the investors. As previously, in space of portfolios (investment projects) $\mathbf{R}^n$ and in space of risks $\mathbf{R}^s$ two Hölder's norms are considered l_p and l_t, $p, t \in [1, \infty]$. The strong stability radius of the problem $Z_s^1(R)$ is denoted by $\rho_s^1(p, t)$. Given $p, t \in [1, \infty]$, $s \in \mathbf{N}$, we denote

$$\widehat{\varphi} = \widehat{\varphi}_s^1(p) = \min_{x \notin E_s^1(R)} \min_{k \in N_s} \max_{x' \in X \setminus \{x\}} \frac{R_k(x - x')}{\|x - x'\|_{p^*}},$$

$$\widehat{\psi} = \widehat{\psi}_s^1(p) = \max_{x' \in E_s^1(R)} \max_{k \in N_s} \min_{x \notin E_s^1(R)} \frac{R_k(x - x')}{\|x - x'\|_{p^*}},$$

$$\widehat{\chi} = \widehat{\chi}_s^1(p, t) = n^{1/p} s^{1/t} \min_{x \notin E_s^1(R)} \max_{k \in N_s} \max_{x' \in E_s^1(R)} \frac{R_k(x - x')}{\|x - x'\|_1},$$

$$\|R\|_{pt} = \|(\|R_1\|_p, \|R_2\|_p, ..., \|R_s\|_p)\|_t.$$

Theorem 2. *For $p, t \in [1, \infty]$, $s \in \mathbf{N}$, the strong stability radius $\rho_s^1(p, t)$ of s-criteria non-trivial Boolean linear problem $Z_s^1(R)$, $R \in \mathbf{R}^{s \times n}$, has the following valid lower and upper bounds:*

$$0 < \max\{\widehat{\varphi}, \widehat{\psi}\} \le \rho_s^1(p, t) \le \min\{\widehat{\chi}, \|R\|_{pt}\}.$$

Proof. From Theorem 1 the validity of upper bounds follows directly. Additionally, since φ, ψ, χ are positive as shown in Theorem 1, the numbers $\widehat{\varphi}$, $\widehat{\psi}$, $\widehat{\chi}$ are also positive.

First we prove inequality

$$\rho_s^1(p,t) \geq \widehat{\varphi}_s^1(p) = \widehat{\varphi}. \tag{18}$$

Let $R^1 \in \mathbf{R}^{s \times n}$ be an arbitrary perturbing matrix with rows $R_k' \in \mathbf{R}^n$, $k \in N_s$, and the norm

$$\|R'\|_{pt} = \|(\|R_1'\|_p, \|R_2'\|_p, ..., \|R_s'\|_p)\|_t < \widehat{\varphi},$$

i.e. $R' \in \Omega_{pt}(\widehat{\varphi})$. Then according to the definition of $\widehat{\varphi}$ and due to (1), we have

$$\forall x \notin E_s^1(R) \quad \forall k \in N_s \quad \exists x^0 \in X \backslash \{x\} \quad \left(\frac{R_k(x - x^0)}{\|x - x^0\|_{p^*}} \geq \widehat{\varphi} > \|R'\|_{pt} \geq \|R_k'\|_p \right).$$

Applying Hölder's inequality (3) for any index $k \in N_s$, we get

$$(R_k + R_k')(x - x^0) = R_k(x - x^0) + R_k'(x - x^0) \geq R_k(x - x^0) - \|R_k'\|_p \|x - x^0\|_{p^*} > 0.$$

So, for any $x \notin E_s^1(R + R')$ we have $x \notin E_s^1(R)$. Thus, any non-optimal portfolio in $Z_s^1(R)$ remains non-optimal in $Z_s^1(R + R')$, i.e. $E_s^1(R + R') \subseteq E_s^1(R)$. Hence, $E_s^1(R) \cap E_s^1(R + R') \neq \emptyset$ for any $R' \in \Omega_{pt}(\widehat{\varphi})$, i.e. (18) holds.

Now we prove inequality

$$\rho_s^1(p,t) \geq \widehat{\psi}_s^1(p) = \widehat{\psi}. \tag{19}$$

Let $R' \in \Omega_{pt}(\widehat{\psi})$ be an arbitrary perturbing matrix with rows R_k', $k \in N_s$. Then according to the definition of $\widehat{\psi}$, there exist $x^0 \in E_s^1(R)$ and $l \in N_s$, such that $x \notin E_s^1(R)$. Using (1) we deduce

$$\frac{R_l(x - x^0)}{\|x - x^0\|_{p^*}} \geq \widehat{\psi} > \|R'\|_{pt} \geq \|R_l'\|_p.$$

Applying Hölder's inequality (3), we get

$$(R_l + R_l')(x - x^0) = R_l(x - x^0) + R_l'(x - x^0) \geq R_l(x - x^0) - \|R_l'\|_p \|x - x^0\|_{p^*} > 0.$$

Thus, $\overline{E_s^1}(R) \cap E_l(x^0, R_l + R_l') = \emptyset$. Then due to Lemma 2, inequality (6), and also inequality (19), holds for any perturbing matrix $R' \in \Omega_{pt}(\widehat{\psi})$. $\square$

From Theorem 2 we obtain the following result specifying the lower and upper bounds on the strong stability radius $\rho_s^1(\infty, \infty)$ for the case of same Hölder's norm l_∞ is used in all the two spaces.

Corollary 2. *Given $p = t = \infty$ and $s \in \mathbf{N}$, the strong stability radius $\rho_s^1(\infty, \infty)$ of non-trivial s-criteria problem of $Z_s^1(R)$, $R \in \mathbf{R}^{s \times n}$, has the following valid lower and upper bounds:*

$$0 < \widehat{\varphi}_s^1(\infty) = \min_{x \notin E_s^1(R)} \min_{k \in N_s} \max_{x' \in X \backslash \{x\}} \frac{R_k(x - x')}{\|x - x'\|_1}$$

$$\leq \rho_s^1(\infty, \infty) \leq \widehat{\chi}_s^1(\infty, \infty) = \min_{x \notin E_s^1(R)} \max_{k \in N_s} \max_{x' \in E_s^1(R)} \frac{R_k(x - x')}{\|x - x'\|_1}.$$

Now we present some results which illustrate attainability of the bounds specified in Theorem 2. First note that from Theorem 2 the well-known formula follows for $p = t = \infty$ and $s = 1$:

$$\rho_1^1(\infty, \infty) = \widehat{\varphi}_1^1(\infty) = \widehat{\chi}_1^1(\infty, \infty) = \min_{x \notin E_1^1(R)} \max_{x' \in E_1^1(R)} \frac{R^T(x - x')}{||x - x'||_1} > 0,$$

where $R \in \mathbf{R}^n$.

The next theorem specifies a class of problems with attainable lower bounds.

Theorem 3. *Given $p, t \in [1, \infty]$ there exists a class of problems $Z_1^1(R)$ such that for the strong stability radius $\rho_1^1(p, t)$ the following formula holds:*

$$\rho_1^1(p, t) = \widehat{\chi}_1^1(p, t) = ||R||_{pt}. \tag{20}$$

Proof. According to Theorem 2, in order to prove (20), it suffices to show that

$$\rho_1^1(p, t) \geq \widehat{\chi}_1^1(p, t) = ||R||_{pt}.$$

Let $X = \{x^0, x^1, x^2, ..., x^n\} \subset \mathbf{E}^n$, where $x^0 = (0, 0, ..., 0)^T \in \mathbf{E}^n$, $x^j = e^j$, $j \in N_n$. Here e^j is a unit column vector of space $\mathbf{R}^n$, i.e. e^j is j-column of identity $n \times n$ matrix. Assume

$$R = (-\alpha, -\alpha, ..., -\alpha) \in \mathbf{R}^n, \quad \alpha > 0.$$

Then we obtain

$$Rx^0 = 0, \; Rx^j = -\alpha, \; j \in N_n,$$

$$x^0 \notin E_1^1(R), \; x^j \in E_1^1(R), \; j \in N_n,$$

$$\widehat{\chi}_1^1(p, t) = ||R||_{pt} = n^{1/p}\alpha. \tag{21}$$

We introduce a perturbing row $R' = (r_1', r_2', ..., r_n')$ that is taken from the set $\Omega_{pt}(n^{1/p}\alpha)$, i.e. $||R'||_{pt} \leq n^{1/p}\alpha$.

Proving by contradiction it is easy to show that there exists at least one index $l \in N_n$ with $|r_l'| < \alpha$. Therefore, we have $(R + R')(x^0 - x^l) = \alpha - r_l' > 0$, i.e. $x^0 \notin E_1^1(R + R')$ for any $R' \in \Omega_{pt}(\widehat{\chi}_1^1(p, t))$. Since, $x^0 \notin E_1^1(R)$ we have $\rho_1^1(p, t) \geq \widehat{\chi}_1^1(p, t)$. Using (21) we finally conclude $\rho_1^1(p, t) = \widehat{\chi}_1^1(p, t) = ||R||_{pt} = n^{1/p}\alpha$. □

Finally, we show that both upper and lower bounds specified in Theorem 2 can be attained in scalar case.

Theorem 4. *Given $p, t \in [1, \infty]$, there exists a class of problems $Z_1^1(R)$ such that for the strong stability radius $\rho_1^1(p, t)$ the following formula holds:*

$$\rho_1^1(p, t) = \widehat{\varphi}_1^1(p) = \widehat{\psi}_1^1(p) = \widehat{\chi}_1^1(p, t) = ||R||_{pt}.$$

Proof. Let $X = \{x^0, x^1\} \in \mathbf{R}^n \subset \mathbf{E}^n$, where $x^0 = (0, 0, ..., 0)^T$, $x^1 = (1, 1, ..., 1)^T$. Assume $R = (1, 1, ..., 1) \in \mathbf{R}^n$. Then we get $x^0 \in E_1^1(R)$, $x^1 \notin E_1^1(R)$. Therefore, $\widehat{\chi}_1^1(p, t) = ||R||_{pt} = n^{1/p}$. Using (2) we deduce $\widehat{\varphi}_1^1(p) = \widehat{\psi}_1^1(p) = n^{1/p}$. Hence, $\rho_1^1(p, t) = \widehat{\varphi}_1^1(p) = \widehat{\psi}_1^1(p) = \widehat{\chi}_1^1(p, t) = ||R||_{pt} = n^{1/p}$. □

6   Conclusion

As a summary, it is worth mentioning that the bounds proven in Theorems 1–4, are theoretical due to their analytical and enumerating structures. Even for a single objective case, the difficulty of exact value calculation for various types of stability radii is a long-standing challenge originally pointed out in [1]. In practical applications, one can try to get reasonable approximation of the bounds using some meta-heuristics, e.g. evolutionary algorithms or Monte-Carlo simulation. This could become a possible subject for future investigations. Another possibility to continue research in this direction is to specify some particular classes of problems where computational burden can be drastically reduced due to a unique structure of the set of efficient outcomes.

Conflict of Interest. The authors declare that they have no conflict of interest.

References

1. Chakavarti, N., Wagelmans, A.: Calculation of stability radius for combinatorial optimization problems. Oper. Res. Lett. **23**, 1–7 (1999)
2. Du, D., Pardalos, P. (eds.): Minimax and Applications. Kluwer, Dordrecht (1995)
3. Frank, J., Fabozzi, C., Markowitz, H. (eds.): The Theory and Practice of Investment Management: Asset Allocation, Valuation, Portfolio Construction, and Strategies. Wiley, New York (2011)
4. Hadamard, J.: Lectures on Cauchys Problem in Linear Partial Differential Equations. Yale University Press, Yale (1923)
5. Howell, L.: Global Risks World Economic Forum Annual Report (2013)
6. Ehrgott, M.: Multicriteria Optimization. Springer, Birkhäuser (2005). https://doi.org/10.1007/3-540-27659-9
7. Emelichev, V., Girlich, E., Nikulin, Yu., Podkopaev, D.: Stability and regularization of vector problem of integer linear programming. Optimization **51**, 645–676 (2002)
8. Emelichev, V., Kotov, V., Kuzmin, K., Lebedeva, N., Semenova, N., Sergienko, T.: Stability and effective algorithms for solving multiobjective discrete optimization problems with incomplete information. J. Autom. Inf. Sci. **26**, 27–41 (2014)
9. Emelichev, V., Korotkov, V., Nikulin, Y.: Post-optimal analysis for Markowitz's multicriteria portfolio optimization problem. J. Multi-Crit. Decis. Analys. **21**, 95–100 (2014)
10. Emelichev, V., Kuzmin, K.: A general approach to studing the stability of a Pareto optimal solutions of a vector integer linear programming problem. Discrete Math. Appl. **17**, 349–354 (2007)
11. Emelichev, V., Kuzmin, K.: On a type of stability of a multicriteria integer linear programming problem in the case of monotonic norm. J. Comput. Syst. Sci. Int. **46**, 714–720 (2007)
12. Emelichev, V., Bukhtoyarov, S., Mychkov, V.: An investment problem under multicriteriality, uncertainty and risk Buletinul Academiei de Stiinte a Republicii Moldova. Matematica **82**, 82–98 (2016)
13. Emelichev, V., Podkopaev, D.: Quantitative stability analysis for vector problems of 0–1 programming. Discrete Optim. **7**, 48–63 (2010)

14. Gordeev, E.: Comparison of three approaches to studying stability of solutions to problems of discrete optimization and computational geometry. J. Appl. Ind. Math. **9**, 358–366 (2015)
15. Hardy, G., Littlewood, J., Polya, G.: Inequalities. Cambridge University Press, Cambridge (1988)
16. Korotkov, V., Emelichev, V., Nikulin, Y.: Multicriteria investment problem with Savage's risk criteria: theoretical aspects of stability and case study. J. Ind. Manag. Optim. **16**(3), 1297–1310 (2020). https://doi.org/10.3934/jimo.2019003
17. Korotkov, V., Wu, D.: Evaluating the quality of solutions in project portfolio selection. Omega **91**, 102117 (2020). https://doi.org/10.1016/j.omega.2019.01.007
18. Kuzmin, K.: A general approach to the calculation of stability radii for the max-cut problem with multiple criteria. J. Appl. Ind. Math. **9**, 527–539 (2015)
19. Kuzmin, K., Haritonova, V.: Estimating the stability radius of an optimal solution to a simple assembly line balancing problem. J. Appl. Ind. Math. **13**, 250–260 (2019)
20. Kuzmin, K., Nikulin, Yu., Mäkelä, M.: On necessary and sufficient conditions of stability and quasistability in combinatorial multicriteria optimization. Control Cybern. **46**, 361–382 (2017)
21. Leontev, V.: Discrete optimization. J. Comput. Phys. Math. **47**, 328–340 (2007)
22. Linura, M., Nikulin, Yu.: Stability and accuracy functions in multicriteria linear combinatorial optimization problem. Ann. Oper. Res. **147**, 255–267 (2006)
23. Libura, M., Van der Port, E.S., Sierksma, G., Van der Veen, J.A.A.: Stability aspects of the traveling salesman problem based on k-best solutions. Discrete Appl. Math. **87**, 159–185 (1998)
24. Markowitz, H.: Portfolio Selection: Efficient Diversification of Investments. Willey, New York (1991)
25. Miettinen, K.: Nonlinear Multiobjective Optimization. Kluwer Academic Publishers, Boston (1999)
26. Nikulin, Yu.: Stability and accuracy functions in a coalition game with bans, linear payoffs and antagonistic strategies. Ann. Oper. Res. **172**, 25–35 (2009)
27. Nikulin, Yu.: Accuracy and stability functions for a problem of minimization a linear form on a set of substitutions. In: Sotskov, Y., Werner, F. (eds.) Sequencing and Scheduling with Inaccurate Data. Nova Science Pub Inc., New York, Chapter 15, pp. 409–426 (2014)
28. Noghin, V.: Reduction of the Pareto Set: An Axiomatic Approach. SSDC, vol. 15. Springer, Cham (2018). https://doi.org/10.1007/978-3-319-67873-3
29. Pareto, V.: Manuel D'economie Politique. V. Giard & E. Briere, Paris (1909)
30. Savage, L.J.: The Foundations of Statistics. Dover Publ., New York (1972)
31. Steuer, R.: Multiple Criteria Optimization: Theory Computation and Application. John Wiley & Sons, New York (1986)

Applications

Production Network Centrality in Connection to Economic Development by the Case of Kazakhstan Statistics

Seilkhan Boranbayev[1] , Nataliia Obrosova[2,3](✉) ,
and Alexander Shananin[2,3]

[1] L.N. Gumilyov Eurasian National University, Satpayev Street 2,
010008 Nur-Sultan, Republic of Kazakhstan
sboranba@yandex.kz
[2] Lomonosov Moscow State University, GSP-1, Leninskie Gory,
119991 Moscow, Russian Federation
{nobrosova,alexshan}@ya.ru
[3] Federal Research Center "Computer Science and Control" of Russian Academy
of Sciences, Vavilov Street 44/2, 119333 Moscow, Russian Federation
https://www.enu.kz
http://www.frccsc.ru/
http://www.msu.ru

Abstract. Analysis of a production network graph allows to determine
the central industries of an economy. According to the well-known concept of economic development, these industries are the main drivers of
economic growth. We discuss this concept in terms of a tractable model.
Our approach is based on the nonlinear input-output balance model that
is developed by authors. The classical method for determining of the
drivers of economic growth is the Leontief's input-output model that
has been widely used since the middle of the XX century. This model
assumes the fixed proportions of material costs for a unit of product output. The nonlinear model is based on the more relevant assumption about
the stability of the structure of financial costs of a production. We use a
Cobb-Douglas production function in order to develop a technology that
allows to analyze the nonlinear input-output balance. The technology is
based on the solution of Fenchel duality problem of resource allocation.
On the base of the obtained results we analyze the concept of centrality and the stability of intersectoral linkages by the case of Kazakhstan
statistics.

Keywords: Production networks · Centrality · Input-output model ·
Cobb-Douglas production function · Fenchel duality problem · Convex
optimization problem · Young transform

This research is funded by the Science Committee of the Ministry of Education and
Science of the Republic of Kazakhstan (Grant No. AP09259435).

N. N. Olenev et al. (Eds.): OPTIMA 2021, LNCS 13078, pp. 321–335, 2021.
https://doi.org/10.1007/978-3-030-91059-4_23

1   Introduction

One of the most actual questions of the development of modern complicate economies is as following. Which industries are more important drivers of the economic growth? This question is especially relevant for the developing economies. The elaboration of government programs of stimulating of economic growth is closely related to the state support of the central industries.

The problem of determining of industry centrality measure in an economy is connected to ambiguity of the concept of centrality in network models. In recent decades, production relations have become much more complicated. Therefore, the approaches related to the analysis of single-product aggregated models of GDP dynamics have lost their relevance. For example, in terms of such approaches, the influence of an industry on the total output of the economy is determined by the amount of revenue of the industry only and does not depend on its position in the supply chain [11]. Crises of the late 19th and early 20th centuries showed that it is important to take into account the intersectoral interactions in supply networks.

In [1] the approach for centrality definition in the case of exchange networks of a general type was developed, that led to the concept of eigenvector centrality (Bonacich centrality). Generalization of this approach led to the concept of influence vector, that allows to analyze the macroeconomical impact of microeconomical shocks in production networks [4–6,12]. The influence vector coincides with the definition of the PageRank vector of a graph and characterizes the industry as more important if it is connected with other important industries [1,13]. Acemouglu et al. in [4–6] considered a connection of centrality measures (including the concept of influence vector) with the problem of the propagation of random shocks in production networks. They considered the economic balance model that describes interactions between different sectors with Cobb-Douglas production technologies.

Other ways of interpretation of industry centrality lead, for example, to the idea of networks clustering, that is based on the allocation of triangles of different types in a given network (for ex., see [14,15]).

A. Shananin [7,8] suggested a generalization of the approach of [4–6], that allows to analyze the stability of intersectoral linkages more completely. Using the Young transform and the Fenchel duality theorem, a method for analyzing of a competitive equilibrium in models of a nonlinear intersectoral balance with concave positively homogeneous production functions and utility function of the final consumer was developed. On the base of such approach the adequate centrality measure of industries can be calculated, which takes into account the influence of a final consumption. That is an advantage of the nonlinear model [7,8]. In this work we use the technique from [7,8] to analyze the centrality in production network according to the statistical data of input-output balances of the economy of Republic of Kazakhstan 2016–2019. We compare the results with centrality measures of the first and second-degrees, evaluated on the base of classical Leontief model of linear inter-industry balance, that was the main stream of intersectoral analysis in the second part of XX century.

2   Model of Nonlinear Intersectoral Balance

Consider a set of m pure industrial sectors connected into a production network. The output of any given sector can be either consumed or used by other sectors as intermediate goods for production. Let X_i^j is the amount of commodity of sector i used in production of a good j (intermediate input). Denote $X^j = \left(X_1^j, \ldots, X_m^j\right)$. Each sector j has a production function $F_j\left(X^j, l^j\right)$ depending on intermediate inputs X^j and n primary production factors $l^j = \left(l_1^j, \ldots, l_n^j\right)$ that are not produced by the considered group of sectors. The production functions of the sectors are assumed to have neoclassical properties, i.e., they are concave, monotonically nondecreasing, and continuous functions on $R_{\geq 0}^{m+n}$, that vanish at the origin $F_j\left(0, 0\right) = 0$. Additionally, the functions $F_j\left(X^j, l^j\right)$ are assumed to be positively homogeneous of degree one. We denote the class of all such functions by Φ_{m+n}.

Let $X^0 = \left(X_1^0, \ldots, X_m^0\right)$ be the vector of final consumption. The demand of the final consumers is described by the utility function $F_0\left(X^0\right) \in \Phi_m$. Assume that the inputs of primary production factors $l^j = \left(l_1^j, \ldots, l_n^j\right)$ are bounded from above by a vector $l = (l_1, \ldots, l_n) \geq 0$. The problem is to find an optimal distribution of these resources between the sectors in order to maximize the utility function of final consumers with balance constraints on the primary production factors and the outputs from the sectors:

$$F_0\left(X^0\right) \to \max \tag{1}$$

$$F_j\left(X^j, l^j\right) \geq \sum_{i=0}^{m} X_j^i, j = 1, \ldots, m \tag{2}$$

$$\sum_{j=1}^{m} l^j \leq l \tag{3}$$

$$X^0 \geq 0, \ X^1 \geq 0, \ldots, X^m \geq 0, \ l^1 \geq 0, \ldots, l^m \geq 0. \tag{4}$$

We assume that the considered set of sectors is productive, i.e., there exist

$$\hat{X}^1 \geq 0, \ldots, \hat{X}^m \geq 0, \ \hat{l}^1 \geq 0, \ldots, \hat{l}^m \geq 0$$

such that

$$F_j\left(\hat{X}^j, \hat{l}^j\right) > \sum_{i=0}^{m} \hat{X}_j^i j = 1, \ldots, m.$$

It is easy to prove that, if the set of sectors is productive and $l = (l_1, \ldots, l_n) > 0$, then the optimization problem (1)–(4) satisfies to the Slater condition.

Since the production functions $F_j\left(X^j, l^j\right) \in \Phi_{m+n}$ are positively homogeneous we need an additional condition on primary production factors inputs to guarantee the existence of the limited optimal solution of (1). Denote

$$A\left(l\right) = \left\{X^0 = \left(X_1^0, \ldots, X_m^0\right) \geq 0 \,\middle|\, X_j^0 \leq F_j\left(X^j, l^j\right) - \sum_{i=1}^m X_j^i, \, j = 1, \ldots, m; \right.$$
$$\left. \sum_{j=1}^m l^j \leq l, \, X^1 \geq 0, \ldots, X^m \geq 0, \, l^1 \geq 0, \ldots, l^m \geq 0 \right\}.$$

We assume that there exists $\hat{l} \in R_{>0}^n$ such that the set $A\left(\hat{l}\right)$ is bounded. Then the set $A\left(l\right)$ is bounded, convex and closed for any $l \in R_{\geq 0}^n$.

Proposition 1 ([9]). *A set of vectors* $\left\{\hat{X}^0, \hat{X}^1, \ldots, \hat{X}^m, \hat{l}^1, \ldots, \hat{l}^m\right\}$, *satisfying constraints (2)–(4) is a solution of the optimization problem (1)–(4) if and only if there exist Lagrange multipliers* $p_0 > 0$, $p = (p_1, \ldots, p_m) \geq 0$ *and* $s = (s_1, \ldots, s_n) \geq 0$ *such that*

$$\left(\hat{X}^j, \hat{l}^j\right) \in Arg\max\{p_j F_j\left(X^j, l^j\right) - pX^j - sl^j \mid X^j \geq 0, l^j \geq 0\}, \, j = 1, \ldots, m \tag{5}$$

$$p_j\left(F_j\left(\hat{X}^j, \hat{l}^j\right) - \hat{X}_j^0 - \sum_{i=1}^m \hat{X}_j^i\right) = 0, \, j = 1, \ldots, m \tag{6}$$

$$s_k\left(l_k - \sum_{j=1}^m \hat{l}_k^j\right) = 0, \, k = 1, \ldots, n \tag{7}$$

$$\hat{X}^0 \in Arg\max\{p_0 F_0(X^0) - pX^0 \mid X^0 \geq 0\}. \tag{8}$$

Thus, the equilibrium market-type mechanisms are the optimal mechanisms of distribution of intermediate goods and primary production factors in production network.

We interpret the Lagrange multipliers $p = (p_1, \ldots, p_m)$ corresponding to the balance constraints on the sectoral outputs (2) as the prices of these outputs, while the Lagrange multipliers $s = (s_1, \ldots, s_n)$ corresponding to the balance constraints on primary production factors (3) as the prices of these factors. Dual description of the technology of industry j is the cost function $q_j(p, s)$, that is the Young transform of the production function $F_j\left(X^j, l^j\right)$

$$q_j(p, s) = \inf\left\{\frac{pX^j + sl^j}{F_j\left(X^j, l^j\right)} \,\middle|\, X^j \geq 0, l^j \geq 0, F_j\left(X^j, l^j\right) > 0\right\}.$$

The cost function $q_j(p, s)$ matches the prices of inputs (p, s) with the production cost of a unit of production in industry j. The dual function to the utility function $F_0\left(X^0\right)$ is the consumer price index

$$q_0(q) = \inf\left\{\frac{qX^0}{F_0(X^0)} \,\middle|\, X^0 \geq 0, F_0\left(X^0\right) > 0\right\}.$$

Note that $q_0(p) \in \Phi_m$, $q_j(p, s) \in \Phi_{m+n}$ $(j = 1, \ldots, m)$. Since the Young transform is involuted (see [10, p.232]), we have

$$F_0\left(X^0\right) = \inf\left\{\frac{qX^0}{q_0(q)} \,\middle|\, q \geq 0, q_0(q) > 0\right\},$$

$$F_j\left(X^j, l^j\right) = \inf\left\{\frac{pX^j + sl^j}{q_j\left(p, s\right)} \mid p \geq 0, s \geq 0, q_j\left(p, s\right) > 0.\right\}.$$

Note that in the case of Cobb-Douglas production function $F_{KD}\left(X_1, \cdots, X_n\right) = AX_1^{\alpha_1}\ldots X_n^{\alpha_n}$, where $A > 0, \alpha_1 > 0, \ldots, \alpha_n > 0, \alpha_1 + \cdots + \alpha_n = 1$, the Young transform gives the following cost function

$$q_{KD}\left(p_1, \ldots p_n\right) = \frac{1}{F_{KD}\left(\alpha_1, \ldots, \alpha_n\right)} p_1^{\alpha_1}\ldots p_n^{\alpha_n}.$$

Optimal value of the functional (1) depending on l (that is the right part of constraint (3)) is called the aggregate production function $F^A\left(l\right)$. Note that $F^A\left(l\right) \in \Phi_n$.

The Young transform of the aggregate production function $F^A\left(l\right)$ determines the aggregate cost function

$$q_A\left(s\right) = \inf\left\{\frac{sl}{F^A\left(l\right)} \mid l \geq 0, F^A\left(l\right) > 0.\right\}.$$

We have $q_A\left(s\right) \in \Phi_n$ and

$$F^A\left(l\right) = \inf\left\{\frac{sl}{q_A\left(s\right)} \mid s \geq 0, q_A\left(s\right) > 0.\right\}.$$

Theorem 1 ([7,8]). *The aggregate cost function $q_A\left(s\right)$ can be written as follows.*

$$q_A\left(s\right) = \sup\left\{q_0\left(p\right) \mid p = \left(p_1, \ldots, p_m\right) \geq 0, q_j\left(s, p\right) \geq p_j, j = 1, \ldots, m\right\}. \quad (9)$$

Moreover, if Lagrange multipliers to the problem (1)–(4) $\hat{p} = \left(\hat{p}_1, \ldots, \hat{p}_m\right) \geq 0$, $\hat{s} = \left(\hat{s}_1, \ldots, \hat{s}_m\right) \geq 0$ satisfy (5)–(8), then $\hat{p} = \left(\hat{p}_1, \ldots, \hat{p}_m\right) \geq 0$ is the solution of the problem (9) for fixed values $\hat{s} = \left(\hat{s}_1, \ldots, \hat{s}_m\right) \geq 0$. Note that if $\hat{p}(\hat{s}) \geq 0$ is the solution of the right hand side of (9), then the value of aggregate cost function $q_A\left(\hat{s}\right)$ equals to the value of consumer price index $q_0\left(\hat{p}(\hat{s})\right)$.

Consider the case of Cobb-Douglas production and utility functions. Let

$$F_0\left(X\right) = \alpha_0 X_1^{a_1^0}\ldots X_m^{a_m^0}, \quad F_j\left(X, l\right) = \alpha_j X_1^{a_1^j}\ldots X_m^{a_m^j} l_1^{b_1^j}\ldots l_n^{b_n^j}, \quad j = 1, \ldots, m,$$

where

$$\alpha_0 > 0, \sum_{i=1}^m a_i^0 = 1, \ a_i^0 \geq 0, \ i = 1, \ldots, m,$$

$$\sum_{i=1}^m a_i^j + \sum_{k=1}^n b_k^j = 1, \ j = 1, \ldots, m; \ \alpha_j > 0, a_i^j \geq 0, \ i = 1, \ldots, m, \ j = 1, \ldots, m,$$

$$b_k^j \geq 0, \ k = 1, \ldots, n, \ j = 1, \ldots, m.$$

Assume that each industry uses at least one primary production factor, i.e. $\sum_{k=1}^{n} b_k^i > 0$, $i = 1, \ldots, m$. Consider the matrices $A = \left\| a_i^j \right\|_{j=1,\ldots,m}^{i=1,\ldots,m}$, $B = \left\| b_k^j \right\|_{k=1,\ldots,n}^{j=1,\ldots,m}$. Denote by E the identity $m \times m$-matrix. Denote

$$a^0 = \left(a_1^0, \ldots, a_m^0 \right)^T, a^j = \left(a_1^j, \ldots, a_m^j \right)^T, b^j = \left(b_1^j, \ldots, b_n^j \right)^T, j = 1, \ldots, m.$$

Since $\sum_{i=1}^{m} a_i^j < 1$, $j = 1, \ldots, m$, it follows that the non-negative matrix A is productive.

Due to the Young transform the consumer price index $q_0(p)$ satisfies

$$q_0(p) = \frac{1}{F_0(a^0)} p_1^{a_1^0} \ldots p_m^{a_m^0}$$

and the cost function of industry j satisfies

$$q_j(p, s) = \frac{1}{F_j(a^j, b^j)} p_1^{a_1^j} \ldots p_m^{a_m^j} s_1^{b_1^j} \ldots s_n^{b_n^j}.$$

Denote

$$(E - A)^{-1} = (\omega_{kj})_{k,\ldots,n}^{j=1,\ldots,n}, \quad d = \left(\ln \left(F_1 \left(a^1, b^1 \right) \right), \ldots, \ln \left(F_m \left(a^m, b^m \right) \right) \right)^T, \tag{10}$$

$$\mu = (\mu_1, \ldots, \mu_m) = -\left(E - A^T \right)^{-1} d, \quad \lambda = q_0 \left(e^{\mu_1}, \ldots, e^{\mu_m} \right). \tag{11}$$

Recall that the matrix A is productive. Then the convex programming problem

$$q_0(p) \to \max \tag{12}$$

$$q_j(p, s) \geq p_j, j = 1, \ldots, m, \tag{13}$$

$$p_j \geq 0, j = 1, \ldots, m, \tag{14}$$

has the solution

$$p_j = e^{\mu_j} s_1^{c_1^j} \ldots s_n^{c_n^j}, j = 1, \ldots, m, \tag{15}$$

where $C = \left\| c_k^j \right\|_{k=1,\ldots,n}^{j=1,\ldots,m} = \left(E - A^T \right)^{-1} B^T$. The aggregate cost function can be written as $q_A(s) = \lambda s_1^{\gamma_1} \ldots s_n^{\gamma_n}$, where $\gamma = (\gamma_1, \ldots, \gamma_n)^T = C^T a^0$ and $\gamma_1 + \cdots + \gamma_n = 1$ (see [7,9]).

Proposition 2. *The elasticity of consumer price index on the productivity* α_j *of industry* j *equals to*

$$-\frac{\alpha_j}{q_A} \frac{\partial q_A}{\partial \alpha_j} = \sum_{k=1}^{n} a_k^0 \omega_{kj}. \tag{16}$$

Proof. Note that

$$\frac{\alpha_j}{q_A} \frac{\partial q_A}{\partial \alpha_j} = \frac{\alpha_j}{\lambda} \frac{\partial \lambda}{\partial \alpha_j} = \frac{\alpha_j}{q_0 \left(e^{\mu_1}, \ldots, e^{\mu_n} \right)} \sum_{k=1}^{n} a_k^0 q_0 \left(e^{\mu_1}, \ldots, e^{\mu_n} \right) \frac{\partial \mu_k}{\partial \alpha_j} = -\sum_{k=1}^{n} a_k^0 \omega_{kj}.$$

3   Analysis of Stability of Inter-industry Balances with Cobb-Douglas Production Functions in the Economy of Republic of Kazakhstan

In this section we apply our nonlinear inter-industry balance model to look at the stability of interconnections implied by the symmetric Input - Output tables of the economy of Kazakhstan 2016–2019 published by the Agency for Strategic planning and reforms of the Republic of Kazakhstan (https://stat.gov. kz). The Kazakhstan input-output matrix we use corresponds to the symmetric table "Supply of goods and services in basic prices" that is the Commodity-by-Commodity matrix, which comprises 68 commodities (pure industries). The nomenclature of the industries remains stable in 2016–2019. This table contains a financial flow data that reflects production and distribution of resources between the various components of intermediate and final demand.

The symmetric input-output table has three quadrants. The first one is $\left\| Z_i^j \right\|$, where $i = 1, \ldots, m$, $j = 1, \ldots, m$. It corresponds to the intermediate domestic demand of industries ($m = 68$). The value Z_i^j denotes the amount of money that industry i received from industry j for the resources produced in industry i and supplied to the industry j. The second quadrant of the table consists of the column vectors of final consumption of the products of pure industries. For the tables of Kazakhstan, final consumption includes the column vectors as follows: final consumption households, final consumption of public administration bodies, final consumption of non-profit organizations serving households (NPOs), gross fixed capital formation, changes in working capital stocks, acquisition minus disposal of valuables, export of goods and services. Let k be the number of columns of the second quadrant. Denote the elements of the second quadrant by $\left\| Z_i^j \right\|$, where $i = 1, \ldots, m$, $j = m+1, \ldots, m+k$. The third quadrant information on the primary production factors used in the economy. In particular, it includes the table contains the data on net taxes on products, on the components of gross value added and import. Recall that the economy uses n primary production factors. Denote the elements of the third quadrant of the table by $\left\| Z_i^j \right\|$, where $i = m, \ldots, m+n$, $j = 1, \ldots, m$. We consider two types of primary production factors: import and labor. Thus, we use only two rows from the third quadrant of initial table: import and gross value added, that is connected to the labor supply, i.e. $n = 2$.

We aggregate the final consumption of the products of all pure industries into one column vector $Z^0 = \left(Z_1^0, \ldots, Z_m^0 \right)$, $Z_i^0 = \sum_{j=m+1}^{m+k} Z_i^j$, $i = 1, \ldots, m$. The sum of elements of Z_0 is

$$A_0 = \sum_{i=1}^{m} \sum_{j=m+1}^{m+k} Z_i^j.$$

Denote the sum of elements of the column j of the obtained table by

$$A_j = \sum_{i=1}^{m+n} Z_i^j, \ j = 1, ..., m.$$

Value A_j equals to the element j of the last row Resources at basic prices of the initial symmetric input-output table, i.e. the total sum of production factors (intermediate inputs and primary factors) consumed by pure industry j including import of product j. Note that at the same time A_j is the total consumption of product j in the economy (the last column of the initial symmetric input-output table). Denote

$$\begin{aligned}
a_i^j &= \frac{Z_i^j}{A_j}, \ i = 1, \ldots, m, \ j = 1, \ldots, m, \\
b_i^j &= \frac{Z_{m+i}^j}{A_j}, \ i = 1, \ldots, n, \ j = 1, \ldots, m, \\
a_i^0 &= \frac{Z_i^0}{A_0}, \ i = 1, \ldots, m.
\end{aligned} \tag{17}$$

Obviously we have

$$\sum_{i=1}^{m} a_i^j + \sum_{t=1}^{n} b_t^j = 1, \ a_i^j \geq 0, \ b_t^j \geq 0, \ j = 1, \ldots, m, \ i = 1, \ldots, m, \ t = 1, \ldots, n, \tag{18}$$

$$\sum_{i=1}^{m} a_i^0 = 1, \ a_i^0 \geq 0, \ i = 1, \ldots, m. \tag{19}$$

We assume that non-negative matrix $\left\| a_i^j \right\|_{i=1,\ldots,m}^{j=1,\ldots,m}$ is productive. That is true if $Z^0 > 0$.

Let the Cobb-Douglas production function of the industry j is defined by

$$F_j\left(X^j, l^j\right) = A_j \left(\prod_{i=1}^{m} \left(\frac{X_i^j}{Z_i^j} \right)^{a_i^j} \right) \left(\prod_{i=1}^{n} \left(\frac{l_i^j}{Z_{m+i}^j} \right)^{b_i^j} \right), \ j = 1, \ldots, m, \tag{20}$$

the Cobb-Douglas utility function of the final consumer is defined by

$$F_0\left(X^0\right) = \prod_{i=1}^{m} \left(X_i^0 \right)^{a_i^0}, \tag{21}$$

the vector of supply of primary production factors is defined by

$$l = (l_1, \ldots, l_n), \ l_i = \sum_{j=1}^{m} Z_{m+i}^j, \ i = 1, \ldots, n.$$

Note that

$$\{\hat{X}_i^0 = Z_i^0, \ i = 1, \ldots, m; \hat{X}_i^j = Z_i^j, \ j = 1, \ldots, m, \ i = 1, \ldots, m; \\ \hat{l}_t^j = Z_{m+t}^j j = 1, \ldots, m, \ t = 1, \ldots, n\}$$

is a solution of convex programming problem (1)–(4) (see [9]). Thus, the constructed problem (1)–(4) explains observed source data (input-output table).

Note that (17)–(20) implies $F_j\left(a^j, b^j\right) = 1$, $j = 1, \ldots, m$, and (21) implies $F_0\left(a^0\right) = 1$. Thus, $d = 0$ and $\mu = 0$ in (10), (11).

Our model of optimal resource allocation allows us to analyze inter-industry financial flows of goods and services under various scenarios of changing of external conditions. In particular, we can analyze the stability of inter-industry balance tables. We select the base year and identify the model of nonlinear inter-industry balance with Cobb-Douglas production functions according to the symmetric input-output table of Kazakhstan for this year. To make a forecast, we use the data on the final consumption of products of industries $\hat{Z}^0 = \left(\hat{Z}^0_1, \ldots, \hat{Z}^0_m\right)$ in the forecast year. The data is available in the symmetric table of Supply of goods and services in basic prices for the forecast year. Then the sum of spending of the final consumer take on a new value $\hat{A}_0 = \sum_{i=1}^{m} \hat{Z}^0_i$, and the utility function of the final consumer is as follows.

$$\hat{F}_0\left(X^0\right) = \sum_{i=1}^{m} \hat{Z}^0_i \left(\frac{X^0_1}{\hat{Z}^0_1}\right)^{\hat{a}^0_1} \cdots \left(\frac{X^0_m}{\hat{Z}^0_m}\right)^{\hat{a}^0_m},$$

where $\hat{a}^0_i = \frac{\hat{Z}^0_i}{\hat{A}_0}$, $i = 1, \ldots, m$.

To analyze the stability of inter-industry connections, we need also a vector of price indices of primary production factors $\hat{s} = \left(\hat{s}^1, \ldots, \hat{s}^n\right)$ in the forecast year in relation to the base year.

Due to the Young transform the new consumer price index $\hat{q}_0\left(p\right)$ has the form

$$\hat{q}_0\left(p\right) = \frac{1}{\hat{F}_0\left(\hat{a}^0\right)} p_1^{\hat{a}^0_1} \ldots p_m^{\hat{a}^0_m}.$$

Then (12)–(15) implies that the vector of price indexes of industries $\hat{p} = \left(\hat{p}_1, \ldots, \hat{p}_m\right)$ equals to $\hat{p}_j = e^{\mu_j} \hat{s}_1^{c_1^j} \ldots \hat{s}_n^{c_n^j}$, $j = 1, \ldots, m$ and gives a solution of the convex programming problem

$$\hat{q}_0\left(p\right) \to \max$$

$$q_j\left(p, \hat{s}\right) \geq p_j \geq 0, \; j = 1, \ldots, m.$$

Proposition 1 implies that the new values of symmetric table of inter-industry balance are equal to

$$\hat{Z}^j_i = a^j_i \hat{Y}^j, \; \hat{Z}^j_{m+t} = b^j_t \hat{Y}^j, \; j = 1, \ldots, m, \; i = 1, \ldots, m; \; t = 1, \ldots, n,$$

where $\hat{Y}^j$ is the output of the industry j. The vector of outputs $\hat{Y} = \left(\hat{Y}^1, \ldots, \hat{Y}^m\right)$ is obtained from the equality $\hat{Y} = (E - A)^{-1} \hat{Z}^0$.

Now we can compare the constructed forecast of the inter-industry balance tables with the known statistical data. Thus, the stability of inter-industry connections can be analyzed. This helps us to understand if it is possible to use the inter-industry balance tables for medium-term analysis of economic development.

Note that the material flows of products in prices against a base year equal to

$$\hat{X}_i^0 = \frac{\hat{Z}_i^0}{\hat{p}_i}, \quad \hat{X}_i^j = \frac{\hat{Z}_i^j}{\hat{p}_i}, \quad \hat{l}_t^j = \frac{\hat{Z}_{m+t}^j}{\hat{s}_t}, \quad i = 1,\dots,m; \quad j = 1,\dots,m; \quad t = 1,\dots,n.$$

Since, we have $d = 0$, $\mu = 0$ in (10) and (11), then by (15) the price indexes vector is $\hat{p}_j = \hat{s}_1^{c_1^j} \dots \hat{s}_n^{c_n^j}$.

We choose 2016 as the base year. The vector of final consumption $\hat{Z}^0 = \left(\hat{Z}_1^0,\dots,\hat{Z}_m^0\right)$ is a column of corresponding initial input-output table of Kazakhstan (2016–2019). As we noted above, we consider two types of primary production factors: import and labor. We identify the corresponding vector of price indexes $\hat{s} = (\hat{s}^1, \hat{s}^2)$ against the base year (2016) on data of dynamics of tenge-dollar exchange rate and the consumer price index that is available at the section "Main socio-economic indicators of the Republic of Kazakhstan" Bureau of National Statistics (https://stat.gov.kz/official/dynamic). The result of identification of $\hat{s}$ is shown in Table 1.

Table 1. Price indexes

	2016	2017	2018	2019
Tenge/$ exchange rate index to the base year	1,000	0,953	1,007	1,119
Consumer price index to the base year	1,000	1,071	1,128	1,189

Based on our nonlinear inter-industry balance model, we calculate a forecast of input-output tables until 2019. We interpret the obtained results by comparing with the statistics of Kazakhstan given by the four vectors of dimension $m = 68$:

- gross output of the economy (**Y**)
- gross value added (**VA**)
- import (**Import**)
- the elasticity of consumer price index on the productivity α_j of industry j, that we calculate from our model in (16) (**Elasticity**).

We consider the angle (in radians) between the corresponding vectors as a measure of proximity between the predicted and actual values of the indicators. On Fig. 1 we show the results of the model forecast to 2019 (from 2016) and statistics 2019 for the first three vectors: Y, VA, Import. The results are given for the first 10 industries with the highest gross output. Figure 2 shows the result of the model forecast and statistics for the elasticity of consumer price index on productivity of the same industries. Table 2 shows the angles (in radians) between the forecast vectors and the corresponding statistics. The conducted estimates show that the nonlinear inter-industry balance model with Cobb-Douglas production functions can be considered as a tool for predicting the average characteristics of the input-output table system. The indicators of gross output, gross value added and the elasticity of consumer price index give approximately the same accuracy of the forecast in the period under review for the statistics of Kazakhstan. A significant weakening of the forecast force for the import vector is probably due to the violation of the hypothesis about the constant structure of producers' inputs in the period. The reason for this is a significant change in the exchange rate of the foreign currency in 2016–2019 (see Table 1).

4   The Analysis of Industries Centrality in Production Network of Kazakhstan

In this section we analyze the centrality of industries of Kazakhstan economy by ranking the elements of linkages vectors of the first- and second-order degrees, which we will compute according to the direct requirements matrix (Leontief matrix) of Kazakhstan. The direct requirements matrix gives the equivalent of the matrix A in our model. We compare the results with ranking of the elasticity of consumer price index on the productivity of industry j, that is calculated in (16). The advantage of our approach to determination of centrality industries (by the elasticity (16) ranking) is that we take into account the influence of the structure of final consumption. The elasticity value allows us to evaluate how the change of productivity in the industry affects the price level in the economy. This approach is close to the mechanism of propagation of productivity shocks from micro- to macro-economic level, proposed in [4]. We use the same input-output matrixes of Kazakhstan 2016–2019 as in the previous section and define a Leontief matrix of direct requirements $A = \|a_i^j\|$ by (17). Below there are the results of the ranking of the first 10 industries by the degree of centrality in accordance with different criteria.

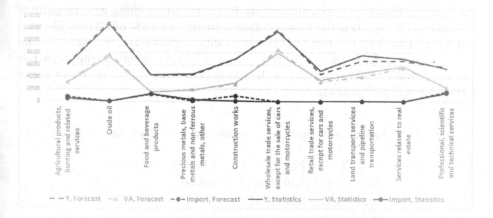

Fig. 1. Forecast and statistics (billions tenge), 2019

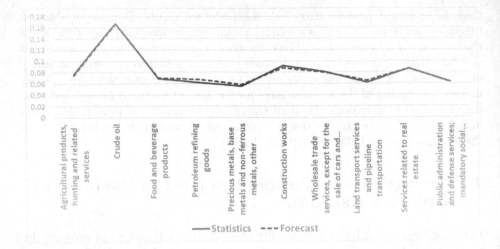

Fig. 2. Forecast and statistics of elasticity, 2019

Figures 3,4 show the results of ranking of industries of Kazakhstan economy by linkages of the first degree (direct costs) $d_i = \sum_{j=1,j\neq i}^{m} a_{ij}$, $i = 1,\ldots,m$ and the second degree (take into account indirect costs) $dd_i = \sum_{j=1}^{m} \sum_{k=1}^{m} a_{ik} a_{kj}$, $i = 1,\ldots,m$, respectively. Figure 5 plots the result of ranking of industries by the elasticity of consumer price index on the productivity evaluated by (16). Ranking of industry centrality is based on 2019. The analysis of the obtained results shows that the rating by the elasticity of industries centrality of the Republic of Kazakhstan is the most stable in the period 2016–2019 (Fig. 5). All the three approaches detect the service sector as one of the key sectors of the economy of Kazakhstan. However, the evaluation method significantly affects the rating of central industries. In the case of centrality measure by linkages of the first-and second-degrees, the central industry is the Wholesale trade. If we consider the elasticity of consumer price index by productivity as a measure of centrality (Fig. 5), then the most central industry is Crude oil. This difference can be explained by the fact that the centrality of elasticity takes into account the structure of final demand in the economy. In this situation, the impact of export volumes on the centrality of the industry is significant. This result seems to be economically justified.

Table 2. Angles (in radians) between forecast vector and statistic vector

	Y	Import	VA	Elasticity centrality
2017	0,0656	0,1025	0,0654	0,0660
2018	0,0653	0,2523	0,0611	0,0693
2019	0,0625	0,2398	0,0745	0,0718

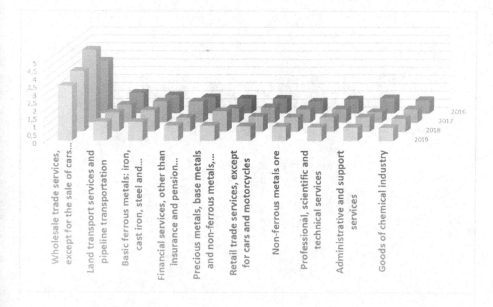

Fig. 3. First-order centrality

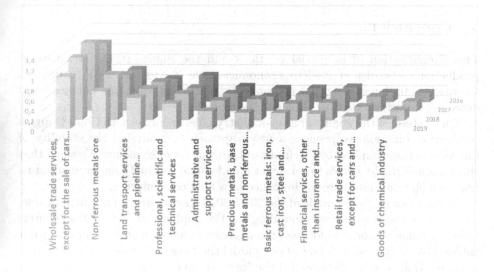

Fig. 4. Second-order centrality

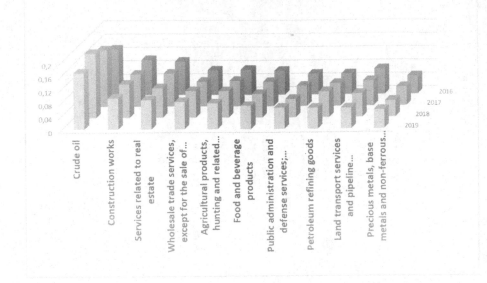

Fig. 5. Elasticity centrality

5   Conclusion

The diversification of industrial relations and the change in the structure of production networks in modern economies have led to the need to develop approaches to forecasting the components of the intersectoral balances and detecting industries that are drivers of economic growth.

On the base of the nonlinear inter-industry balance model with Cobb-Douglas production function we propose a new approach to analyzing the stability of interconnections implied by the input-output matrix of economy. We show that the nonlinear model allows to construct an adequate centrality measure of industries, which takes into account influence of a final consumption.

In our work on the example of the 2016–2019 Kazakhstan input-output table, we obtain a rough empirical grounding for the applying of our nonlinear inter-industry balance model to forecast the aggregate characteristics of input-output tables. Due to the assumptions of our model the forecast is valid if the structure of production costs is stable and if the forecast values of final consumption and price indices for primary production factors are given.

The model forecast of industry centrality by the elasticity of consumer price index on the productivity of the industry is more adequate in conditions of variability of finite demand.

On the base of Kazakhstan input-output matrix we show that the approach to detecting of the industry centrality based on the classical Leontief model (linkages of the first-and second-degrees) has a less accurate result compared with the centrality measure by the elasticity of consumer price index on the productivity of the industry that we evaluate in terms of our model.

A nonlinear model of inter-industry balance with Cobb-Douglas production functions assumes the constancy of the structure of financial costs of producers. The evolution of production networks leads to the emergence of new connections in production networks as a result of changes in the productivity of central industries. As a result, the proportions of financial costs of producers may change and the systems of the central sectors of the economy may be rebuilt. Therefore, the transition in nonlinear inter-industry models from Cobb-Douglas production functions to more general CES functions allows to increase the predictive power of such type of models. That is a potential area for our future research.

References

1. Bonacich, P.: Power and centrality: a family of measures. Am. J. Sociol. **92**(5), 1170–1182 (1987)
2. Barro, R.J., Sala-i-Martin, X.: Economic Growth. MIT Press, Cambridge Mass (2004). see https://mitpress.mit.edu/books/economic-growth-second-edition for details.
3. Chamberlin, E.H.: The Theory of Monopolistic Competition, 8th edn. Harvard University Press, Cambridge (1962)
4. Acemouglu, D., Ozdaglar, A., Tahbaz-Salehi, A.: The network origins of aggregate fluctuations. Econometrica **80**(5), 1977–2016 (2012)
5. Acemouglu, D., Ozdaglar, A., Tahbaz-Salehi, A.: Networks, shocks, and systemic risk. In The Oxford Handbook of the Economics of Networks, pp. 569–607. Oxford University Press, New York (2016)
6. Acemouglu, D., Ozdaglar, A., Tahbaz-Salehi, A.: Microeconomic origins of macroeconomic tail risks. Am. Econ. Rev. **107**(1), 54–108 (2017)
7. Shananin, A.A.: Young duality and aggregation of balances. Doklady Math. **102**(1), 330–333 (2020)
8. Shananin, A.A.: Problem of aggregating of an input-output model and duality. Comput. Math. Math. Phys. **61**(1), 162–176 (2021)
9. Rassokha, A.V., Shananin, A.A.: Inverse problems of the analysis of input-output balances. Math. Model. **33**(3), 39–58 (2021)
10. Ashmanov, S.A.: Introduction to Mathematical Economics. Fizmatlit, Moscow (1984)
11. Hulten, C.R.: Growth accounting with intermediate inputs. Rev. Econ. Stud. **45**, 511–518 (1978)
12. Baqaee, D.R., Farhi, E.: The macroeconomical impact of microeconomical shocks: beyond Hulten's Theory. Econometrica **87**(4), 1155–1203 (2019)
13. Newman, M.E.J.: Networks: An Introduction. Oxford University Press, Oxford (2010)
14. Fagiolo, G.: Clustering in complex networks. Phys. Rev. E-Stat. Nonlinear Soft Matter Phys. **76**(2), 1–16, 0612169 (2007)
15. McAssey, M.P., Bijma, F.: A clustering coefficient for complete weighted networks. Netw. Sci. **3**(02), 1–13, 183195 (2015)

Techniques for Speeding up *H*-Core Protein Fitting

Andrei Ignatov[1,2]([⊠]) [iD] and Mikhail Posypkin[1,2] [iD]

[1] HSE University, Moscow, Russia
[2] Federal Research Center Computer Science and Control of the Russian Academy
of Sciences, Moscow, Russia

Abstract. Restoration of the 3D structure of a protein from the sequence
of its amino acids ("folding") is one of the most important and challenging
problems in computational biology. The most accurate methods require
enormous computational resources due to the large number of variables
determining a protein's shape. Coarse-grained models combining several
protein atoms into one unified globule partially mitigate this issue. The
paper studies one of these models where globules are located in the nodes
of the two-dimensional triangular lattice. In this model, folding is reduced
to the discrete optimization problem: find positions of protein's globules to
maximize the number of contacts between them. We consider a standard
procedure that finds an exact solution to this problem. It first generates an
H-core—a set of positions for hydrophobic globules, which is followed by
mapping of protein's hydrophobic globules to these positions by the constraint satisfaction techniques. We propose a way to avoid unnecessary
enumeration by skipping infeasible *H*-cores prior to mapping. Another
contribution of our paper is a procedure that automatically generates constraints to simplify finding the feasible mapping of proteins globules to
the lattice nodes. Experiments show that the proposed techniques tremendously accelerate the problem's solving process.

Keywords: Protein · HP-models · *H*-core · Constraint satisfaction ·
Discrete optimization

1 Introduction

Protein folding is the physical process by which a protein chain acquires its
native 3-dimensional structure (conformation). The folding process has been
studied for the last 50 years by numerous researchers worldwide. Modeling of
the folding process implies answering a question: "Is it possible to predict a
protein's conformation having only its formula?".

A protein is a chain of amino acid residues. There are only 20 types of amino
acids from which a protein can be built. An amino acid has a backbone (three

Partially supported by HSE University project "Development, research and application
of optimization methods".

N. N. Olenev et al. (Eds.): OPTIMA 2021, LNCS 13078, pp. 336–350, 2021.
https://doi.org/10.1007/978-3-030-91059-4_24

main atoms – one nitrogen and two carbons) and a side-chain connected to the first carbon of the backbone (see Fig. 1). For clarity, this carbon is named C_α. Amino acids differ by their side-chains, while their backbones are the same.

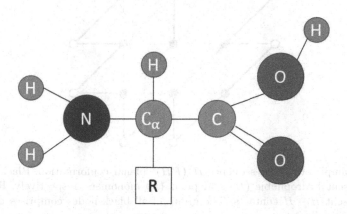

Fig. 1. The basic structure of an amino acid. N, C_α, and C atoms are the main backbone atoms. R is the side-chain.

The most accurate way of predicting protein conformation is minimization of an all-atom potential [1]. The idea behind this approach implies that the target protein acquires the conformation with the minimal potential energy value. Several different force fields and potentials exist. The major disadvantage of this approach is a large number of degrees of freedom and complex non-linear multiextremal objective function. As a result, the problem is known to be very challenging [2,3].

One possible way to mitigate this issue is provided by coarse-grained models [4–6]. These models are based on simplified geometry, consisting in grouping several atoms into one particle. Accuracy of these models is usually limited by the atom grouping resolution level. They can provide a good initial approximation of the native conformation, but further steps are required to obtain the target conformation.

The highest level of geometry simplification is achieved in HP-model [7]. In this model, residues are considered as single particles ("unified atoms"). All protein residues are split into two groups—hydrophobic (H) and polar (P). Hydropathy is one of the key aspects that drive the folding process. Hydrophobic side-chains tend to 'hide' inside the protein, while polar (hydrophilic) side-chains stay on its surface. This mechanism leads to the creation of a robust core with all hydrophobic side-chains inside (see Fig. 2).

The initial protein sequence of amino acid codes can be converted to a sequence of 'H' and 'P' letters. This is called an HP-*sequence*. In the paper, long HP-sequences are presented in a diminished form, e.g. $H^2(PH)^2$ instead of $HHPHPH$. The number over a single monomer or a subsequence in brackets implies the number of successive repeats of the corresponding base.

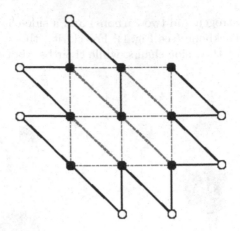

Fig. 2. Example of an HP-sequence $H^2(PH)^7$ planar conformation. Black and white dots represent hydrophobic (H) and polar (P) monomers, respectively. Red dashed lines represent $H-H$ contacts. The group of all black nodes comprises the H-core. (Color figure online)

In HP-model, positions of all monomers are constrained to nodes of some 2- or 3-dimensional lattice. In what follows, we consider only planar structures and thus 2-dimensional lattices. The goal function is simply the number of $H-H$ contacts. This number should be maximized to achieve a maximally dense H-core. Positions of remaining atoms are not regarded as important in this model. A protein conformation that provides the maximal possible number of $H-H$ contacts is called optimal.

The standard approach to constructing optimal conformations [8] in HP-models is a two-stage process. At the first stage, H-monomers in a protein sequence are assembled into a maximally dense H-core structure without accounting for any constraints related to the protein sequence. Then, at a second phase, an attempt to fit the whole protein chain into this structure is performed. If the attempt is successful, the algorithm terminates with the found conformation as an output. Otherwise, the next H-core is tried. If all maximally dense H-cores are examined without success, configurations with less number of $H-H$ contacts are considered. The process is continued until the first successful fitting attempt.

Normally, there exist many different H-cores with the same number of $H-H$ contacts. The complete enumeration of all such configurations is very time-consuming. In this paper, we propose a way of avoiding unnecessary enumeration by skipping infeasible H-cores prior to fitting. We have developed a set of rules that help identify such configuration by a fast checking procedure.

Another contribution of our paper is a procedure that automatically generates constraints that significantly simplify the fitting phase of the folding process. We showed that the proposed techniques may result in a reasonable fitting time decrease.

The paper is organized as follows. Section 2 provides the problem statement. In Sect. 3, H-Core filtering criteria are discussed. Section 4 introduces a list of constraints and several sets of domains for H- and P-monomer's coordinates. In Sects. 5 and 6, experimental results are presented and discussed.

2   Basic Definitions and Problem Statement

In HP-model, a protein is represented as a sequence of hydrophobic (H) and polar (P) monomers. In a sequel, we use the following notation:

- S is the HP-sequence under consideration (target HP-sequence);
- n_S is the length of S;
- n_H is the number of H-monomers in sequence S;
- $S_i,\ i \in 1 \ldots, n_S$ is the ith monomer in this sequence.

Monomers are located in the nodes of a 2-D lattice. A **2-D lattice** is a countable set L of points on a plane such that $u + v \in L$ for any $u, v \in L$. A simple way of representing a lattice is using a minimal set of vectors N_L that encode all its points [8]. Every node u of the lattice can be represented as a linear combination with non-negative integral coefficients of vectors from N_L:

$$L = \{u \in \mathbb{R}^2 | u = \sum_{v \in N_L} c_v \cdot v, c_v \in \mathbb{Z}^+\}.$$

N_L sets for squared and triangular lattices are listed in Table 1.

Table 1. N_L sets for squared and triangular 2D lattices

Lattice	Square	Triangular
N_L vectors	(0, 1), (0, −1), (1, 0), (−1, 0)	(1, 0), (−1, 0), (0, 1), (0, −1), (−1, 1), (1,−1)

In this paper, the triangular lattice is used as it provides a more accurate approximation (see [9] for a detailed explanation). The conformation of the target sequence is defined by **mapping** $M = \{M_1, \ldots, M_{n_S}\}$, where $M_i = (x_i, y_i)$ is the position (lattice node) of monomer S_i. This mapping should preserve the connectivity of the protein, i.e. the successive monomers should be connected by the vectors from N_L:

$$M_{i+1} - M_i \in N_L,\ i = 1, \ldots, n_S - 1.$$

Such mappings are called *valid*.

In what follows we use the notion of a *distance between two lattice nodes*. The distance $\rho(c, c')$ between two lattice nodes c and c' is defined as follows:

$$\rho(c, c') = \min \left\{ t | c + \sum_{i=1}^{t} v_i = c', \, v_i \in N_L, \, i = 1, \ldots, t \right\}.$$

Informally, $\rho(c, c')$ is the minimal number of steps along the lattice nodes required to reach c' from c.

Two nonsuccessive monomers are *in contact* if they are located in the neighboring nodes of the lattice. Formally, a contact between monomers $i, j = 1, 2, \ldots, n_S$ is defined as $M_i - M_j \in N_L$. It should be noticed that in a valid mapping, two monomers that are successive along the HP-sequence are always in contact.

A goal function in HP-model is the number of $H - H$ contacts. It can be formally written as follows:

$$E_S(M) = \sum_{1 \le i+1 < j \le n_S} \Delta(M_i, M_j) \cdot e(S_i, S_j), \text{ where} \tag{1}$$

$$\Delta(M_i, M_j) = \begin{cases} 1, \text{if } M_i - M_j \in N_L, \\ 0, \text{ otherwise,} \end{cases}$$

$$e(S_i, S_j) = \begin{cases} -1, \text{ if } S_i = \text{`}H\text{'} \text{ and } S_j = \text{`}H\text{'}, \\ 0, \text{ otherwise.} \end{cases}$$

Having the goal function, we can formulate the main problem as follows:

$$\begin{cases} E_S(M) \longrightarrow \min, \\ M_i - M_{i-1} \in N_L, i = 2, \ldots, n_S, \\ M_i \ne M_j, i, j = 1, \ldots, n_S, \, i \ne j. \end{cases}$$

Constraints in the general problem definition imply that two monomers cannot be located in the same lattice node, and monomers that are successive along the sequence must be located in neighboring nodes.

Due to an enormous number of possible mappings of the target HP-sequence to the lattice, the attempts to directly find an optimal conformation are impractical. In [8], it was observed that only positions of H-monomers influence the number of contacts while the positions of P-monomers are irrelevant. Thus, the authors have proposed a natural two-stage approach. At the first stage, the positions of H-monomers in a lattice are determined thereby forming a so-called H-core. Then, a valid mapping M is constructed in such a way that H-monomers are placed to the positions of the H-core. The latter procedure is called *fitting*.

To maximize the number of contacts, the maximally dense H-cores are tried first. If the fitting attempts fail, H-cores with the less number of contacts are tried.

Formally, an **H-core H** is a subset of n_H lattice points. We restrict our consideration only to connected H-cores. An H-core **H** is *connected* if any of its nodes can be reached by a path along the lattice having all intermediate nodes belonging to **H**. In other words, for any nodes $c, c' \in$ **H** there exists a finite path $c_1, \ldots, c_k$, where $c_1 = c, c_k = c'$, $c_i \in$ **H**, $i = 1, \ldots, k$ and $c_{i+1} - c_i \in N_L$, $i = 1, \ldots, k-1$. Focusing only on connected H-cores makes sense because connection of several distinct components may only increase the number of $H - H$ contacts, which is preferable for the problem.

Based on the discussed definitions, we can formulate two subproblems that naturally appear when the approach from [8] is applied. The first subproblem is to develop an efficient algorithm for enumerating H-cores with a given number of contacts. Not all H-cores with the same number of contacts may be feasible for the target HP-sequence. Avoiding principally infeasible H-cores tremendously reduces the number of generated H-cores. This can dramatically decrease the whole fitting time.

According to the approach under consideration, H-cores generated at the first stage are submitted to the second stage to find positions of the protein monomers. The commonly adopted technique to do the latter is constraint satisfaction [10,11]. Constraint satisfaction is a problem of finding a solution which is a set of variable values subject to a given set of constraints. The efficiency of constraint satisfaction largely depends on the tightness of the imposed constraints. This entails the second subproblem: introducing a set of additional constraints to accelerate the constraint satisfaction algorithm.

Below we address both subproblems. We propose an approach that significantly reduces the amount of generated H-cores by skipping infeasible configurations and introduce additional constraints for accelerating the fitting phase.

3   Efficient H-Core Enumeration

3.1 H-Core Node Depth

In this section, a classification of H-core nodes is suggested. In further discussion, let us consider only connected H-cores.

An H-core can be treated as a graph with N vertices (points) and K edges (contacts). According to the lattice constraints, contact is possible only between neighboring nodes, thus, its corresponding graph can only be planar in a 2-D lattice.

Nodes of an H-core may be classified by their buriedness (distance to the closest lattice point that is outside the H-core). Let us define the H-core node depth.

Definition 1. Depth $d(c)$ *of a node* $c \in$ ***H*** *is the minimal distance to a lattice node* $\tilde{c}$ *such that* $\tilde{c} \notin$ ***H***, *i.e.*

$$d(c) = \min \{\rho(c, \tilde{c}) \mid \tilde{c} \in L, \tilde{c} \notin \boldsymbol{H}\}.$$

Nodes of such H-core can be split into two groups by the following definition.

Definition 2. *Let H be an H-core. A node $c \in H$ is **inner** if its depth $d(c) > 1$. Otherwise, c is called **boundary**.*

As a corollary from Definition 2, degrees of inner vertices of the corresponding graph are equal to $|N_L|$, which is the cardinality of the set N_L. Degrees of boundary vertices are smaller than $|N_L|$.

Depth computing is performed with the procedure outlined in Algorithm 1. Initially, the depths of all nodes are set to infinity. In the algorithm, the list of node depths is denoted as *depths*. At the first step of the algorithm, degrees of all nodes are computed and placed to *degrees* variable. A node with the minimal degree c_s is selected. Its depth is set equal to 1. After that, the function from Algorithm 1 is executed: $UpdateDepths(c_s)$.

Function UpdateDepths(c_0):
    $neighbors := \{c \in \mathbf{H}, c - c_0 \in N_L\}$
    $updated\_nodes := [\,]$
    for $\tilde{c} \in neighbors$ **do**
        if $degrees[\tilde{c}] < |N_L|$ **then**
            if $depths[\tilde{c}] > 1$ **then**
                $depths[\tilde{c}] := 1$
                $updated\_nodes := updated\_nodes + [\tilde{c}]$
        else
            if $depths[\tilde{c}] > depths[c_0] + 1$ **then**
                $depths[\tilde{c}] := depths[c_0] + 1$
                $updated\_nodes := updated\_nodes + [\tilde{c}]$
        end
    end
    for $c \in updated\_nodes$ **do**
        $UpdateDepths(c)$
    end

Algorithm 1: H-core node depth computation

The algorithm is considered to be fast enough for applications. In the most widely used benchmark for 2-D triangular lattice [12], the greatest number of H-monomers is only 42. As shown in Fig. 3 (right), depth computation for 100 nodes takes less than $0.1\,s$, which is obviously a satisfying result.

Figure 3 (left) presents an example of an H-core for $N = 36$ H-monomers. $H - H$ contacts of this H-core are shown as red dashed lines.

Let us denote the ***set of indices of H-monomers*** in an HP-*sequence* S as $h_S = \{i \text{ for } i \in 1, \ldots, n_S \text{ if } S_i = \text{'}H\text{'}\}$. The respective ***set of P-monomers' indices*** is denoted as $p_S = \{i \text{ for } i \in 1, \ldots, n_S \text{ if } S_i = \text{'}P\text{'}\}$. Having that, let us denote the ***distance from an H-monomer S_i to the closest P-monomer*** along the HP-sequence as $d_P(i) = \min_{j \in p_S} |i - j|$. For a P-monomer S_j, the corresponding ***distance to the closest H-monomer*** is denoted as $d_H(j)$ Having Definition 1, we can formulate the following theorem:

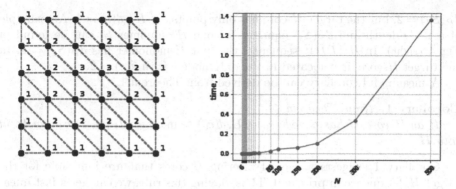

Fig. 3. Sample H-core for $N = 36$ (left). Depths are annotated, dashed lines are contacts. Time scaling of the node depth computation algorithm (right).

Theorem 1. *Let S be an HP-sequence and $\boldsymbol{H}$ be an H-core. An H-monomer S_i can be mapped only to nodes $c \in \boldsymbol{H}$ such that $d(c) \leq d_P(i)$.*

Proof. Assume that there exists an H-monomer S_j located in a point $c_0 \in \boldsymbol{H}, d(c_0) > d_P(j)$. Possible lattice nodes for the closest P-monomer along the chain are located not farther than $d_P(j)$ steps from c_0 along the lattice. According to the assumption, these nodes belong to $\boldsymbol{H}$. But P-monomers cannot be located in H-core nodes. Thus, we obtained a contradiction. The theorem is proved.

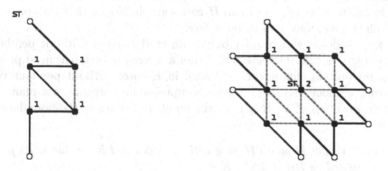

Fig. 4. Fittings for sequences PH^4P (left), $H^2(PH)^7$ (right). Black points represent H-monomers, white points are P-monomers. Depths are annotated, dashed lines are contacts. The first monomer of the chain is marked as 'ST'. (Color figure online)

Theorem 1 provides a correct estimation for depths of feasible H-core nodes. For instance, if S has k adjacent H-monomers, they may be located in nodes whose depth is less than their respective $d_P(i)$'s. An example of such S fitting can be found in Fig. 4 (left). PH^4P sequence has two H-monomers with $d_P(S_3) =$

$d_P(S_4) = 2$, but the target H-core has only points $c \in H : d(c) = 1$. An example of a node depth and $d_P(S_i)$ equality for one H-monomer S_i can be found in Fig. 4 (right). In $H^2(PH)^7$ sequence, the first H-monomer has $d_P(S_1) = 2$. In the target H-core, it is located in the only node c with $d(c) = 2$.

A meaningful corollary can be derived from Theorem 1.

Corollary 1. *(from Theorem 1)*
If an H-core $\boldsymbol{H}$ has a node $c_0 \in \boldsymbol{H}, d(c_0) > \max_{i \in h_S} d_P(i)$, S cannot be fit into $\boldsymbol{H}$.

Corollary 1 suggests a way of filtering H-cores that are infeasible for the target HP-sequence in principal. Thus, having this rule, we can get a fast infeasibility verdict without applying constraint satisfaction techniques.

The ideas of Theorem 1 can be applied to P-monomers. Let $\mathbf{H}_S$ be the *surface* of an H-core $\mathbf{H}$, i.e. $\mathbf{H}_S = \{c \,| c \in \mathbf{H}, d(c) = 1\}$.

Theorem 2. *Let S be an HP-sequence, $\boldsymbol{H}$ be an H-core. A P-monomer S_i can be mapped only to lattice nodes c such that $c \notin \boldsymbol{H}$ and $\min_{h \in \mathbf{H}_S}(\rho(c, h)) \leq d_H(i)$.*

Proof. (The proof is identical to the proof of Theorem 1)

3.2   H-Core Perimeter

The ideas suggested in Sect. 3.1 can be generalized for all H-cores with N nodes and K contacts.

It is possible to build a graph on H-core nodes, considering contacts as edges. According to lattice requirements, this graph is always planar as it has no crossing edges. Thus, its edges and vertices split the plane into regions called *faces*. Let F be the number of faces in an H-core's graph. Notice that the space outside the graph is also a face called *outer face*.

In what follows, we consider only connected H-cores without pendant vertices, i.e. vertices with degree 1. This is not a severe limitation as most practically meaningful protein chains can be folded in H-cores without pendant vertices. An H-core without pendant vertices occupies some polygon on a plane. Define the *perimeter* R_H of an H-core as the number of edges in the boundary of this polygon.

Statement 1. *Let $\boldsymbol{H}$ be an H-core with N nodes and K contacts. Its perimeter can be computed as $R_H = 3N - K - 3$.*

Proof. Since an H-core graph is planar, due to Euler's formula for planar graphs [13]:

$$N - K + F = 2. \tag{2}$$

The boundary of the outer face of the H-core graph has R_H edges. All other faces have 3 edges in their boundaries as we consider a triangular lattice. Since each edge (contact) belongs to exactly two faces we have:

$$2K = 3(F - 1) + R_H.$$

From (2) we easily get an expression for F:

$$F = 2 - N + K.$$

Thus,

$$R_H = 2K - 3(F - 1) = 2K - 3(1 - N + K) = 3N - K - 3.$$

This completes the proof.

Having the formula for H-core perimeter, we can calculate the number of its inner nodes.

Statement 2. *Let H be an H-core without pendant vertices. Let it have N nodes and K contacts. The number of its inner nodes can be computed as $N_{in} = K - 2N + 3$.*

Proof. Let H be an H-core with N nodes and K contacts. Obviously, all outer nodes of H contribute in its perimeter R_H. Then, all other nodes are inner:

$$N_{in} = N - R_H$$

Using formula from Statement 1, we get:

$$N_{in} = N - R_H = N - (3N - K - 3) = K - 2N + 3.$$

This completes the proof.

Having formula from Statement 2, we can compute the exact number of inner nodes for any H-core with N nodes and K contacts, assuming it has no pendant vertices. This can be done analytically, without full enumeration of these H-cores and looping over their nodes. That provides another H-core filtering criterion.

Corollary 2. *(from Theorem 1)*
   Let S be an HP-sequence, H be an H-core. Let H have N_{in} inner nodes. Then, if $N_{in} > |\{i \mid d_P(S_i) > 1, \; i \in h_S\}|$, S cannot be fit into H.

The condition in Corollary 2 implies that the number of nodes $c \in H, d(c) > 1$ is greater than the number of H-monomers located farther than 1 from the respective closest P-monomers. In this case, some inner H-core positions would be unreachable by H-monomers.

Let a set of H-cores with the same N and K be denoted as (N, K) **flavor**. Then, if the condition in the Corollary 2 is true, we can avoid checking all H-cores in (N, K) flavor.

4 HP-Sequence Fitting Constraints

In the folding method under consideration, after an H-core is generated, the HP-sequence is fit into it with constraint satisfaction techniques. The first set of constraints is derived from obvious geometric observations:

1. Different monomers cannot be mapped into the same position: $M_i \neq M_j$, $i, j = 1, \ldots, n_S, i \neq j$.
2. Monomers that are adjacent along the HP-sequence must be located in neighboring nodes of the lattice: $M_i - M_{i-1} \in N_L$, $i = 2, \ldots, n_S$.
3. H-monomers can be located only in nodes of the target H-core $\mathbf{H}$, while P-monomers must be located outside it:

$$\begin{cases} M_i \in \mathbf{H}, & \text{if } S_i = H \\ M_i \notin \mathbf{H}, & \text{otherwise} \end{cases}, \ i = 1, \ldots, n_S.$$

Theorems 1 and 2 allow us to inroduce additional constraints:

4. According to Theorem 1, an H-monomer S_i, $i \in h_S$ cannot be located in H-core nodes that are deeper than $d_P(i)$: $M_i \in \{c \ | c \in \mathbf{H}, d(c) \leq d_P(i)\}$.
5. According to Theorem 2, a P-monomer S_i, $i \in p_S$ cannot be located in nodes of lattice L that are not farther from H-core surface $\mathbf{H}_S$ than $d_H(i)$: $M_i \in \{c \ | c \in L, c \notin \mathbf{H}, \min_{h \in \mathbf{H}_S} \rho(c, h) \leq d_H(i)\}$.

Besides constraints, constraint satisfaction techniques require setting domains for problem variables. Narrowing these domains usually accelerates the solution process and therefore is desirable. Monomer positions M_i are represented as pairs of two variables (x_i, y_i). Below we outline four possible ways to define these domains.

Domain Set 1
If S_i is hydrophobic it is mapped to the positions of an H-core. Thus, $x_i \in \{x | (x, y) \in \mathbf{H}\}$, $i \in h_S$. Similarly, $y_i \in \{y, \ (x, y) \in \mathbf{H}\}$, $i \in h_S$. P-monomers are restricted to nodes that are not farther than n_S from the H-core $\mathbf{H}$: $x_i \in \{x \ | (x, y) \in L, \ (x, y) \notin \mathbf{H}, \ \min_{h \in \mathbf{H}} \rho((x, y), h) \leq n_S\}$, $i \in p_S$. Domains for y_i, $i \in p_S$ are set in the same way.

Domain set 1 accounts only for basic geometric observations. Tighter bounds can be obtained if a more deep analysis of monomers' positions is performed.

Domain Set 2
The maximal possible offset of a P-monomer from $\mathbf{H}$ is computed as $d_H^{max} = \max_{i \in p_S} d_H(i)$. Therefore, domains of M_i, $i \in p_S$, of P-monomers are bound to lattice points with

$$x_i \in \{x \ | (x, y) \in L \setminus \mathbf{H}, \min_{h \in \mathbf{H}} \rho((x, y), h) \leq d_H^{max}\}, \ i \in p_S.$$

Domains for y_i, $i \in p_S$ are defined on the same way.

Further domain contraction is obtained if we consider each monomer individually and apply Theorems 1, 2.

Domain Set 3

For each H-monomer S_i, $i \in h_S$, its domain is set as follows:

$x_i \in \{x \,|(x, y) \in \mathbf{H} : d((x, y)) \leq d_P(i)\}$,

$y_i \in \{y \,|(x, y) \in \mathbf{H} : d((x, y)) \leq d_P(i)\}$.

Accordingly, domains for P-monomers S_i, $i \in p_S$, are:

$x_i \in \{x \,|(x, y) \in L \setminus \mathbf{H}, \min_{h \in \mathbf{H}_S} \rho((x, y), h) \leq d_H(i)\}$,

$y_i \in \{y \,|(x, y) \in L \setminus \mathbf{H}, \min_{h \in \mathbf{H}_S} \rho((x, y), h) \leq d_H(i)\}$.

Some CSP solvers admit vector variables. For such solvers, a pair of point coordinates is considered as a single variable and different domains for x_i and y_i coordinates can be replaced with sets of possible lattice nodes for M_i, $i = 1, \ldots, n_S$. Such domain definitions make constraints 3—5 redundant. Fortunately, the solver *python-constraint* [14] used in our study enables vector variables, and we add the following domain set to a comparison.

Domain Set 4

Domain for every H-monomer S_i, $i \in h_S$, is set as follows.

$M_i \in \{c \,|c \in \mathbf{H} : d(c) \leq d_P(i)\}$.

Accordingly, domains for P-monomers S_i, $i \in p_S$, are:

$M_i \in \{c \,|c \in L \setminus \mathbf{H}, \min_{h \in \mathbf{H}_S} \rho(c, h) \leq d_H(i)\}$.

5   Results

The main contribution of the paper are two criteria for filtering H-cores that are principally infeasible for the target HP-sequence. Additionally, several sets of intelligent constraints for this problem were suggested. Utilizing them leads to a significant speed-up while solving the Constraint Satisfaction Problem (CSP).

The developed methodology was implemented in Python programming language. A third-party Python module *python-constraint* [14] was used for solving Constraint Satisfaction Problem (CSP). Experiments were run on MacBook Pro 2020 machine, equipped with the 2 GHz Intel Core i5-1038NG7 CPU and 16 Gb of RAM.

Table 2. Configurations for CSP solver.

Configuration	Constraints	Domain set
1	1—5	1
2	1—5	2
3	1—5	3
4	1, 2	4

The two-stage algorithm under consideration was tested in several configurations outlined in Table 2. Two HP-sequences were taken for comparing these configurations:

1. $((HP)^2PH)^2PH^2P^2$;
2. $H^2(PH)^7$.

Running times for configurations 1—4 are listed in Table 3. Optimal configurations for both sequences are shown in Fig. 5.

Table 3. Running times for test HP-sequences and solver configurations, seconds

Sequence	Configuration 1	Configuration 2	Configuration 3	Configuration 4
$((HP)^2PH)^2PH^2P^2$	15.34	1.42	1.05	0.01
$H^2(PH)^7$	–	680	283	0.01

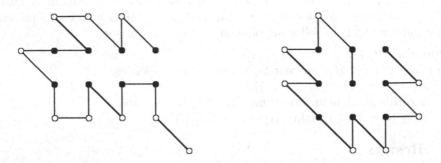

Fig. 5. Resulting optimal H-cores for sequences $((HP)^2PH)^2PH^2P^2$ (left) and $H^2(PH)^7$ (right).

Time evaluation exceeded the limit for the second sequence and solver configuration 1. In Table 3, the corresponding cell is left empty. Data in the table demonstrates that utilizing solver configuration 4 leads to a tremendous decrease in the fitting time. This is likely to be a result of avoiding checks of constraints 3—5, which are always true for the selected variable domains.

H-core filtering criteria were tested at a longer sequence: $((HP)^2PH)^2PH^2P(PH)^2$ [12]. This sequence has $N = 10$ H-monomers, which results in $N_{in} = 1$ for the most dense H-core. At the same time, this node cannot be reached by any H-monomer of this sequence (according to Corollary 2). Thus, the number of contacts must be reduced. Avoiding all H-cores with the maximal number of contacts might save nearly half of the whole time needed to find a solution. The distribution of time for this problem is presented in Fig. 6 (left). In this experiment, solver configuration 4 was used. This experiment demonstrates that usage of the developed filtering criteria can significantly decrease the overall fitting time.

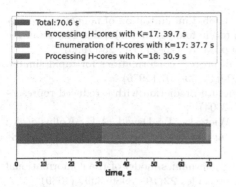
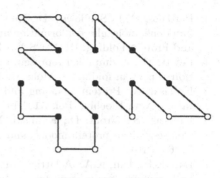

Fig. 6. Breakdown of time spent on searching an optimal H-core for sequence $((HP)^2PH)^2PH^2P(PH)^2$ (left). The resulting configuration for H-core with 17 contacts (right). Black and white dots represent H- and P-monomers, respectively.

6 Conclusion

The paper addresses a highly important and very challenging computational problem: protein folding. We considered a coarse-grained 2-D lattice protein model. The folding is treated as a combinatorial optimization problem with the number of contacts between hydrophobic monomers as an objective. We focused on one of the possible approaches to solving this problem when the structure of the folded protein is obtained in two stages. At the first stage, an H-core is generated. At the second stage, H-monomers from the sequence are mapped to nodes of the H-core preserving the integrity of the protein chain. The latter is done via constraint satisfaction techniques.

We developed methods to accelerate both stages. A methodology for avoiding principally infeasible H-cores that significantly reduced the H-core enumeration phase was proposed. Also, we introduced a set of sophisticated constraints that entail a significant speedup of the fitting stage. Experiments showed a remarkable efficiency of the proposed approach.

In the future, a more deep investigation of the connection between a protein sequence and its H-core structure is planned. Based on discovered H-core properties, new efficient exact and approximate folding algorithms can be developed. With such algorithms at hand, long peptide chains can be folded in a reasonable time.

References

1. Jorgensen, W.L., Maxwell, D.S., Tirado-Rives, J.: Development and testing of the OPLS all-atom force field on conformational energetics and properties of organic liquids. J. Am. Chem. Soc. **118**(45), 11225–11236 (1996)
2. Pardalos, P.M., Shalloway, D., Xue, G.: Optimization methods for computing global minima of nonconvex potential energy functions. J. Glob. Optim. **4**(2), 117–133 (1994)

3. Pardalos, P.M., Shalloway, D., Xue, G.: Global minimization of nonconvex energy functions: molecular conformation and protein folding. In: Molecular Conformation and Protein Folding: DIMACS Workshop, March 20–21, vol. 23 (1996)

4. Levitt, M.: A simplified representation of protein conformations for rapid simulation of protein folding. J. Mole. Biol. 104(1), 59–107 (1976)

5. Koliński, A.: Protein modeling and structure prediction with a reduced representation. Acta Biochim. Pol. 51, 349–371(2004)

6. Kmiecik, S., Gront, D., Kolinski, M., Wieteska, L., Dawid, A.E., Kolinski, A.: Coarse-grained protein models and their applications. Chem. Rev. 116(14), 7898–7936 (2016)

7. Lau, K.F., Dill, K.A.: A lattice statistical mechanics model of the conformational and sequence spaces of proteins. Macromolecules 22(10), 3986–3997 (1989)

8. Mann, M., Backofen, R.: Exact methods for lattice protein models. Bio-Algorith. Med-Syst. 10(4), 213–225 (2014)

9. Böckenhauer, H.J., Ullah, A.Z. M.D., Kapsokalivas, L., Steinhöfel, K.: A local move set for protein folding in triangular lattice models. In: International Workshop on Algorithms in Bioinformatics, pp. 369–381 (2008)

10. Dal Palu, A., Dovier, A., Fogolari, F.: Constraint logic programming approach to protein structure prediction. BMC Bioinform. 5(1), 1–12 (2004)

11. Dal Palu, A., Dovier, A., Pontelli, E.: A constraint solver for discrete lattices, its parallelization, and application to protein structure prediction. Softw. Pract. Exp. 37(13), 1405–1449 (2007)

12. Liu, J., Song, B., Liu, Z., Huang, W., Sun, Y., Liu, W.: Energy-landscape paving for prediction of face-centered-cubic hydrophobic-hydrophilic lattice model proteins. Phys. Rev. E 88(5), 052704 (2013)

13. Agnarsson, G., Greenlaw, R.: Graph theory: Modeling, Applications, and Algorithms. Prentice-Hall, Inc., Englewood Cliffs (2006)

14. PyPI page for python-constraint module, https://pypi.org/project/python-constraint. Accessed 24 June 2021

Application of Second-Order Optimization Methods to Solving the Inverse Coefficient Problems

Alla Albu[✉][iD] and Vladimir Zubov[iD]

Federal Research Center Computer Science and Control of the Russian Academy
of Sciences, Moscow, Russia

Abstract. The inverse problem of determining the thermal conductivity coefficient depending on temperature is considered and investigated. The consideration is based on the initial boundary value problem for the non-stationary heat equation. The work is devoted to obtaining the necessary conditions for non-uniqueness of the considered inverse problem solution in the n-dimension case, and also to examine the possibility of applying the Fast Automatic Differentiation Technique to solve this problem by second-order methods. The examples of solving the inverse coefficient problem confirm the accuracy and efficiency of the proposed algorithm.

Keywords: Inverse coefficient problems · Heat equation · Numerical algorithm · Fast automatic differentiation · Levenberg-Marquardt algorithm

1 Introduction

Inverse coefficient problems are of great interest and have been considered for a long time (see, e.g., [1–7]). At the same time, much attention is paid not only to the theoretical study of these problems, but also to the development of numerical methods for solving them.

An algorithm for the numerical solution of the thermal conductivity coefficient identification problem for one-dimensional and two-dimensional unsteady heat equation was proposed by the authors in previous papers publicised in journal Comp. Math. and Math. Phys. (56 (10); 58 (10); and 58 (12)). It is based on the Fast Automatic Differentiation technique (FAD-technique, see [8]), which allowed to successfully solve a number of complex optimal control problems for dynamic systems.

In [9] we compare three approaches for calculation the gradient of a complex function of many variables. Comparison was based on a complex function that represents the energy of a system of atoms whose interaction potential is the

---

This work was partially supported by the Russian Foundation for Basic Research (project no. 19-01-00666 A).

N. N. Olenev et al. (Eds.): OPTIMA 2021, LNCS 13078, pp. 351–364, 2021.
https://doi.org/10.1007/978-3-030-91059-4_25

Tersoff potential. The results of that work show the superiority of the FAD-technique in comparison to the approach that is based on analytical formulas. In the present paper we compare the gradient method and a second-order optimization method, which uses formulas that have been obtained in the current work for the first time and by means of the FAD-technique.

In the developed algorithm the inverse coefficient problem was reduced to the following variational problem: it is required to find such dependence of the thermal conductivity coefficient on the temperature, at which the temperature field and heat fluxes at the boundary of the object, obtained as a result of solving the primal problem, differ little from the data obtained experimentally.

The gradient descent method was usually used for numerical solution of the obtained optimization problems. It is well known that for gradient methods to work effectively, one needs to know the exact value of the gradient of the cost function. The use of the FAD-technique in the proposed algorithm made it possible to achieve this goal: the gradient of the cost function was calculated with machine accuracy. However, in the neighborhood of the solution, the gradient method converges slowly. Taking into account this fact, in the present paper an algorithm for solving inverse coefficient problems by second-order methods is proposed. It is also based on the FAD-technique.

The most popular second-order method is Newton's method. It requires calculating and reversing the Hessian at each iteration, which is often quite complex and requires a lot of machine resources.

One of the varieties of the Newton method is the Newton-Gauss method. This iterative method is intended for solving the least squares problem. There is no need to calculate and reverse the matrix of second derivatives here, but a Jacobi-type matrix is built. Its elements are the gradient components of each of the quadratic terms of the cost functional. In this case, the convergence rate of the Newton-Gauss method is close to the quadratic one.

The advantage of the Newton-Gauss method is its simplicity of implementation. However, its application to solving specific problems has revealed a number of problems associated with incorrect operation and slowing down convergence. The study of these problems led to a modification of the Newton-Gauss method—the Levenberg-Marquardt algorithm. This algorithm differs from the Newton-Gauss method by introducing a special regularization parameter. The Jacobi-type matrix mentioned above, as well as the gradient, requires high accuracy in determining its elements.

The accuracy of calculating the elements of a Jacobi-type matrix has an essential effect on the convergence of the Levenberg-Marquardt method.

The purpose of this paper is to study the possibility of applying and efficiency of the Levenberg-Marquardt algorithm to the solution of inverse coefficient problems. It is essential that in the proposed approach, the elements of the Jacobi-type matrix are calculated with machine accuracy due to the use of the FAD-technique.

This paper is also devoted to the proving of necessary conditions for non-uniqueness of the inverse problem solutions.

2   Mathematical Formulation of the Problem

We consider a restricted domain $Q \subset R^n$ with piecewise-smooth boundary $S = \partial Q$. This domain is filled with the substance being investigated. The distribution of the temperature field in Q at each time moment is described by the following initial boundary value (mixed) problem:

$$C(x)\frac{\partial T(x,t)}{\partial t} = div_x(K(T(x,t))\nabla_x T(x,t)), \qquad x \in Q, \quad 0 < t \leq \Theta, \quad (1)$$

$$T(x,0) = w_0(x), \qquad x \in Q, \qquad (2)$$

$$T(x,t) = w_s(x,t), \qquad x \in S, \quad 0 \leq t \leq \Theta. \qquad (3)$$

Here $x = (x_1, ..., x_n)$ are the Cartesian coordinates; t is time; $T(x,t)$ is the temperature of the material at the point with the coordinates x at time t; $C(x)$ is the volumetric heat capacity of the material; $K(T)$ is the thermal conductivity; $w_0(x)$ is the given temperature at the initial time $t = 0$; $w_s(x,t)$ is the given temperature on the boundary of the object. The volumetric heat capacity $C(x)$ of a substance is considered as known function of the coordinates.

If the dependence of the thermal conductivity $K(T)$ on the temperature T is known, then we can solve the mixed problem (1)–(3) to find the temperature distribution $T(x,t)$ in $G = Q \times (0, \Theta]$. We will call problem (1)–(3) the direct problem. The inverse coefficient problem is reduced to the following variational problem: find the dependence $K(T)$ on T under which the temperature field $T(x,t)$, obtained by solving the mixed problem (1)–(3), is close to the field $Y(x,t)$ obtained experimentally, and the heat flux $\left(-K(T(x,t))\frac{\partial T(x,t)}{\partial n}\right)$ on the boundary of the domain is close to the experimental data $P(x,t)$. The quantity

$$\Phi(K(T)) = \int_0^\Theta \int_Q [T(x,t) - Y(x,t)]^2 \cdot \mu(x,t)dx\,dt$$

$$+ \int_0^\Theta \int_S \beta(x,t)\left[-K(T(x,t))\frac{\partial T(x,t)}{\partial n} - P(x,t)\right]^2 ds\,dt + \varepsilon \int_a^b (K'(T))^2 dT \quad (4)$$

can be used as the measure of difference between these functions. Here, $\varepsilon \geq 0$, $\beta(x,t) \geq 0$, $\mu(x,t) \geq 0$ are given weight parameters; $Y(x,t)$ is a given temperature field; $P(x,t)$ is a given heat flux at the boundary S of the domain Q, $\frac{\partial T}{\partial n}$ is the derivative of the temperature along the outer normal to the boundary of the domain; $[a;b]$ is the interval where the function $K(T)$ will be restored. The last term in functional (4) is used to obtain a smooth solution of variational problem with moderate perturbations of experimental data.

3   Non-uniqueness of the Solution of Inverse Problem

An analysis of results obtained by solving a large number of formulated inverse coefficient problems has shown that these problems can have a non-unique solution. Whether or not the solution of the inverse problem is unique depends substantially on the given experimental field $Y(x,t)$. The following result presents the necessary condition for $Y(x,t)$ when the formulated inverse coefficient problem has a non-unique solution.

Theorem 1. *Let $Y(x,t) \in C_{x,t}^{2,1}(G) \cap C^1(\overline{G})$ be a solution of the direct problem (1)–(3) for two admissible thermal conductivities $K_1(T) \in C^1([a,b])$ and $K_2(T) \in C^1([a,b])$, $T \in [a,b]$. Then the following assertions are true:*

(a) There exists a function $R(T) \in C([a,b])$ such that

$$\Delta_x Y(x,t) = R(Y(x,t)) \cdot |\nabla_x Y(x,t)|^2,$$

(b) For such a function $Y(x,t)$, the inverse problem has infinitely many solutions $K(T)$.

Proof. Let $K_1(T) \in C^1([a,b])$ and $K_2(T) \in C^1([a,b])$ be two different functions, and let $Y(x,t) \in C_{x,t}^{2,1}(G) \cap C^1(\overline{G})$ be a function in the reachability domain that is a solution to the direct problem (1)–(3) for $K(T) = K_1(T)$ and $K(T) = K_2(T)$.

The function $Y(x,t)$ simultaneously satisfies two equations

$$C(x)\frac{\partial Y(x,t)}{\partial t} = div_x(K_1(Y(x,t)) \cdot \nabla_x Y(x,t)), \qquad (x,t) \in G,$$

$$C(x)\frac{\partial Y(x,t)}{\partial t} = div_x(K_2(Y(x,t)) \cdot \nabla_x Y(x,t)), \qquad (x,t) \in G.$$

By subtracting the first equation from the second one, we see that the function $w(T) = K_2(T) - K_1(T)$ satisfies the condition (5)

$$div_x(w(Y(x,t)) \cdot \nabla_x Y(x,t)) = 0, \qquad (x,t) \in G. \qquad (5)$$

Assuming that the field $Y(x,t)$ and the sought function $w(T)$ are smooth enough, we obtain an equation satisfied by $w(T)$, namely,

$$w'(Y) \cdot \sum_{k=1}^{n} \left(\frac{\partial Y(x,t)}{\partial x_k}\right)^2 + w(Y) \cdot \sum_{k=1}^{n}\left(\frac{\partial^2 Y(x,t)}{\partial x_k^2}\right) = 0, \qquad (x,t) \in G. \qquad (6)$$

A function $w(Y)$ which is not identically zero can be determined by (6) only in the case when the function

$$B(x,t) = \frac{\sum_{k=1}^{n}\left(\frac{\partial^2 Y(x,t)}{\partial x_k^2}\right)}{\sum_{k=1}^{n}\left(\frac{\partial Y(x,t)}{\partial x_k}\right)^2} = \frac{\Delta_x Y(x,t)}{|\nabla_x Y(x,t)|^2}$$

can be represented as a composition of the one-variable function $R(z)$ and the function $Y(x,t)$, i.e., $B(x,t) = R(Y(x,t))$. It is under this condition that the solution of the formulated inverse coefficient problem (1)–(4) is not unique. Moreover, under this condition, Eq. (6) has an infinite number of solutions.

To illustrate what was said above, we consider the case when the experimental temperature field $Y(x,t)$ is a sufficiently smooth function and monotonically depends on a linear combination of spatial coordinates and time, i.e.,

$$Y(x,t) = f\left(\sum_{k=1}^{n} \alpha_k x_k + \gamma t\right) = f(\xi),$$

where $\xi = \sum_{k=1}^{n} \alpha_k x_k + \gamma t$ and α_k, γ are given constants such that $\sum_{k=1}^{n} \alpha_k^2 > 0$ (otherwise, the field $Y(x,t) \equiv Const$ is of no interest). In this case, we obtain

$$B(x,t) = B(\xi) = f''(\xi) \cdot (f'(\xi))^{-2}$$

and Eq. (6) becomes

$$w'(Y) \cdot (f'(\xi))^2 + w(Y) \cdot f''(\xi) = 0.$$

After introducing the new function $\varphi(\xi) = w(Y) = w(f(\xi))$, this equation is transformed into

$$\varphi'(\xi)f'(\xi) + \varphi(\xi)f''(\xi) = 0.$$

Integrating this equation, we obtain

$$\varphi(\xi)f'(\xi) = Const. \tag{7}$$

Determining function $\varphi(\xi)$ from Eq. (7) and taking the inverse transform $\xi = \eta(Y)$ yields the sought functions $w(Y)$.

4 Application of the FAD-Technique for Solving Optimal Control Problem by Second-Order Methods

The optimal control problem formulated above was solved numerically. One of the main elements of the proposed numerical method for solving inverse coefficient problem is the solution of the mixed problem (1)–(3). Spatial and time grids (generally non-uniform) were introduced for numerical solution of the problem. In each node of the calculation area $\overline{Q} \times [0, \Theta]$ all the functions are determined by their point values. In previous works two finite-difference schemes (a two-layer implicit scheme with weights, and implicit alternating directions scheme) were used to approximate the thermal conductivity equation.

The temperature interval $[a, b]$ on which the function $K(T)$ will be restored is defined as the set of values of the given functions $w_0(x)$ and $w_s(x,t)$, i.e. the boundaries of the segment $[a, b]$ were assigned as the minimum and maximum values of the indicated functions. This interval is partitioned by the points $\tilde{T}_0 = a, \tilde{T}_1, \tilde{T}_2, \ldots, \tilde{T}_M = b$ into M parts (they can be equal or of different lengths). Each point $\tilde{T}_m$ ($m = 0, \ldots, M$) is connected with a number $k_m = K(\tilde{T}_m)$. The function $K(T)$ to be found is approximated by a continuous piecewise linear function with the nodes at the points $\left\{(\tilde{T}_m, k_m)\right\}_{m=0}^{M}$, so that

$$K(T) = k_{m-1} + \frac{k_m - k_{m-1}}{\tilde{T}_m - \tilde{T}_{m-1}} (T - \tilde{T}_{m-1}) \quad for \quad \tilde{T}_{m-1} \le T \le \tilde{T}_m, \quad m = 1, \dots M.$$

If the temperature at the point fell outside the boundaries of the interval $[a, b]$, then the linear extrapolation was used to determine the function $K(T)$.

We illustrate the application of the FAD-technique for solving the optimization problem by second-order methods using the example of the inverse coefficient problem for the one-dimensional heat equation ($Q \subset R^1$). In [10] this problem was considered under the assumption that the heat flux was known only on the left boundary of the domain. Here this problem will be considered in a more general case.

A layer filled with material of width L is considered. The distribution of the temperature field at each instant of time is described by the initial boundary value problem (1)–(3), where $Q = (0, L)$, and the boundary S consists of two points: $x = 0$ and $x = L$.

To solve the problem numerically, the domain $[0, L] \times [0, \Theta]$ was decomposed by the grid lines $\{\tilde{x}_i\}_{i=0}^{I}$ and $\{\tilde{t}^j\}_{j=0}^{J}$ into rectangles. At each node $(\tilde{x}_i, \tilde{t}^j)$ of G all the functions are determined by their point values (e.g., $T(\tilde{x}_i, \tilde{t}^j) = T_i^j$). To approximate the heat equation a two-layer implicit scheme with weights was used and resulting system of nonlinear algebraic equations is solved iteratively using the Gaussian elimination (see [10]).

Here, we give only the canonical form of these finite-difference equations, which will be needed further for using the FAD-technique:

$$T_i^j = b_i^j \left(K(T_i^j) + K(T_{i+1}^j)\right)(T_{i+1}^j - T_i^j) - a_i^j \left(K(T_i^j) + K(T_{i-1}^j)\right)(T_i^j - T_{i-1}^j)$$

$$+ T_i^{j-1} + c_i^j \left(K(T_i^{j-1}) + K(T_{i+1}^{j-1})\right)(T_{i+1}^{j-1} - T_i^{j-1})$$

$$- d_i^j \left(K(T_i^{j-1}) + K(T_{i-1}^{j-1})\right)(T_i^{j-1} - T_{i-1}^{j-1}) \equiv \Psi_i^j,$$

$$a_i^j = \sigma \tau^j / (C_i h_{i-1}(h_i + h_{i-1})), \qquad b_i^j = \sigma \tau^j / (C_i h_i (h_i + h_{i-1})),$$

$$c_i^j = (1 - \sigma)\tau^j / (C_i h_i (h_i + h_{i-1})), \qquad d_i^j = (1 - \sigma)\tau^j / (C_i h_{i-1}(h_i + h_{i-1})),$$

σ – weight parameter, $h_i = \tilde{x}_{i+1} - \tilde{x}_i$, $\tau^j = \tilde{t}^j - \tilde{t}^{j-1}$, $i = \overline{1, I-1}$, $j = \overline{1, J}$.
The cost functional (4) was approximated by a function $F(k_0, k_1, \dots, k_M)$ of the finite number of variables using the rectangles method:

$$\Phi(K(T)) \approx F = \sum_{j=1}^{J} \sum_{i=1}^{I-1} \left((T_i^j - Y_i^j)^2 \mu_i^j h_i \tau^j\right)$$

$$+ \sum_{j=1}^{J} \left(\beta_0^j \cdot \tau^j \left[\frac{\sigma}{2h_0} \left(K(T_0^j) + K(T_1^j)\right)(T_1^j - T_0^j)\right.\right.$$

$$+ \frac{1-\sigma}{2h_0} \left(K(T_0^{j-1}) + K(T_1^{j-1}) \right) (T_1^{j-1} - T_0^{j-1}) - \frac{C_0 h_0}{2\tau^j} (T_0^j - T_0^{j-1}) - P_0^j \Bigg]^2 \Bigg)$$

$$+ \sum_{j=1}^{J} \left(\beta_I^j \cdot \tau^j \left[\frac{\sigma}{2h_{I-1}} \left(K(T_I^j) + K(T_{I-1}^j) \right) (T_{I-1}^j - T_I^j) \right. \right.$$

$$+ \frac{1-\sigma}{2h_{I-1}} \left(K(T_I^{j-1}) + K(T_{I-1}^{j-1}) \right) (T_{I-1}^{j-1} - T_I^{j-1}) - \frac{C_I h_{I-1}}{2\tau^j} (T_I^j - T_{I-1}^{j-1}) - P_I^j \Bigg]^2 \Bigg)$$

$$+ \varepsilon \sum_{m=1}^{M} \frac{(k_m - k_{m-1})^2}{(\tilde{T}_m - \tilde{T}_{m-1})}. \tag{8}$$

The second-order iterative methods converge well in the neighborhood of the solution, but at the beginning of the iterative process it is recommended to use some gradient method. Taking into account this fact, we present a formula for calculating the gradient of the cost function $F(k_0, k_1, \ldots, k_M)$, which here is more complex than in [10].

The adjoint problem for computation the conjugate variables p_i^j ($i = \overline{1, I-1}$, $j = \overline{1, J}$) was similar in form to that presented in [10]. The only difference was that the derivatives $\partial F / \partial T_i^j$ ($i = \overline{1, I-1}, j = \overline{1, J}$) were computed in a different manner, since in [10] another simpler cost functional was used. In the present case, these derivatives were calculated using the formulas:

$$\frac{\partial F}{\partial T_1^j} = G_1^j + \frac{D^j}{h_0} \left(\beta_0^j \tau^j A^j \sigma + \beta_0^{j+1} \tau^{j+1} A^{j+1} (1-\sigma) \right),$$

$$\frac{\partial F}{\partial T_i^j} = G_i^j, \qquad (i = \overline{2, I-2},$$

$$\frac{\partial F}{\partial T_{I-1}^j} = G_{I-1}^j + \frac{E^j}{h_{I-1}} \left(\beta_I^j \tau^j B^j \sigma + \beta_I^{j+1} \tau^{j+1} B^{j+1} (1-\sigma) \right),$$

where

$$G_i^j = 2(T_i^j - Y_i^j) \mu_i^j h_i \tau^j,$$

$$D_i^j = K(T_0^j) + K(T_1^j) + K'(T_1^j) T_1^j - K'(T_1^j) T_0^j,$$

$$E_i^j = K(T_I^j) + K(T_{I-1}^j) + K'(T_{I-1}^j) T_{I-1}^j - K'(T_{I-1}^j) T_I^j,$$

$K'(T) = (k_m - k_{m-1})/(\tilde{T}_m - \tilde{T}_{m-1})$ (index m is determined by the condition $\tilde{T}_{m-1} \leq T \leq \tilde{T}_m$),

$$\beta_0^{J+1} = \beta_I^{J+1} = \tau^{J+1} = A^{J+1} = B^{J+1} = 0,$$

$$A^j = \left[\frac{\sigma}{2h_0} \left(K(T_0^j) + K(T_1^j) \right) (T_1^j - T_0^j) \right.$$

$$+ \frac{1-\sigma}{2h_0} \left(K(T_0^{j-1}) + K(T_1^{j-1}) \right) (T_1^{j-1} - T_0^{j-1}) - \frac{C_0 h_0}{2\tau^j} (T_0^j - T_0^{j-1}) - P_0^j \Bigg],$$

$$B^j = \left[\frac{\sigma}{2h_{I-1}} \left(K(T_I^j) + K(T_{I-1}^j) \right) (T_{I-1}^j - T_I^j) \right.$$

$$\left. + \frac{1-\sigma}{2h_{I-1}} \left(K(T_I^{j-1}) + K(T_{I-1}^{j-1}) \right) (T_{I-1}^{j-1} - T_I^{j-1}) - \frac{C_I h_{I-1}}{2\tau^j} (T_I^j - T_I^{j-1}) - P_I^j \right].$$

According to the FAD-technique, the components of function gradient (8) with respect to the components $\{k_m\}_{m=0}^M$ are calculated by formula:

$$\frac{\partial F}{\partial k_m} = \sum_{n=0}^{J} \sum_{l=0}^{I} \left(\sum_{j=1}^{J} \sum_{i=1}^{I-1} \frac{\partial \Psi_i^j}{\partial K(T_l^n)} \cdot p_i^j \right) \cdot \frac{\partial K(T_l^n)}{\partial k_m}$$

$$+ \sum_{j=0}^{J} \left(\frac{\partial F}{\partial K(T_0^j)} \frac{\partial K(T_0^j)}{\partial k_m} + \frac{\partial F}{\partial K(T_1^j)} \frac{\partial K(T_1^j)}{\partial k_m} \right)$$

$$+ \sum_{j=0}^{J} \left(\frac{\partial F}{\partial K(T_I^j)} \frac{\partial K(T_I^j)}{\partial k_m} + \frac{\partial F}{\partial K(T_{I-1}^j)} \frac{\partial K(T_{I-1}^j)}{\partial k_m} \right)$$

$$+ 2\lambda\varepsilon \frac{(k_m - k_{m-1})}{(\tilde{T}_m - \tilde{T}_{m-1})} - 2\nu\varepsilon \frac{(k_{m+1} - k_m)}{(\tilde{T}_{m+1} - \tilde{T}_m)}, \qquad m = \overline{0, M}. \tag{9}$$

where

$$\lambda = \begin{cases} 0, & m = 0, \\ 1, & m = \overline{1, M}, \end{cases} \qquad \nu = \begin{cases} 0, & m = M, \\ 1, & m = \overline{0, M-1}, \end{cases}$$

Ψ_i^j—right part of the canonical form of equations for calculation temperature field,

$$\frac{\partial K(T_l^n)}{\partial k_{m-1}} = 1 - \frac{T_l^n - \tilde{T}_{m-1}}{\tilde{T}_m - \tilde{T}_{m-1}}, \qquad \frac{\partial K(T_l^n)}{\partial k_m} = \frac{T_l^n - \tilde{T}_{m-1}}{\tilde{T}_m - \tilde{T}_{m-1}}$$

(index m is determined by the condition $\tilde{T}_{m-1} \le T_l^n \le \tilde{T}_m$),

$$\frac{\partial F}{\partial K(T_0^j)} = \frac{\partial F}{\partial K(T_1^j)} = \frac{T_1^j - T_0^j}{h_0} \left(\beta_0^j \tau^j A^j \sigma + \beta_0^{j+1} \tau^{j+1} A^{j+1} (1 - \sigma) \right),$$

$$\frac{\partial F}{\partial K(T_I^j)} = \frac{\partial F}{\partial K(T_{I-1}^j)} = \frac{T_{I-1}^j - T_I^j}{h_{I-1}} \left(\beta_I^j \tau^j B^j \sigma + \beta_I^{j+1} \tau^{j+1} B^{j+1} (1 - \sigma) \right).$$

The sums $H_l^n = \sum_{j=1}^{J} \sum_{i=1}^{I-1} \frac{\partial \Psi_i^j}{\partial K(T_l^n)} p_i^j$ was calculated in the same way to that presented in [10].

Further, to use second-order methods to solve the optimization problem, note that the cost function (8) is the sum of N squares of scalar functions, i.e.

$\Phi(K(T)) \approx F = \sum\limits_{n=1}^{N} W_n^2$, where $N = (I-1)J + 2J + M$. Scalar functions W_n, $n = \overline{1,N}$ allow one to build a vector-function with components

$$
W_n = \begin{cases}
\widetilde{W}_n^{ij} = (T_i^j - Y_i^j) \cdot \sqrt{\mu_i^j h_i \tau^j}, & n = i + (j-1)(I-1), \ i = \overline{1,I-1}, j = \overline{1,J}, \\
\widetilde{W}_n^j = A^j \cdot \sqrt{\beta_0^j \tau^j}, & n = j + (I-1)J, \ j = \overline{1,J}, \\
\widetilde{W}_n^j = B^j \cdot \sqrt{\beta_I^j \tau^j}, & n = j + I \cdot J, \ j = \overline{1,J}, \\
\widetilde{W}_n^m = \dfrac{\sqrt{\varepsilon}(k_m - k_{m-1})}{\sqrt{(\widetilde{T}_m - \widetilde{T}_{m-1})}}, & n = (I+1)J + M, \ m = \overline{1,M}.
\end{cases}
$$

Note, that functions $W_n, (n = \overline{1,N})$, are complex functions, which in the general case depend on the components of the control vector $(k_0, k_1, \ldots, k_M)^T$ and on the phase variables T_i^j. The control vector components $\{k_m\}_{m=0}^{M}$ and phase variables T_i^j $(i = \overline{0,I}, j = \overline{0,J})$ are connected between themselves by the system of equations (3.1), which is given in [10]. With the help of the FAD-technique for each complex function $W_n, (n = \overline{1,N})$ we will find its gradient with respect to the components $k_0, k_1, \ldots, k_M$.

The adjoint problem for computation of the conjugate variables p_i^j, $(i = \overline{1,I-1}, \ j = \overline{1,J})$ corresponding to the cost function $W_n, (n = \overline{1,N})$ was similar in form to that presented in [10]. The only difference was that the derivatives $\partial F / \partial T_i^j$ must be replaced by derivatives $\partial W_n / \partial T_i^j$, that are calculated by the formulas:

for all $n = 1, 2, \ldots, (I-1)J$

$$
\frac{\partial W_n}{\partial T_p^l} = \frac{\partial \widetilde{W}_n^{ij}}{\partial T_p^l} = \begin{cases} \sqrt{\mu_p^l h_p \tau^l}, & p = i, l = j, \\ 0, & otherwise, \end{cases}
$$

for all $n = (I-1)J + 1, \ldots, (I-1)J + J$

$$
\frac{\partial W_n}{\partial T_p^l} = \frac{\partial \widetilde{W}_n^j}{\partial T_i^l} = \begin{cases} \sqrt{\beta_0^l \tau^l} \cdot \sigma D^l / (2h_0), & i = 1, \ l = j, \\ \sqrt{\beta_0^l \tau^l} \cdot (1-\sigma) D^l / (2h_0), & i = 1, \ l = j-1, \\ 0, & otherwise, \end{cases}
$$

for all $n = (I-1)J + J + 1, \ldots, (I-1)J + 2J$

$$
\frac{\partial W_n}{\partial T_p^l} = \frac{\partial \widetilde{W}_n^j}{\partial T_i^l} = \begin{cases} \sqrt{\beta_I^l \tau^l} \cdot \sigma E^l / (2h_{I-1}), & i = I-1, \ l = j, \\ \sqrt{\beta_I^l \tau^l} \cdot (1-\sigma) E^l / (2h_{I-1}), & i = I-1, \ l = j-1, \\ 0, & otherwise, \end{cases}
$$

for all $n = (I-1)J + 2J + 1, \ldots, (I-1)J + 2J + M$

$$
\frac{\partial W_n}{\partial T_i^j} = \frac{\partial \widetilde{W}_n^m}{\partial T_i^j} = 0, \qquad i = \overline{1,I-1}, \ j = \overline{1,J}.
$$

According to the FAD-technique, the components of the gradient of functions W_n $(n = \overline{1, N})$ with respect to the components $\{k_m\}_{m=0}^M$ are calculated by formulas:

for all $n = 1, 2, ..., (I-1)J$

$$\frac{\partial W_n}{\partial k_m} = \frac{\partial \widetilde{W}_n^{rs}}{\partial k_m} = \Lambda_m \sqrt{\mu_r^s h_r \tau^s},$$

for all $n = (I-1)J + 1, ..., (I-1)J + J$

$$\frac{\partial W_n}{\partial k_m} = \frac{\partial \widetilde{W}_n^s}{\partial k_m} = \Lambda_m \sqrt{\beta_0^s \tau^s} + \sqrt{\beta_0^s \tau^s} \frac{\sigma(T_1^s - T_0^s)}{2h_0} \left(\frac{\partial K(T_0^s)}{\partial k_m} + \frac{\partial K(T_1^s)}{\partial k_m} \right)$$

$$+ \sqrt{\beta_0^s \tau^s} \frac{(1-\sigma)(T_1^{s-1} - T_0^{s-1})}{2h_0} \left(\frac{\partial K(T_0^{s-1})}{\partial k_m} + \frac{\partial K(T_1^{s-1})}{\partial k_m} \right),$$

for all $n = (I-1)J + J + 1, ..., (I-1)J + 2J$ the derivatives $\widetilde{W}_n^s/k_m$ are calculated in a similar way,

for all $n = (I-1)J + 2J + 1, ..., (I-1)J + 2J + M$

$$\frac{\partial W_n}{\partial k_m} = \frac{\partial \widetilde{W}_n^s}{\partial k_m} = \begin{cases} \Lambda_m + \sqrt{\dfrac{\varepsilon}{\overline{T}_m - \overline{T}_{m-1}}}, & s = m, \\ \Lambda_m, & otherwise, \end{cases}$$

$$\frac{\partial W_n}{\partial k_{m-1}} = \frac{\partial \widetilde{W}_n^s}{\partial k_{m-1}} = \begin{cases} \Lambda_{m-1} - \sqrt{\dfrac{\varepsilon}{\overline{T}_m - \overline{T}_{m-1}}}, & s = m, \\ \Lambda_{m-1}, & otherwise, \end{cases}$$

$$\Lambda_m = \sum_{l=0}^J \sum_{p=0}^I \left(\sum_{j=1}^J \sum_{i=1}^{I-1} \frac{\partial \Psi_i^j}{\partial K(T_p^l)} \cdot p_i^j \right) \cdot \frac{\partial K(T_p^l)}{\partial k_m}.$$

It should be noted that the conjugate variables p_i^j that appear in these formulas are different for each $n = \overline{1, N}$, and the corresponding adjoint problem is solved in each case separately for all $n = \overline{1, N}$.

Finally, it is possible to build the Jacobian $\Omega = \{\Omega_{nm}\}$ of a vector function $\overrightarrow{W}$. The matrix Ω consists of N rows and $(M+1)$ columns. Each n-th row of this matrix consists of $(M+1)$ elements: $\Omega_{nm} = \frac{\partial W_n}{\partial k_m}$, $(n = \overline{1, N}, m = \overline{0, M})$. This matrix was obtained for the most general case: temperature field and the heat fluxes on both ends of the layer are experimentally obtained. If any weight parameters in the cost functional are zero, then in vector function $\{W\}_{n=1}^N$ the corresponding rows are also missing.

It is important to note that the elements of the Jacobian thus obtained are calculated with machine accuracy. For solving the optimization problem the Levenberg–Marquardt algorithm was used as the method of the second order of convergence. According to this algorithm the control vector $\overrightarrow{k} = \{k\}_{m=0}^M$

changes at each subsequent iteration $(s + 1)$ by the formula: $\vec{k}_{s+1} = \vec{k}_s + \vec{r}_s$. The direction of descent $\vec{r}_s$ is determined from the following system of linear algebraic equations:

$$\left[\Omega^T(\vec{k}_s)\Omega(\vec{k}_s) + \lambda_s \cdot diag\left(\Omega^T(\vec{k}_s)\Omega(\vec{k}_s)\right)\right]\vec{r}_s = -\Omega^T(\vec{k}_s)\vec{W}(\vec{k}_s).$$

The studies related to the choice of constants λ_s at each step of descent were carried out. For the considered here inverse problems they were calculated by the formula:

$$\lambda_s = \alpha \cdot \max_{0 \leq m \leq M}\left(diag\left(\Omega^T(\vec{k}_s)\Omega(\vec{k}_s)\right)\right),$$

where α is some non-negative constant, which was selected for each example individually.

5   Numerical Results

The effectiveness of proposed method was tested on all examples that were considered in [10]. Initially, until the functional fell by several orders, they were solved using the gradient method. Furthermore, the above indicated algorithm was used, with the help of which it was enough to perform just several iterations to obtain a solution of the optimization problem. The thermal conductivity coefficient in all considered examples was restored with precision 10^{-12} in norm C.

A comparative analysis of the results of solving the problem using the gradient method and the second-order method mentioned above is presented here for the following two examples. In the studies it was assumed that $L = 1$, $\Theta = 1$, $C(x) \equiv 1$, the parameter ε in the cost functional (4) was equal to zero. When solving primal and conjugate problems, a uniform grid with parameters $I = 150$ (number of intervals along the x-axis), $J = 3000$ (number of intervals along the t-axis) was used, which provided sufficient accuracy for calculating the temperature field and the field of conjugate variables. Parameter σ was equal to 0.6. The initial control for the methods used was the same.

1. The problem of finding the thermal conductivity coefficient is considered with the following input data

$$w_0(x) = \sqrt{2(1.5 - x)}, \qquad\qquad\qquad 0 \leq x \leq 1,$$
$$w_1(t) = \sqrt{2(1.5 + t)}, \qquad w_2(t) = \sqrt{2(0.5 + t)}, \qquad 0 \leq t \leq 1,$$
$$Y(x, t) = \sqrt{2(1.5 + t - x)}, \qquad 0 \leq x \leq 1, \qquad\qquad 0 \leq t \leq 1,$$
$$a = 1, \qquad\qquad b = \sqrt{5}.$$

The inverse problem with this input data has an analytical solution, since the function $Y(x, t) = \sqrt{2(1.5 + t - x)}$ is a solution of the mixed problem (1)–(3) with the indicated above parameters and $K(T) = T^2$. It was assumed that in the cost functional (4) the weight function $\mu(x, t)$ is equal to zero, and $\beta(x, t) = 1$ (the thermal conductivity coefficient is restored by the heat flux on the boundary). The segment $[a, b]$ was divided into $M = 40$ intervals. The function $K(T) = 4.0$ was selected as the initial control. The cost functional at the initial control was equal to $F_0 = 1.284$.

When solving the problem with the help of the gradient method, about 20000 iterations were required. The cost functional decreased to $F_{opt} = 1.216 \cdot 10^{-21}$, and the gradient value at optimal control reached the value 10^{-14} in the norm C. The thermal conductivity coefficient was determined with accuracy 10^{-10}.

When solving the problem using the Levenberg-Marquardt method, only 12 iterations were required. In this case, the problem was solved twice as fast. The constant α that appears in the Levenberg-Marquardt method was equal to 10^{-6}. The cost functional was reduced to the value $F_{opt} = 1.767 \cdot 10^{-25}$, the gradient value at optimal control reached the value 10^{-15} in the norm C. The thermal conductivity coefficient was determined with accuracy 10^{-12}.

2. The effectiveness of the Levenberg-Marquardt method was also tested using the example with the following parameters:

$$w_0(x) = 2x, \qquad w_1(t) = 0, \qquad w_2(t) = 2, \qquad a = 0, \qquad b = 2.$$

The "experimental" temperature field was defined as the solution of the direct problem (1)–(3) with

$$K(T) = \begin{cases} 1, & T < 1, \\ 3, & T \geq 1. \end{cases}$$

On the first q subintervals it was assumed $K(T) = 1$, on the subintervals $(q + 2), (q + 3), ..., (2q + 1)$ it was assumed $K(T) = 3$, and on the subinterval $(q + 1)$ the function $K(T)$ is a linear function varying from 1 to 3. The entire segment $[a, b]$ was divided into 79 intervals ($M = 2q + 1 = 79$).

It was assumed that in the cost functional (4) the weight function $\mu(x, t) = 1$, and $\beta(x, t) = 0$ (the coefficient of thermal conductivity is restored by a given temperature field). The function $K(T) = T$ was selected as the initial control. The cost functional at the initial control was equal to $F_0 = 9.504 \cdot 10^{-3}$.

When solving the problem using the gradient method, about 550000 iterations were performed. The cost functional was reduced to the value $F_{opt} = 7.338 \cdot 10^{-24}$, the gradient value at optimal control reached the value $5 \cdot 10^{-16}$ in the norm C. The thermal conductivity coefficient was determined with accuracy 10^{-8}.

When solving the problem using the Levenberg-Marquardt method, only 11 iterations were performed. The problem was solved 8.5 times faster than using the gradient method. The constant α that appears in the method Levenberg-Marquardt was equal to 10^{-3}. The cost functional was reduced to the value $F_{opt} = 5.953 \cdot 10^{-29}$, the gradient value at optimal control reached the value $2 \cdot 10^{-16}$ in the norm C. The thermal conductivity coefficient was determined with accuracy 10^{-13}.

Solutions of this optimization problem were additionally performed in such a way that the solution obtained by the gradient method was used as the initial approximation for the Levenberg-Marquardt method. With the help of the

Levenberg-Marquardt method it was possible to carry out 3 iterations. The cost functional decreased to $F_{opt} = 6.021 \cdot 10^{-29}$ and the gradient value at optimal control was equal to $2 \cdot 10^{-16}$. The thermal conductivity coefficient was determined with accuracy 10^{-13}.

6 Conclusions

Based on results obtained in this paper, the following conclusions can be made.

On the one hand, using the Levenberg-Marquardt method requires additional effort related to obtaining formulas for calculating the Jacobi matrix, additional memory for working with this matrix, and additional machine time. In this respect, gradient methods look preferable.

On the other hand, these additional costs are more than repaid. First, it should be noted that obtaining the formulas using the FAD-technique for calculating the elements of the Jacobi matrix is much simpler than obtaining the formulas for the gradient of the cost function. Second, the machine time required to obtain a solution of optimization problem is significantly reduced when using the Levenberg-Marquardt method. Finally, using second-order methods, it is possible to obtain a more accurate solution than using first-order methods.

The best results, as shown by our experience in solving optimization problems, can be obtained by combining the first and second order methods, namely: at the first stage, when the functional is significantly reduced and the gradient of the functional is large, it is advisable to use gradient methods, and then, when the functional gradient becomes close to zero, to continue the minimization of the functional using second-order methods.

References

1. Kozdoba, L.A., Krukovskii, P.G.: Methods for Solving Inverse Thermal Transfer Problems. Naukova Dumka, Kiev (1982).[in Russian]
2. Alifanov, O.M.: Inverse Heat Transfer Problems. Springer, Berlin (2011). https://doi.org/10.1007/978-3-642-76436-3. [in Russian]
3. Vabishchevich, P.N., Denisenko, A.Yu.: Numerical methods for solving inverse coefficient problems. In: Method of Mathematical Simulation and Computational Diagnostics, pp. 35–45. Mosk. Gos. Univ., Moscow (1990). [in Russian]
4. Samarskii, A.A., Vabishchevich, P.N.: Difference methods for solving inverse problems of mathematical physics. In: Fundamentals of Mathematical Simulation, pp. 5–97. Nauka, Moscow (1997). [in Russian]
5. Samarskii, A.A., Vabishchevich, P.N.: Computational Heat Transfer. Editorial URSS, Moscow (2003).[in Russian]
6. Czel, B., Grof, G.: Inverse identification of temperature-dependent thermal conductivity via genetic algorithm with cost function-based rearrangement of genes. Int. J. Heat Mass Trans. **55**(15), 4254–4263 (2012)
7. Matsevityi, Y.M., Alekhina, S.V., Borukhov, V.T., Zayats, G.M., Kostikova, A.O.: Identification of the thermal conductivity coefficient for quasi-stationary two-dimensional heat conduction equations. J. Eng. Phys. Thermophy. **90**(6), 1295–1301 (2017)

8. Evtushenko, Yu.G.: Computation of exact gradients in distributed dynamic systems. Optim. Methods Softw. **9**, 45–75 (1998). https://doi.org/10.1080/10556789808805686

9. Albu, A.F., Gorchakov, A.Yu., Zubov, V.I.: On the effectiveness of the fast automatic differentiation methodology. In: Optimization and Applications. OPTIMA 2018. Communications in Computer and Information Science, vol. 974, pp. 264–276. Springer, Cham (2019). https://doi.org/10.1007/978-3-030-10934-9-19

10. Zubov, V.I.: Application of fast automatic differentiation for solving the inverse coefficient problem for the heat equation. Comp. Math. and Math. Phys. **56**(10), 1743–1757 (2016). https://doi.org/10.7868/S0044466916100148

Author Index

Printed in the United States
by Baker & Taylor Publisher Services